D1793751

Polymer Molecular Weight Methods

Myer Ezrin, *Editor*

A symposium co-sponsored by the Division of Analytical Chemistry and the Division of Polymer Chemistry at the 162nd Meeting of the American Chemical Society, Washington, D. C., Sept. 14–16, 1971.

ADVANCES IN CHEMISTRY SERIES **125**

AMERICAN CHEMICAL SOCIETY

WASHINGTON, D. C. 1973

ADCSAJ 125 1-350 (1973)

Library of Congress Catalog Card 73-89047

ISBN 8412-0187-0

PRINTED IN THE UNITED STATES OF AMERICA

Advances in Chemistry Series

Robert F. Gould, *Editor*

FOREWORD

ADVANCES IN CHEMISTRY SERIES was founded in 1949 by the American Chemical Society as an outlet for symposia and collections of data in special areas of topical interest that could not be accommodated in the Society's journals. It provides a medium for symposia that would otherwise be fragmented, their papers distributed among several journals or not published at all. Papers are refereed critically according to ACS editorial standards and receive the careful attention and processing characteristic of ACS publications. Papers published in ADVANCES IN CHEMISTRY SERIES are original contributions not published elsewhere in whole or major part and include reports of research as well as reviews since symposia may embrace both types of presentation.

CONTENTS

PREFACE

This book is the outgrowth of a joint Analytical and Polymer Division Symposium, entitled "Recent Trends in the Determination of Molecular Weight." The papers deal primarily with molecular weight methods for polymers, but some are also applicable to non-polymeric materials.

With the advent of gel permeation chromatography (GPC) during the middle of the last decade, molecular weight determination has been so strongly dominated by this outstanding new method that it has tended to overshadow older methods and other developing new methods. In the present book on molecular weight methods, approximately one-third of the 26 papers are on GPC, indicating its continuing strong position in this field. However, there are also papers on four exciting new methods that promise to extend the determination of molecular weight into new techniques that seem far removed from conventional methods. These include self-beat spectroscopy, thin layer chromatography, mass chromatography, and electrospray mass spectroscopy. The papers on these provide an excellent introduction and progress report on the very newest molecular weight methods; these will bear close watching in the years ahead as they develop further. Taken together with other papers on recent developments in older methods, such as ebulliometry and ultracentrifugation, this collection of papers represents a significant sourcebook for the scientist concerned with molecular weight in the broadest sense.

Generally, the molecular weight literature on synthetic polymers and on biochemical systems has been separate, as though they were unrelated. The two fields, of course, share common methods of molecular weight. There is much to be gained by an open exchange of ideas and experience between the biochemist and polymer physical chemist. This volume contains six papers by outstanding biochemical scientists, two of which deal with sedimentation equilibrium. Here, too, the emphasis is on recent trends in molecular weight determination.

A very important aspect of recent molecular weight work is the effort by the National Bureau of Standards to provide well documented, certified standard samples of known molecular weight. One of the papers included here is a progress report on this work. NBS is anxious to provide the type of standards that are needed on as broad a basis as possible. Dr. Wagner and his associates at NBS welcome suggestions from scientists in all fields as to standards that may be needed and the usefulness of the present standards.

Another recent trend in molecular weight work that was discussed at the ACS Symposium but is not covered formally in the book is the work of the American Society for Testing and Materials; ASTM is in the process of preparing and testing standard methods of measuring molecular weight. Those who are interested or have suggestions to make may write to ASTM, 1916 Race Street, Philadelphia, Pa. 19103. The committee numbers are D20.70.04 and .05.

Enfield, Conn.
September 1973

MYER EZRIN

1

Determination of Number-Average Molecular Weights by Ebulliometry

CLYDE A. GLOVER

Research Laboratories, Tennessee Eastman Co., Division of Eastman Kodak Co., Kingsport, Tenn. 37662

This paper describes an ebulliometric system for routine and special determinations of molecular weights. The system uses a simple ebulliometer, an immersion heater, and a Cottrell-type pump. Temperature sensing is by differential thermopile. Precision varies from about 1 to 6%, and values compare well with those from other laboratories and those from other methods. Values as high as 170,000 have been successfully measured. Some problems encountered in using the ebulliometric method are: selection and effect of reference temperature, limitations of the vapor lift pump and a possible substitute for it, measurement of equilibrium concentrations within the operating ebulliometer, and the experimentally determined ebulliometric constant and some factors which influence its value.

Ebulliometry, one of the classical methods for determining molecular weights, has undergone great improvement in recent years. The requirements of a successful ebulliometric system are thermal stability, temperature and concentration equilibrium, and temperature sensing. These requirements have been met by a number of investigators in various ways. The systems of each are reported to have met the need for which they were designed; in spite of this, however, the method does not now appear to be used widely either for routine determinations or in special problems. Here we present the case for what appears to be a neglected technique which is felt to have great potential. We describe a system which has been in use for several years for both routine and special molecular weight measurements, discuss some of the results obtained, and finally consider some of the problems and unanswered questions which have been encountered.

The Ebulliometric System

The system consists of a simple ebulliometer, a temperature sensing element, and an electrical recording device. The ebulliometer is essentially the same as that described by Glover and Stanley (*1*) and is shown in Figure 1. It consists of a platinum immersion heater, a vapor lift pump of the Cottrell type, and a recycle arrangement patterned after that described earlier by Ray (*2*). There are other successful designs, such as those of Ray (*2*), Lehrle (*3*), and others, which differ in detail from the one shown, and there are different types, such as the shaking apparatus of Schultz (*4*) which was improved by Ezrin (*5*). Temperature sensing is an important part of ebulliometry. In the work being described, thermopiles are used. They are very stable in operation and change little, if any, with age. One advantage is the leveling effect of the multiple

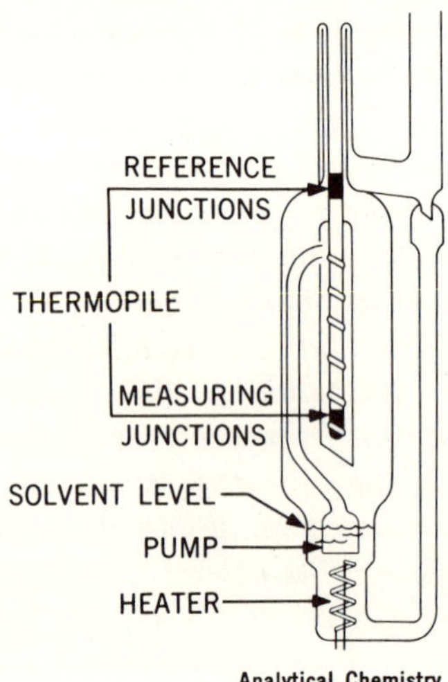

Figure 1. Ebulliometric apparatus (1)

temperature sensing points of the thermopile on rapid temperature fluctuations caused by uneven boiling or pumping. Thermistors have been used, with equal success, for temperature sensing by many workers. Also, devices such as the quartz crystal thermometer might be adapted to this application. Finally, the signal from the thermopile is amplified in this system by a Leeds and Northrup stabilized microvolt amplifier No. 9835A and is recorded on a strip chart recorder. Thermal stability is obtained with a vapor jacket. This jacket normally contains the solvent to be used in the determination and is covered with aluminum foil which serves as a light shield as well as a thermal barrier. The jacket is connected directly to the ebulliometer to prevent pressure differences, and the system is vented to ambient pressure through a surge tank. With a thermopile having 80 measuring junctions in the ebulliometer, temperature differences of 2×10^{-5} °C can be detected with satisfactory accuracy.

Operation of the system has been simplified to permit routine determinations by a laboratory technician with no active supervision. The procedure differs from those published by Dimbat and Stross (*6*) and others chiefly in that the ebulliometer is neither washed nor dried between determinations. However, after fresh solvent is added, the output of the thermopile ("zero line") for the solvent is recorded. Further, a decreasing heat input program is followed to eliminate the effects of foaming and superheating. Normally three or four weighed portions of the sample are added successively, and the elevation in boiling temperature is recorded for each portion. The molecular weight is obtained from these data by the limiting slopes calculation, or by any of the established calculation procedures (*6, 7*), with the aid of an IBM 1130 computer programmed to give the molecular weight and an "error" indication for each data point.

About 8000 determinations involving a range of solvents and solutes have been made with the system as described. The most recent and perhaps most interesting development in this respect is the use of hexafluoro-2-propanol (HFIP) as an ebulliometric solvent for the study of polyesters and polyamides. HFIP has an ebulliometric constant (K_b) of 3.0, compared with a value of 3.3 for toluene, and behaves satisfactorily as an ebulliometric solvent. Precision of these measurements varies with molecular weight level, solvent, and to some extent, solute. With a series of polyglycols, as shown in Table I, the standard deviation was about 1% at a molecular weight level of about 4100. For polyethylene, as shown in Table II, precision does vary with molecular weight and is obviously influenced by other factors, such as homogeneity and purity of the sample.

Table I. Precision in Molecular Weight Determination: Polyglycol in Toluene

Determination	*Molecular Weight*
1	4189
2	4138
3	4112
4	4112
5	4164
Mean	4143
Standard Deviation	36 or 0.85%

Table III shows some surprising results obtained from the NBS (National Bureau of Standards) polystyrene sample 705. This material has a very narrow molecular weight distribution; since it is a specially prepared

Table II. Precision in Molecular Weight Determination: Polyethylene in Toluene[a]

Av. Mol. Wt.	18,170
n	7
Standard Dev.	756 or 4.2%
Av. Mol. Wt.	24,620
n	5
Standard Dev.	1886 or 5.6%
Av. Mol. Wt.	34,280
n	5
Standard Dev.	942 or 2.7%

[a] Data from *Anal. Chem.* (1961) **31,** 449.

Table III. Precision in Molecular Weight Determination: Polystyrene in Toluene

NBS Reference Material 705	
Mol. Wt. Found	171,900
	168,400
	169,700
Av.	170,000
Standard Dev.	1756 or 1.03%

reference material, it probably is more homogeneous than most samples and thus gives more precise data.

Assessment of accuracy presents a problem since few authentic reference materials exist in the molecular weight range of interest. However, our results have been compared with those obtained by other laboratories and those obtained by other methods. One comparison is shown in Table IV. The accuracy of the method is also shown by the data obtained from the previously mentioned NBS sample. These data are shown in Table V.

Effect of Experimental Variables

In spite of the apparently successful performance of the ebulliometric system in routine use, a number of variables exist in connection with the operation of the system as it has been described. Many of these variables, as many as possible, are overcome by standardization in routine operation. However, they must be recognized if the ebulliometric method is to be understood and applied to special problems.

First, the use of the condensing solvent vapors as a reference temperature can lead to considerable uncertainty, particularly with nonpolymeric solutes, because changes in the reference temperature, which result from the presence of impurities in the solvent or the solute or from solute volatility, may go undetected. Such changes have actually been measured with special thermopile circuits and are real and significant. If a boiling liquid reference (twin ebulliometer) is used, the problem is less acute. An example is shown in Table VI; the results were obtained when both types of reference temperatures were used to determine the molecular weight of biphenyl (bp, 256°C) in toluene (bp, 110°C). These results may explain some of the confusion found in the literature concerning the required difference between the boiling temperatures of a solute and a solvent necessary for valid use of the ebulliometric method.

Table IV. Accuracy in Molecular Weight Determination: Polyethylene in Toluene[a]

	Ebulliometry				
Sample	*Lab A*	*Lab B*	*Lab C*	*Cryoscopy*	*Vapor Pressure Osmometry*
1	11,500	11,500	11,500	10,700	10,900
2	18,400	19,200	—	19,100	18,800

[a] "Advances in Analytical Chemistry and Instrumentation," Vol. 5, Table XII, Chapter 1, p. 63, Wiley, New York 1966.

Table V. Accuracy in Molecular Weight Determination: Polystyrene in Toluene

NBS Reference Material 705

Mol. Wt. (Elbulliometry)	170,000
Mol. Wt. (Osmometry—NBS)	170,900

Table VI. Molecular Weight of Biphenyl in Toluene

Reference Type	*Mol. Wt. Found*	*Theory*
Condensing Vapor	512	154
Boiling Liquid	171	154

The Cottrell-type pump presents an additional variable. With it the rate and the heat input to the boiling solution cannot be varied independently. For this reason, the possibility of superheating cannot be rigorously eliminated. To overcome superheating, a no-dead-space mechanical pump (Figure 2) was designed, constructed, and installed in an experimental ebulliometer as shown in Figure 3. The unique feature of this pump is the loose fit of the piston which permits continuous flow of liquid through the piston chamber during operation. The problems of superheating and equilibration are currently being reexamined with this apparatus in a twin configuration.

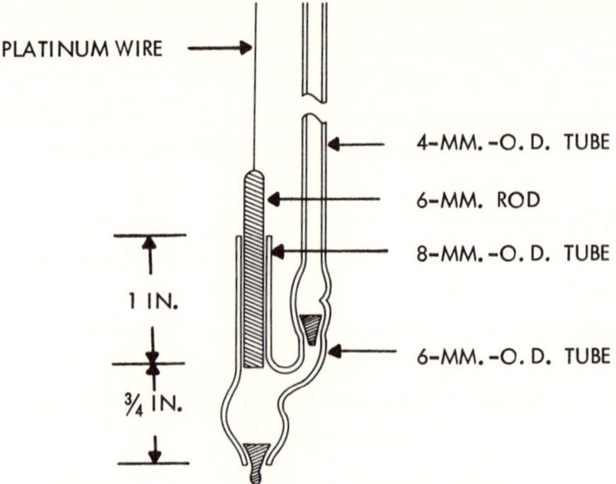

Figure 2. Mechanical pump for ebulliometer

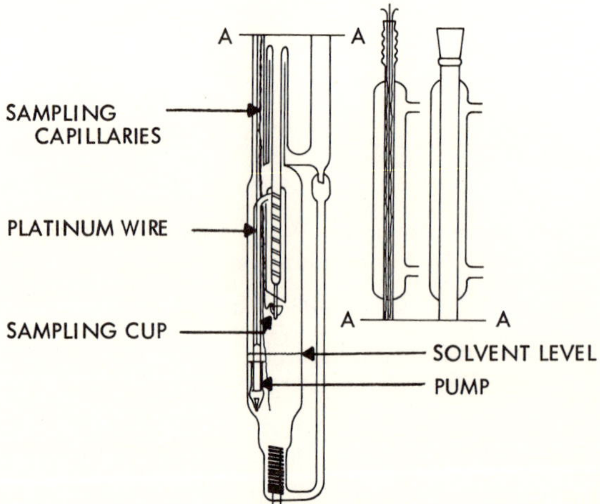

Figure 3. Experimental ebulliometer

A further need is to know the actual solute concentration at the temperature sensing point in the ebulliometer at operating equilibrium. Amounts of solvent and solute added to the ebulliometer can be accurately measured and controlled; however, their distribution within the apparatus at equilibrium is difficult to establish. As shown in Figure 3, provisions have been made for removing samples during operation. The problem of solvent–solute distribution and the accompanying problem of analysis are being studied, but so far the studies have been less than completely successful.

The last variable to be discussed is that of the experimentally determined value for the ebulliometric constant, K_b. Equation 1 indicates that theoretically ΔT_b varies with solute molecular weight to make K_b constant.

$$K_b = \frac{\text{Solute mol wt (grams)} \times \Delta T_b\,(^\circ\text{C}) \times \text{solvent wt (grams)}}{1000 \times \text{solute wt (grams)}} \quad (1)$$

In the work described here, this has not been found to be true. Considerable effort has been expended to eliminate this apparent inconsistency or to explain it as instrumentally induced. To date neither has been successful, and the observations remain unexplained. It appears from this work that the experimentally determined value of K_b is influenced by both the chemical nature and the molecular weight of the solute.

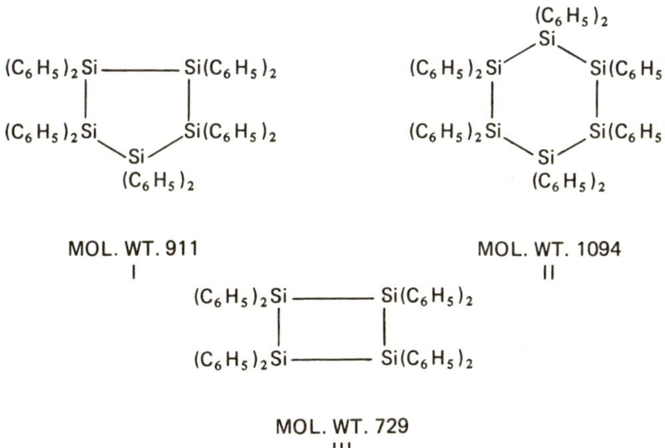

Figure 4. Organosilicon compounds

This is demonstrated by a problem in which an effort was being made to establish the structure of an unidentified organosilicon compound prepared by Dr. Gilman of Iowa State University. Other data and the history of the compound indicated that the structure was either I or II as shown in Figure 4. The boiling point elevation of the unknown compound was obtained in toluene with the apparatus shown in Figure 1. A K_b which had been established with tristearin (molecular weight 891) was used to calculate the molecular weight of the unknown, and an inconclusive value of 973 was obtained. A new K_b was then established with an authentic compound of structure III, Figure 4. The new K_b was used to recalculate the molecular weight of the unknown, and a value of 921 was obtained. The structure was later confirmed by x-ray diffraction as I, Figure 4.

This paper has attempted to show that the ebulliometric method can be used successfully for the routine determination of molecular weights and in special problems. It still presents some worthy challenges and some opportunities of a theoretical nature.

Literature Cited

1. Glover, C. A., Stanley, R. R., *Anal. Chem.* (1961) **33,** 477.
2. Ray, N. H., *Trans. Faraday Soc.* (1952) **48,** 809.
3. Lehrle, R. S., Majury, T. G., *J. Polym. Sci.* (1958) **29,** 219.
4. Schön, K. G., Schultz, G. V., *Z. Physik. Chem. (Frankfurt am Main)* (1954) **2,** 197.
5. Ezrin, M., Eastern Analytical Symposium, Symposium on Molecular Weight Measurements, New York, 1962.
6. Dimbat, M., Stross, F. H., *Anal. Chem.* (1957) **29,** 1517.
7. Lehrle, R. S., "Progress in High Polymers," Robb, J. C., Peaker, F. W., Eds., pp. 57–61, 1961, Academic Press, New York.

Received January 17, 1972.

2

Application of the Summative-Fractionation Method to the Determination of $\overline{M}_w/\overline{M}_n$ for Narrow-Distribution Polymers

FRED W. BILLMEYER, JR. and LEONARD R. SIEBERT[a]

Rensselaer Polytechnic Institute, Troy, N. Y. 12181

The summative-fractionation method was extended to apply to narrow-distribution polymers with polydispersity ($\overline{M}_w/\overline{M}_n$) *less than 1.12. A fractionation parameter* H, *previously defined and calculated for theoretical molecular weight distributions for normal polymers, was computed for narrow-distribution polymers. The calculations were made both with and without correction for fractionation errors, using the Flory–Huggins treatment. The method was applied to a well-characterized anionic polystyrene with* $\overline{M}_w$ = *97,000, for which the polydispersity was estimated by this technique to be 1.02 (in the range 1.014–1.027, 95% confidence limits).*

Precise measurement of some parameter describing the breadth of the molecular weight distribution (MWD) has become particularly important for narrow-distribution polymers in recent years. The commercial availability (*1*) of narrow-distribution anionic polystyrenes (*2*, *3*) has provided valuable standards for the calibration of gel permeation chromatography (GPC) and other characterization techniques. Although it is generally accepted that these polymers have approximately the Poisson distribution (*4*) of molecular weights, the measurement of the breadth of their MWD remains a difficult task.

In terms of the ratio $\overline{M}_w/\overline{M}_n$, hereafter called the polydispersity, taken as the common measure of distribution breadth, these polymers usually have nominal polydispersities below 1.05 or so. Classical methods of determining polydispersity are not accurate in this range. In the separate determinations of $\overline{M}_w$ and $\overline{M}_n$, combined experimental errors are

[a] Present address: Rochester Products Division, General Motors Corp., Rochester, N. Y.

estimated to range between ±5% and ±15%, with the smaller figure likely too optimistic in our opinion. Fractionation by conventional solubility-based techniques and ultracentrifugation is likewise subject to significant uncertainties in absolute results and is relatively tedious, difficult, or expensive to perform. While one might have thought that GPC would provide an easy solution to the problem, the errors derived from peak broadening (*5*) are comparable with the broadening caused by polydispersity in this range; in fact, accurate independent knowledge of the polydispersity of these calibrating standards would greatly alleviate the difficulty of correcting GPC results (*6*). Only the recycle technique proposed by Waters (*7*) but not yet widely adopted offers promise of overcoming this problem.

For these reasons, we sought a fresh approach to this problem through the application of the summative fractionation technique as described by Billmeyer and Stockmayer in 1950 (*8*). (In this paper we follow a conservative treatment (*8*), feeling that less conservative treatments (*9*) can easily lead to overinterpretation of the data.) This method is relatively rapid and inexpensive and would seem to be particularly suitable for determining polydispersities in the range of interest since the experimentally determined parameter H experiences its greatest fractional change per unit change in polydispersity as the polydispersity approaches unity. Countering this advantage are increased experimental error in H at low polydispersities and the importance of the correction for imperfect fractionation. This paper presents a progress report on the application of the summative technique to determining the polydispersity of narrow-distribution anionic polystyrenes.

Theory

The Summative-Fractionation Method. To review briefly (*8*), in the summative-fractionation method one performs a series of small-scale single-step fractional precipitations in which increasing percentages of the total polymer are precipitated. The weight fractions x and average molecular weights $\overline{M}_w$ of all the precipitates are determined. One calculates and plots against x a parameter

$$z = x(\overline{M}_x - \overline{M}_w)/\overline{M}_w = (1/\overline{M}_w)\int_0^\infty (M - \overline{M}_w)f(M)w(M)dM \qquad (1)$$

in which $w(M)$ is the MWD of the sample and $f(M)$ is the fraction of the polymer of molecular weight M appearing in the precipitate. For an infinitely sharp fraction, z is always zero while for a polydisperse polymer z is finite and positive (except at $x = 0$ and $x = 1$, where $z = 0$) and is a measure of the MWD.

If fractionation were perfect, $f(M)$ would be zero below some value of M representing the transition between supernatant and precipitate and unity for all higher values of M. It is shown (*8*) that the most readily determined value of the $z(x)$ curve is its maximum, H, occurring at the point where this transition value of M is $\overline{M}_w$. Adopting the use of an asterisk to denote quantities calculated with the above assumption of perfect fractionation

$$H^* = (1/\overline{M}_w) \int_{\overline{M}_w}^{\infty} (M - \overline{M}_w) w(M)\, dM \tag{2}$$

$$H = (1/\overline{M}_w) \int_{\overline{M}_w}^{\infty} (M - \overline{M}_w) f(M) W(M)\, dM \tag{3}$$

If the form of $w(M)$ is known, H^* can be related explicitly to the polydispersity $\overline{M}_w/\overline{M}_n$. It was shown that the shape of the $z^*(x^*)$ curve is virtually independent of the nature of the distribution, so that no further information about the MWD, beyond the polydispersity, can be obtained from the experiment. The magnitude of the error from imperfect fractionation, expressed as the difference between H^* and H, was also explored and shown to be small for polydispersities of the order of 2 or greater.

The Poisson Distribution. We have extended the earlier work (*8*) to obtain the relation between H^* and $\overline{M}_w/\overline{M}_n$ for the Poisson distribution.

$$w(M) = \frac{u}{u+1} \frac{Me^{-u}\, u^{M-2}}{(M-1)!} \tag{4}$$

$$\overline{M}_w/\overline{M}_n = 1 + \frac{u}{(u+1)^2} \tag{5}$$

The calculations were performed by numerical integration, after replacing the factorial by a lower end Stirling approximation

$$M! = (2\pi M)^{1/2}(M/e)^M \tag{6}$$

which is within 1% of the true value at $M = 10$ and closer for all higher values of M. The results are summarized in Table I and are plotted in Figure 1. The calculations were carried only up to a polydispersity of 1.122, corresponding to $u = 6$; lower values of u are not realistic since any sample with this high polydispersity can probably be described better by a distribution other than the Poisson.

Other Distributions. We have recalculated earlier results (*8*) for several distributions useful for describing samples with higher polydispersity. These include the Schulz–Zimm form of the familiar exponential distribution (*10, 11*)

$$w(M)dM = \frac{p^n e^{-p} dp}{\Gamma(n+1)} \qquad p = (n+1)M/\overline{M}_w \qquad n > 0 \tag{7}$$

Table I. Summative-Fractionation Parameter *H* for the Poisson Distribution

$\overline{M}_w/\overline{M}_n$	H*	$H_{0.001}$	$H_{0.01}$
1.001	0.0126	0.0071	0.0068
1.002	0.0178	0.0105	0.0100
1.003	0.0220	0.0135	0.0127
1.005	0.0280	0.0179	0.0167
1.008	0.0352	0.0236	0.0218
1.010	0.0392	0.0270	0.0248
1.014	0.0465	0.0333	0.0304
1.019	0.0546	0.0405	0.0368
1.024	0.0605	0.0460	0.0418
1.031	0.0691	0.0541	0.0490
1.045	0.0827	0.0673	0.0611
1.083	0.1071	0.0918	0.0840
1.099	0.1165	0.1013	0.0928
1.122	0.1291	0.1139	0.1048

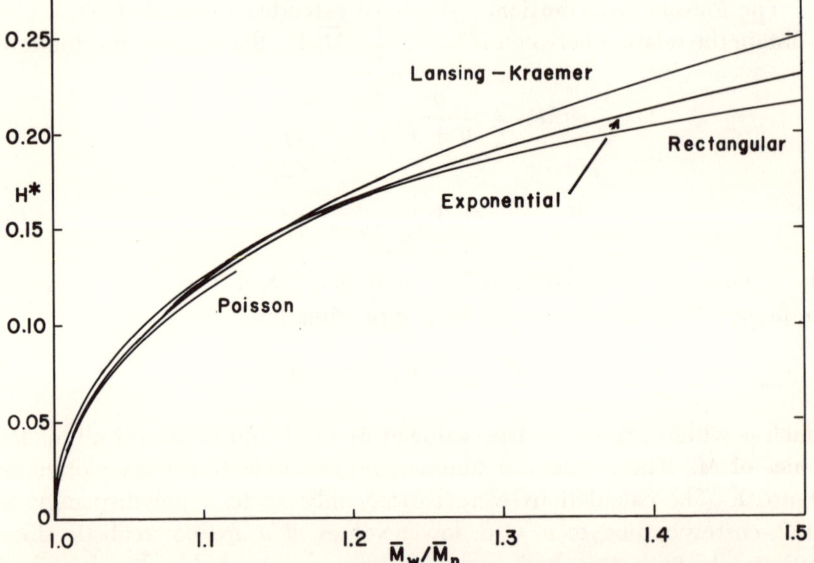

Figure 1. The summative-fractionation parameter H* *(ideal-fractionation assumption) as a function of polydispersity* $\overline{M}_w/\overline{M}_n$ *for the Poisson, Schulz–Zimm exponential, Lansing–Kraemer logarithmic-normal, and rectangular box distributions*

and the Lansing–Kraemer logarithmic normal distribution (*12*)

$$w(M) = Ke^{-y^2} \qquad y = (1/\beta) \ln (M/M_0) \tag{8}$$
$$1/K = \beta M_0(\pi)^{1/2} e^{\beta^2/4}$$

In addition, to demonstrate the lack of dependence of H^* on the form of the distribution function, we included the calculation for a highly artificial rectangular box distribution

$$w(M) = 1/(2a\overline{M}_w) \text{ for } \overline{M}_w (1 - a) < M < \overline{M}_w (1 + a) \tag{9}$$
$$w(M) = 0 \text{ elsewhere}$$

The results of these recalculations are listed in Table II and plotted in Figure 1. The curves for all the distributions studied differ insignificantly at polydispersities below, say, 1.2.

Table II. Summative-Fractionation Parameter *H* for Other Distribution Functions

Exponential		*Lansing-Kraemer*		*Rectangular*	
$\overline{M}_w/\overline{M}_n$	H*	$\overline{M}_w/\overline{M}_n$	H*	$\overline{M}_w/\overline{M}_n$	H*
1.001	0.0126	1.005	0.0282	1.001	0.015
1.002	0.0178	1.020	0.0564	1.002	0.020
1.005	0.0281	1.046	0.0846	1.005	0.030
1.010	0.0397	1.083	0.113	1.015	0.052
1.020	0.0599	1.133	0.140	1.020	0.060
1.050	0.0871	1.197	0.168	1.050	0.0925
1.10	0.120	1.277	0.196	1.10	0.126
1.20	0.163	1.377	0.223	1.20	0.165
1.33	0.199	1.499	0.250	1.33	0.194
1.50	0.230			1.50	0.215

Fractionation Error. The effect of imperfect fractionation was studied by integrating Equation 3 numerically with $f(M)$ given by the Flory–Huggins theory

$$f(M) = Re^{aM}/(1 + Re^{aM}) \tag{10}$$

where R is the ratio of the volumes of precipitate and solution, and a can be eliminated by noting that at $M = \overline{M}_w$, $f(M) = 0.5$ when H is calculated. These calculations were performed with the Poisson distribution for $w(M)$ and for $R = 0.01$ and $R = 0.001$, covering the range expected in application of the summative-fractionation technique. These results are also listed in Table I and are plotted together with values of H^* for the Poisson distribution in Figure 2. The differences, though not insignificant, are relatively small and decrease with decreasing polydispersity.

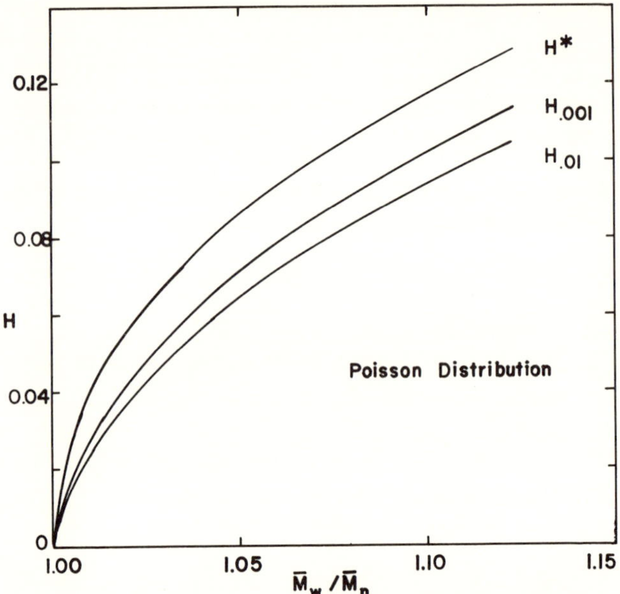

Figure 2. The summative-fractionation parameter H *as a function of polydispersity for the Poisson distribution.* H* *is calculated for the ideal-fractionation assumption while* $H_{0.001}$ *and* $H_{0.01}$ *are calculated including the Flory–Huggins correction for imperfect fractionation, with the volume ratio* R *equal to 0.001 and 0.01, respectively.*

Experimental

The summative technique was applied to the fractionation of a polystyrene standard, catalog No. 500-16, molecular weight 97,200, ArRo Laboratories. This sample was prepared by Pressure Chemicals Co. and is identified by them as Polymer 4A. As measured at the Mellon Institute and reported by Pressure Chemicals it is characterized as follows: light-scattering $\overline{M}_w = 96{,}200 \pm 2\%$; osmometry $\overline{M}_n = 97{,}600 \pm 5\%$; viscosity ($\theta$ cyclohexane) $\overline{M}_v = 98{,}200 \pm 3\%$; MWD by 20-cut fractionation, including maximum opposite error (plus for $\overline{M}_w$ and minus for $\overline{M}_n$) $<$ 1.06:1. The fractionation was carried out in general as described in Ref. 8. The solvent was benzene (reagent, redistilled), and the precipitant was hexane with mixed isomers (Fisher H-291, ACS reagent grade, bp, 68.7°C). The fractionation flasks were roughly conical with ground glass stoppers and an 8 × 60 mm tube extending down from the narrow end into which the precipitated phase settled. Capacity was approximately 40 ml. Polymer concentration at precipitation was about 0.07 gram per liter. Final equilibration was for 24 hr at 25.0°C. The supernatant phase was drawn off with a hypodermic syringe, and the polymer was recovered from both phases by freeze drying. Molecular weights were determined from intrinsic viscosities in benzene using the relation

$$[\eta] = 8.5 \times 10^{-5} \overline{M}_v{}^{0.75} \quad (11)$$

for benzene at 25°C (*3*).

In replicate series of experiments, the following values of *H* were obtained: 0.0425, 0.0423, 0.0415. The corresponding values of polydispersity are in the range 1.014–1.027, or an average value of $\overline{M}_w/\overline{M}_n = 1.02$.

Table III. Upper and Lower 95% Confidence Limits (C.L.) on *H* and on the Polydispersity

H	*Lower C. L.*	*Upper C. L.*
0.028	0.021	0.036
0.040	0.032	0.048
0.056	0.047	0.065
0.078	0.068	0.089
0.087	0.076	0.098
$\overline{M}_w/\overline{M}_n$	*Lower C. L.*	*Upper C. L.*
1.005	1.0029	1.0080
1.01	1.0065	1.0146
1.02	1.014	1.027
1.05	1.038	1.065
1.10	1.078	1.13

Independent of these results, an error analysis was carried out using data obtained for the reproducibility of the precipitation step, determining $\overline{M}_x$. The results, summarized in Table III, were calculated using the exponential MWD, but should apply equally well to the Poisson distribution at these low values of polydispersity. The experimental results fall close to the calculated 95% confidence limits for *H* at the appropriate level.

Discussion

The argument for using the summative-distribution method for determining the polydispersity of narrow-distribution polymers is far from complete, but it has been demonstrated for a single sample that a result with usable precision can be obtained. In fact, no other method known to us combines this order of precision with low cost and short time of the analyses. More work is needed, however, in several areas, including (1) improvement of experimental techniques, (2) application to samples at different levels of molecular weight (clearly, poorer precision can be expected for high molecular weight samples), and (3) comparison with recycle GPC results (*7*) on the same sample.

Acknowledgments

This research was taken in part from the M.Sc. Thesis of L. R. Siebert, Materials Division, Rensselaer Polytechnic Institute. The work was supported by the General Motors Institute and carried out in Rensselaer's Materials Research Center, a facility supported by the National Aeronautics and Space Administration. We thank Howard Loy for experimental assistance, ArRo Laboratories for the donation of the standard polystyrene, and Robert T. Marcus for assistance in programming the numerical integrations for digital computer solution.

Literature Cited

1. ArRo Laboratories, Joliet, Ill.; Pressure Chemicals Co., Pittsburgh, Pa.; Waters Associates, Inc., Framingham, Mass.
2. Altares, T., Jr., Wyman, D. P., Allen, V. R., *J. Polym. Sci., Part A* (1964) **2,** 4533.
3. Wyman, D. P., Elyash, L. J., Frazer, W. J., *J. Polym. Sci., Part A* (1965) **3,** 681.
4. Flory, P. J., *J. Amer. Chem. Soc.* (1940) **62,** 1561.
5. Kelley, R. N., Billmeyer, F. W., Jr., *Sep. Sci.* (1970) **5,** 291.
6. Duerksen, J. H., Hamilec, A. E., *J. Polym. Sci., Part C* (1968) **21,** 83.
7. Waters, J. L., *J. Polym. Sci., Part A-2* (1970) **8,** 411.
8. Billmeyer, F. W., Jr., Stockmayer, W. H., *J. Polym. Sci.* (1950) **5,** 121.
9. Coppick, S., Battista, O. A., Lytton, M. R., *Ind. Eng. Chem.* (1950) **42,** 2533; *see also* Battista, O. A., *in* "Polymer Fractionation," Chapter C.4, Academic, New York, 1967.
10. Schulz, G. V., *Z. Phys. Chem., Abt. B* (1939) **44,** 227.
11. Zimm, B. H., *J. Chem. Phys.* (1948) **16,** 1099.
12. Lansing, W. D., Kraemer, E. O., *J. Amer. Chem. Soc.* (1935) **57,** 1369.

RECEIVED January 17, 1972.

3

The Polymer Standard Reference Materials Program at the National Bureau of Standards

HERMAN L. WAGNER

Institute for Materials Research, National Bureau of Standards,
Washington, D. C. 20234

The National Bureau of Standards now distributes four polymer Standard Reference Materials designed for use in the calibration of instruments employed in polymer characterization. Polystyrene is available in narrow (SRM 705) and broad (SRM 706) distributions and polyethylene in high-density linear (SRM 1475) and low-density branched (SRM 1476) whole polymer. These materials are characterized with respect to many but not necessarily all of the following properties: weight and number-average molecular weight, limiting viscosity number in several solvents, ASTM density, and ASTM melt flow rate. In addition the molecular-weight distribution of the linear polyethylene is given, making it suitable for the calibration of gel-permeation chromatographs at high temperatures.

The National Bureau of Standards has for more than 60 years provided Standard Reference Materials which have assumed a vital role in many areas of production, commerce, and research. These samples, which provide a universal basis for comparison and standardization of many material properties, are characterized with great care before they are certified and released for use. They also have the advantage of being continuously available, unchanged, for long periods of time. In 1963 the first polymer Standard Reference Materials were issued, polystyrenes SRM 705 and SRM 706.

The need for improving the reliability of characterization of polymers has long been evident. Other efforts in this direction include the round-robin testing conducted at various times, beginning in 1950, under the auspices of the Commission of Macromolecules of IUPAC (*1, 2, 3*), as well as the work now in progress in ASTM concerned with establishing standardized methods of characterization.

A recent survey has shown that our first polymer standard samples are widely distributed and used in research and industry for the calibration of a variety of characterization instruments, particularly gel-permeation chromatographs. They also serve as materials with well defined properties for research in many areas. These properties should become better defined with time as results accrue in the literature.

Polystyrene Standard Reference Materials

The choice of the first polymer Standard Reference Material (*4*) was governed by some fairly simple requirements—that it be reasonably stable, that it be soluble in solvents ordinarily used in characterization work, and that it be readily obtained in the molecular weight range from 40,000 to 500,000. This is the range most readily measured by conventional molecular-weight techniques. Polystyrene easily meets these criteria and is of special importance because it is probably one of the most studied amorphous polymers. In addition, an attractive feature of polystyrene is the fact that it can be polymerized anionically to give polymers of very narrow molecular-weight distribution. This is particularly important in characterization work because of the problems caused by diffusion of the low molecular-weight species of broad molecular-weight-distribution polymers across the semipermeable membranes during osmotic pressure measurements.

SRM 705 is the narrow molecular-weight-distribution sample; the other SRM 706, by contrast, is a broad molecular-weight-distribution polymer, the result of a thermal polymerization which more closely resembles commercially available polymers. Both of these were supplied by the Dow Chemical Co. (Certain commercial equipment, instruments, or materials are identified in this paper in order to specify adequately the experimental procedure. In no case does such identification imply recommendation or endorsement by the National Bureau of Standards, nor does it imply that the material or equipment identified is necessarily the best available for the purpose.)

Because sample homogeneity is particularly important for a Standard Reference Material, it was carefully assessed for both polystyrenes using solution viscosity, a measure of molecular weight, as an index. There is essentially no variation with location within the lot, or from pellet to pellet, within the limits of error of the viscosity measurements. Viscosity measurements may be made with a standard deviation of a single determination of about 0.3%.

The details of the measurements made on these samples are discussed elsewhere (*4, 5*). The certified values from the certificates for all the samples are shown in the Appendix. For SRM 705 the number-average molecular weight by osmometry, 170,900, and the weight-average molecular weight by light scattering, 179,300, are, as expected, very close to each other, giving an $\overline{M}_w/\overline{M}_n$ of 1.05. The value of $\overline{M}_z/\overline{M}_w/\overline{M}_n$ was determined by fractionation into 36 fractions giving an $\overline{M}_w/\overline{M}_n$ of 1.07. Weight-average molecular weight by sedimentation equilibrium was about 190,000. Since both the light-scattering and the sedimentation-equilibrium results are based on completely separate absolute calibrations, the 6% difference between the two values is small and certainly well within the expected range of uncertainty for such measurements, which is usually considered to be ±10% at best. In addition to the molecular-weight values, limiting viscosity numbers in benzene at 25° and 35°C as well as in cyclohexane at 35°C are certified.

For SRM 706, the broad molecular-weight-distribution polystyrene, the certificate provides weight-average molecular weight both by light scattering and sedimentation equilibrium, with the latter method again showing the higher value. Limiting viscosity numbers in benzene at 25°C and in cyclohexane at 35°C are given as well. Number-average molecular weight by osmometry could not be determined because of the diffusion problems referred to above. However, from the ratio of molecular weights found by fractionation into 44 fractions, $\overline{M}_z/\overline{M}_w/\overline{M}_n$ = 2.9/2.1/1.0, and the number average is estimated to be 123,000.

Weight- and number-average molecular weights were recently redetermined for SRM 705. No significant difference could be found between the recent and earlier measurements. The same holds true for the viscosity numbers. The results of this more recent work are given in a Special Publication available from the National Bureau of Standards (5).

Polyethylene Standard Reference Polymers

The next polymer Standard Reference Materials, issued in 1970, were a linear and a branched polyethylene. They are representative of crystalline olefin polymers, which have assumed great commercial and scientific importance. The linear material was kindly provided by the Dupont Co. and the branched by Union Carbide Corp.

Both samples are in the form of pellets containing antioxidants as specified on the certificates. They have also been examined for sample homogeneity by dilute-solution viscosity. In the case of the linear material, SRM 1475, although the lot is uniform with respect to location, a measurable pellet-to-pellet variation exists. To make certain that a uniform sample is obtained, at least one gram of pellets should be blended to reduce the expectation of error from pellet variability to less than 0.5%. The branched material on the other hand is quite uniform and does not show pellet-to-pellet variation. Both materials have low ash and volatile content.

There is no evidence of branching in the linear material, SRM 1475. Infrared measurements show that there is about 0.15 methyl group for 100 carbon atoms. This is consistent with the value expected for a number-average molecular weight of 18,000, assuming two methyl end groups per chain and no short-chain branching. The absence of long-chain branching is indicated by the linearity of the log limiting viscosity number–log molecular-weight relationship for the fractions (*6*).

The weight-average molecular weight of SRM 1475 determined by light scattering in 1-chloronaphthalene at 135°C is 52,000 with an estimated standard deviation of about 4%. Again, number-average molecular weight could not be determined by osmometry for this whole polymer. It was possible, however, to determine the number-average molecular weight and, in fact, the entire distribution by gel-permeation chromatography. The columns of the gel-permeation chromatograph were first calibrated with 9 fractions of polyethylene obtained by column elution. The number- and weight-average molecular weights were determined by osmometry and light-scattering measurements, respectively, and ranged in weight-average molecular weight from 19,000 to 688,000. Two zone-refined linear hydrocarbons, C_{94} and C_{36}, were also employed to help anchor the lower end of the calibration curve, which is shown in Figure 1. The distribution is given in tabular form on the certificate (*see* Appendix). Hence this linear whole polymer may be conveniently used for the calibration of gel-permeation chromatographs without the necessity of resorting to a whole series of narrow-distribution polymers in the range

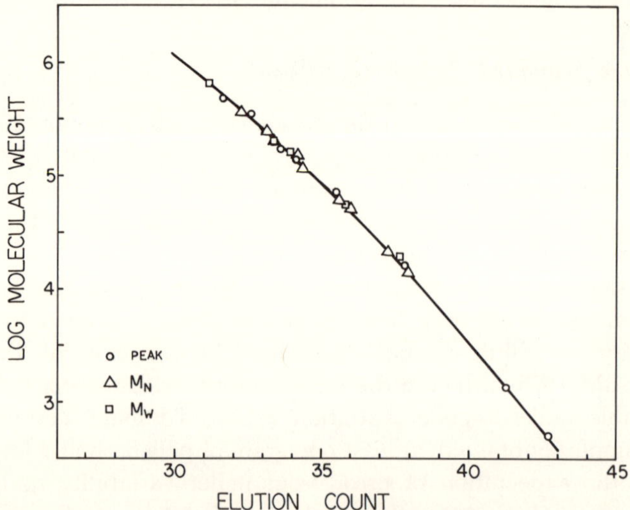

Figure 1. Calibration curve for the gel-permeation chromatographic tracing obtained with polyethylene fractions, and n-$C_{94}H_{190}$ *and* n-$C_{36}H_{74}$

of a thousand to a million. This is the range of molecular weights most readily measured by conventional methods.

The certificate also gives limiting viscosity numbers in 1-chloronaphthalene, 1,2,4-trichlorobenzene, and decahydronaphthalene. Density and melt flow viscosity under specified ASTM conditions are also included. Hence SRM 1475 may be utilized in the calibration of a variety of instruments used for characterizing polyethylene. A series of papers describing the methods used to characterize this sample have been published (7).

The low density-branched polyethylene, SRM 1476, has been characterized with respect to limiting viscosity number in the same solvents as the linear material, namely 1-chloronaphthalene, 1,2,4-trichlorobenzene, and decahydronaphthalene. Melt index and density were also determined using ASTM procedures indicated in the certificate. Because of the complications arising from branched polyethylene in light scattering and gel-permeation chromatography, meaningful molecular weights could not be obtained by these methods for the whole polymer. The material has been passed through an elution column and the fractions are being characterized for molecular weight and branching.

In addition to these studies on the branched polyethylene, fractions of linear polyethylene prepared by large-scale gel-permeation chromatography are being characterized for certification in the near future. These should be useful for gel-permeation chromatography calibration. We also expect them to be particularly valuable in dilute solution, crystallization, and rheological studies.

Appendix

Standard Reference Material 705, Polystyrene (Narrow Molecular-Weight Distribution)

Specification	*No. of Determinations*	*Average*	*Standard Deviation of Average*
Number-average molecular weight (measured by osmotic pressure)	12	170,900	580
Weight-average molecular weight (measured by light scattering)	9	179,300	740
Weight-average molecular weight (measured by sedimentation equilibrium)	22	189,800	2,100
Limiting viscosity number (ml/gram) (intrinsic viscosity)			
Benzene, 25°C	5	74.4	0.18
Benzene, 35°C	13	74.5	0.23
Cyclohexane, 35°C	6	35.4	0.24
Ratios of molecular weight (based on fractionation)	$\overline{M}_z/\overline{M}_w/\overline{M}_n = 1.12{:}1.07{:}1$		

Standard Reference Material 706, Polystyrene (Broad Molecular-Weight Distribution)

Specification	*No. of Determinations*	*Average*	*Standard Deviation of Average*
Weight-average molecular weight (measured by light scattering)	12	257,800	930
Weight-average molecular weight (measured by sedimentation equilibrium)	4	288,100	9.600
Limiting viscosity number (ml/gram) (intrinsic viscosity)			
Benzene, 25°C	17	93.7	0.19
Cyclohexane, 35°C	4	39.5	0.10
Ratios of molecular weight (based on fractionation)	$\overline{M}_z/\overline{M}_w/\overline{M}_n = 2.9{:}2.1{:}1$		

Standard Reference Material 1475, Linear Polyethylene (Whole Polymer)

Specification	*Average Value*	*Estimated Standard Deviation of Average*
Molecular weight		
Weight-average molecular weight[a]	52,000	2,000
Number-average molecular weight	18,310	360
Weight-average molecular weight	53,070	620
Z-average molecular weight	138,000	3,700
Ratio of molecular weights $\overline{M}_z/\overline{M}_w/\overline{M}_n$	7.54:2.90:1	
	See Table I below	
Limiting viscosity number (dl/gram)		
1-Chloronaphthalene, 130°C	0.890	0.0032
1,2,4-Trichlorobenzene 130°C	1,010	0.0086
Decahydronaphthalene,[c] 130°C	1.180	0.0032
Melt flow rate (gram/10 min)[d]	2.07	0.0062
Density (gram/cm³)[e]	0.97844	0.00004

[a] By light scattering in 1-chloronaphthalene at 135°C.

[b] By gel-permeation chromatography.

[c] "Technical" grade, which assayed at approximately equal proportions of *cis*- and *trans*-decahydronaphthalenes.

[a] By a procedure similar to Procedure A, ASTM Method D1238-65T, Test Condition D, 190°C, load 325 gram.

[e] By ASTM Method D1505-67; sample prepared by Procedure A, ASTM Method D1928-68.

Table I. Cumulative Molecular Weight Distribution by Gel-Permeation Chromatography

log M	*Wt. %*	*log M*	*Wt. %*	*log M*	*Wt. %*
2.800	0.0	4.014	15.2	5.065	90.7
2.865	0.005	4.070	18.1	5.113	92.2
2.929	0.020	4.126	21.5	5.161	93.7
2.992	0.052	4.182	25.2	5.209	94.8
3.056	0.105	4.237	29.3	5,256	95.8
3.119	0.185	4.292	33.7	5,303	96.6
3.181	0.343	4,346	38.5	5.349	97.3
3.243	0.475	4.400	43.4	5.395	97.9
3.305	0.706	4.454	48.5	5.440	98.4
3.366	0.999	4.507	53.5	5.485	98.7
3,427	1.38	4.560	58.3	5.530	99.1
3.488	1.88	4.612	62.9	5.574	99.3
3.548	2.51	4.664	67.3	5.618	99.5
3.607	3.30	4.715	71.4	5.662	99.7
3.667	4.28	4.766	75.1	5.705	99.8
3.725	5.46	4.817	78.5	5.789	99.9
3.784	6.87	4.868	81.6	5.87	100.0
3.842	8.56	4.918	84.4		
3.900	10.50	4.967	86.7		
3.957	12.7	5.016	88.9		

Standard Reference Material 1476, Branched Polyethylene (Whole Polymer)

Specification	*Average Value*	*Estimate of Precision*	*No. of Points or Measurements*
Limiting viscosity number (dl/gram)			
1-Chloronaphthalene, 130°C	0.8132	0.0033 [d]	14
1,2,4-Trichlorobenzene, 130°C	0.9024	0.0034 [d]	30
Decalin,[a] 130°C	1.042	0.0022 [d]	5
Melt index (gram/10 min) [b]	1.19	0.010 [e]	35
Density (gram/cm^3) at 23°C [c]	0.9312	0.0006 [e]	10

[a] "Technical" grade, a mixture of *cis-* and *trans*-decahydronaphthalene.
[b] By Procedure A, ASTM Method D1238-65T, Test Condition E, 190°C, load 2160 gram.
[c] By ASTM Method D1505-67T; Sample prepared by Procedure A, ASTM Method D1928-68.
[d] Standard deviation of the intercept of the least-squares line.
[e] Standard deviation of a single determination.

Literature Cited

1. Frank, H. P., Mark, H., *J. Polym. Sci.* (1953) **10,** 129.
2. Atlas, S. M., Mark, H. P., Report, presented to the Commission on Macromolecules, International Union of Pure and Applied Chemistry, Montreal, July 28, 1961.

3. Strazielle, C., Benoit, H., *Rept. Pure Appl. Chem.* (1971) **26**, 453.
4. McIntyre, D. J., *J. Res. Nat. Bur. Stand.* (1967) **71A,** 43.
5. Wagner, H. L., *Nat. Bur. Stand. Special Publ.* **260-33** (1972).
6. Wagner, H. L., Hoeve, C. A. J., to be published.
7. Hoeve, C. A. J., Wagner, H. L., Verdier, P. H., and the following papers, *J. Res. Nat. Bur. Stand.* (1972) **76A,** 137. Also available as *Nat. Bur. Stand. Special Publ.* **260-42.**

RECEIVED January 17, 1972.

4

Self-Beat Spectroscopy and Molecular Weight

N. C. FORD, JR., R. GABLER, and F. E. KARASZ

University of Massachusetts, Amherst, Mass. 01002

Self-beat spectroscopy is a newly developed optical technique in which the spectral distribution of light scattered from solute molecules undergoing Brownian motion is analyzed to allow calculation of the diffusion coefficient of the molecules. A brief review of the literature shows that this technique has been applied to many types of solutions containing molecules whose molecular weights range from 10^4 to 10^8. Using proper experimental procedure and solutions in thermal equilibrium, diffusion coefficients can be obtained in several minutes with accuracies typically on the order of 1%. Diffusion coefficients can be related to molecular weights by the Stokes–Einstein equation, by combination with sedimentation velocity data, or by the consideration of homologous polymer solutions.

The interaction of light with matter has long provided important clues into the mysteries of molecular structure and behavior. The information obtained from such interactions is almost as varied as are the different parts of the electromagnetic spectrum itself. The phenomena of light scattering, absorption, fluorescence, rotation, reflection, and refraction have been used to add to our knowledge of molecular systems. Techniques utilizing these phenomena can yield information on a gross level in which parameters such as the general shape, conformation, and molecular weight are important; they can also help elucidate details of microstructure in which the nature and strengths of chemical and physical forces are important. Each technique gives a unique type of information about the system being studied because of the nature of the interaction process with it. In this paper we describe a recently developed technique utilizing the interaction of light with matter—*i.e.*, self-beat spectroscopy. In particular, we consider the light scattered from a solution of molecules which are in a state of macroscopic equilibrium and show that diffusion coefficients of these molecules may be readily obtained by studying the spectral distribution of this scattered light.

When light traverses a perfectly homogeneous transparent solution, it will not be scattered although it may be refracted at the entrance and exit interfaces. If, however, the solution contains inhomogeneities, which all real solutions do because of thermal excitations, the incident light will also be scattered. Lord Rayleigh (*1*) was the first to obtain a quantitative expression relating the intensity of the scattered light of incident wavelength λ_0 to the scattering angle θ. His results pertained to dilute gases containing molecules small compared with λ_0, and his work also provided a theoretical explanation for the blue color of the sky. (It has also been reported (*2*) that Leonardo da Vinci around 1500 thought the color of the sky was caused by scattered light.) In 1935, Putzeys and Brosteaux (*3*) reasoned that Rayleigh's equations would also describe the excess light scattered by a dilute solution of macromolecules over that of the solvent, and they used this fact to examine the molecular weights of several small proteins. Thus began the use of light scattering as an analytical tool in physics and chemistry. Analysis also showed that the light scattered from particles whose dimensions were on the order of the wavelength of the incident light contained information regarding their size and shape (*4–6*). This result becomes plausible when one considers that the light scattered from different parts of a particle with a fixed geometry will interfere constructively and destructively in a manner dependent on that geometry, thus giving rise to an angular distribution of intensity. Zimm (*7*) utilized these concepts to develop a technique which would give both the molecular weight of a scattering molecule and its radius of gyration R_G. This method has since become a standard way to determine molecular weights.

However, Rayleigh's work failed to describe adequately the effects of scattering from concentrated solutions. Experimentally, one observes a diminution of scattering intensity from the expected Rayleigh intensity as the solution concentration increases. This is caused by stronger interactions between scattering molecules and a loss of the spatial domain in which the molecules may move. Concentrated solutions have neighboring molecules which scatter light whose phases are related, and the Rayleigh assumption of independent (or incoherent) scatterers is no longer valid. A theory describing all situations was advanced by Einstein (*8*) who asserted that the scattering was caused by fluctuations in the local dielectric constant. Debye (*9, 10*) then interpreted these fluctuations of the dielectric constant in solutions to be caused by fluctuations in the local concentration of the scattering molecules; it is this approach which has laid the theoretical foundation for macromolecular light-scattering experiments. The classical technique of light scattering, where the scattered intensity is measured as a function of scattering angle, can provide the molecular weight, information about the gross shape, and, although it is not discussed here, information pertaining to the thermodynamic interaction with the solvent in which the molecules are dispersed (virial coefficients). There are several reviews concerning this type of scattering and its applications (*6, 11, 12, 13*).

In the application of Einstein's theory to conventional light scattering, only the time-averaged dielectric constant fluctuation $<(\delta\epsilon)>$ was considered and not its time dependence. The time dependence of these fluctuations would be expected to contain additional information. Microscopic fluctuations in concentration would not only cause the dielectric constant to deviate from an average value but would also shift the frequency of the scattered light in a manner directly related to the time dependence of the fluctuations. Because these fluctuations grow and decay in time, the scattered light is modulated, and consequently the frequency of the scattered light is shifted from that of the incident. Alternatively, it is perfectly correct to consider the frequency shifts of the scattered light to be a Doppler effect initiated by the Brownian motion of the individual molecules. The light scattered from a macroscopic solution illuminated by a monochromatic beam then would display a spectrum centered around the incident frequency. The width of this spectrum would contain information related to the statistical motion of the molecules. This implies that the light scattered by concentration fluctuations in a solution exhibits a spectrum characteristic of the time dependence of these concentration fluctuations. The width of the expected spectrum, however, is much too small to be observed by conventional spectrometers. However, in 1947, Forrester, Parkins, and Gerjouy (*14*) proposed a technique (optical mixing) where the beat frequencies of the interfering optical waves might be measured. Using this idea, Forrester, Gudmundsen, and Johnson (*15*) did observe beat frequencies between the split Zeeman lines of a mercury light source although their signal-to-noise ratios were low; because of this, the method was not directly applicable to molecular light scattering. With the development of the laser, a sufficiently powerful light source with precise monochromaticity became available, and measurements of the beat frequencies contained in the scattered light became possible.

Both Pecora (*16*) and Komarov and Fisher (*17*) adapted van Hove's space–time correlation function approach for neutron scattering (*18*) to the light-scattering problem to calculate the spectral distribution of the light scattered from a solution. Using a molecular analysis, Pecora assumed the scattering particles to be undergoing Brownian motion, and predicted a Lorentzian line shape for the spectral distribution of the scattered light. He also predicted that the width of the Lorentzian would be directly related not only to the translational diffusion coefficient but also in some cases to the rotational diffusion coefficient and the frequencies of the normal modes of vibration for flexible molecules. In later work he considered the effects of solutions containing polymers and other types of molecules with a variety of different conformations and flexibilities on the spectral profile (*19–21*).

In 1964, Cummins, Knable, and Yeh (*22*) observed the spectral broadening of the scattered light from the Brownian motion of solute molecules for the first time. Using dilute solutions of polystyrene latex spheres (PSL) and the technique of optical mixing, they achieved a resolution ($\omega/\Delta\omega$) on the order of 10^{14} which was much greater than any other form of optical spectroscopy. It remained, however, for Dubin, Lunacek, and Benedek (*23*) and Arecchi, Giglio, and Tartari (*24*) to make quantitative measurements on solutions of PSL to determine that the observed spectral profile agreed with Pecora's theory. In their work, Dubin *et al.* (*23*) also showed the scattered profile to be Lorentzian for solutions of several small proteins and non-Lorentzian for scattering from solutions of calf thymus deoxyribonucleic acid (DNA) and tobacco mosaic virus (TMV), respectively. With these initial successes, the technique of optical mixing rapidly became a means of measuring molecular parameters not easily observed with other methods.

A survey of the literature shows that the technique of optical mixing has been used to obtain the spectrum of the scattered light from many types of solutions containing biologically interesting molecules whose molecular weights have ranged from 10^4 to 10^8 daltons. As stated before, initial measurements on solutions of TMV showed a non-Lorentzian line shape (*23*) although later analysis (*25*) demonstrated that the observed shape was a sum of superimposed Lorentzians whose individual widths were related to the translational diffusion coefficient (D_t) and the rotational diffusion coefficient (D_R). Wada, Suda, Tsuda, and Soda (*26*) also measured the scattered depolarized spectral profile of TMV solutions to determine D_R while Dubin, Clark, and Benedek (*27*) performed a similar measurement on solutions of the protein lysozyme. Other macromolecules whose spectral widths have been measured are the coliphages T4, T5, T7, and λ (*28*); the blood protein haemocyamin (*29*); the poly(α-amino acid) poly(γ-benzyl L-glutamate) (*30*); polystyrene in 2-butanone (*31*); the copolymer poly-d(AT) (*32*); the enzyme RNase (*33*); the viruses MS2 (*34*) and R17 (*35*); the proteins lysozyme (*23, 36*), bovine serum albumin (*23, 37*), ovalbumin (*23*), and casein (*38*); and the contractile muscle protein myosin (*39, 40*). The effect of motile organisms such as sperm and bacteria on the scattered spectrum has also been considered (*41–43*). Finally, it has been reported that the spectral profile of light scattered from solutions may be used as a probe to investigate the kinetics of chemical reactions which are occurring within a solution (*32, 44–48*). A comprehensive treatment of many phases of light scattering is found in the text by Fabelinskii (*49*). A treatment of the specific topic of optical mixing techniques may be found in reviews by Chu (*50*), Cummins and Swinney (*51*), Benedek (*52a*), and Ford (*52b*).

From this brief review of light-scattering techniques, it can be seen that this approach is one of the most powerful methods for molecular investigation. Not only is it profitable to measure the distribution of scattered intensities as a function of scattering angle, but it is also practical and advantageous to measure the spectral distribution as well. Within the last several years, investigators using optical mixing techniques have progressed from considering simple systems such as solutions of PSL spheres to pursuing the complicated dynamics of myosin and probing the kinetics of the helix–coil transformation in polymers. From a biochemical viewpoint, optical mixing techniques represent a powerful investigative tool, both for the type of information obtained and for the rapidity with which information is obtained. When used in conjunction with already established physicochemical techniques, this should provide a more detailed description of molecular systems.

Theory

When an electromagnetic wave of angular frequency ω_0 is incident upon a dilute solution, it causes electrons to oscillate with a frequency equal to that of the initial wave. The magnitude of these displacements depends on the polarizability of the media, which we will assume to be isotropic, and the magnitude of the electric field. Since it is a fundamental principle of electrodynamics that an oscillating electric charge will radiate energy, the vibrating electrons reradiate energy in the form of spherical electromagnetic waves of frequency ω_0. These waves combine through interference to form the resultant wave in the liquid. This then, is a brief qualitative sketch of the phenomenon responsible for light scattering as well as refraction. We now present a mathematical description of the process and later consider the statistical nature of the light scattered from a group of randomly moving particles.

Consider a planar electromagnetic wave in a medium with index of refraction n, having the form

$$\mathbf{E}_L = \mathbf{E} \exp[i(\mathbf{k} \cdot \mathbf{r} - \omega_0 t)] \tag{1}$$

where $\mathbf{E}$ is the vector magnitude, $\mathbf{k}$ is the wave vector with $|\mathbf{k}| = 2\pi n/\lambda_0$, ω_0 is the angular frequency, and $\mathbf{r}$ is the position vector. The medium is then polarized according to the relation

$$\mathbf{P}(\mathbf{r}, t) = \alpha(\mathbf{r}, t)\mathbf{E}_L \tag{2}$$

where $\alpha(\mathbf{r}, t)$ is local polarizability. On a microscopic level $\alpha(\mathbf{r}, t)$ will not actually be a constant but will vary because of thermal fluctuations; hence we can write

$$\alpha(\mathbf{r}, t) = \alpha_0 + \delta\alpha(\mathbf{r}, t) \tag{3}$$

with α_0 being the time-averaged polarizability. The first term on the RHS of Equation 3 is responsible for refraction although we will consider only the second term and its relation to the light-scattering phenomenon. This fluctuation of local polarization then gives rise to a fluctuating polarization vector of the form

$$\delta\mathbf{P}(\mathbf{r}, t) = \delta\alpha(\mathbf{r}, t)\mathbf{E}_L \tag{4}$$

It is these fluctuations that cause the inhomogeneities in the fluid and the reradiated waves. The total amplitude of the electric field at a distance R from the scattering volume is given by the summation of the contributions of the infinitesimal scattering elements and can be expressed as

$$\mathbf{E}_s(\mathbf{R}, t) = \frac{1}{c^2}\int_V \frac{1}{|\mathbf{R} - \mathbf{r}|}\, \mathbf{k}_s\mathbf{x}\left[\mathbf{k}_s\mathbf{x}\,\frac{d^2\delta\mathbf{P}(\mathbf{r}, t_R)}{dt^2}\right]dV \tag{5}$$

where V is the illuminated volume, $t_R = t - (|\mathbf{R} - \mathbf{r}|/c_L)$ is the retarded time, c_L and c are the velocity of light in the liquid and *in vacuo*, respectively, and $\mathbf{k}_s = (\mathbf{R} - \mathbf{r})/|\mathbf{R} - \mathbf{r}|$ is a unit vector in the direction of the scattered light. If it is assumed that the magnitude of the first two time derivatives of $\delta\alpha(\mathbf{r}, t)$ is negligible compared with the frequency of the incident radiation ω_0, then after some manipulation and, remembering that for a nonmagnetic media

$$\varepsilon = 4\pi\alpha + 1 \tag{6}$$

where ϵ is the dielectric constant, Equation 5 reduces to

$$\mathbf{E}_s(\mathbf{R}, t) = A\mathbf{k}_s\mathbf{x}(\mathbf{k}_s\mathbf{x}\mathbf{E})\exp[i(\mathbf{k}_s \cdot \mathbf{R} - \omega_0 t)]\int \delta\varepsilon(\mathbf{r}, t) \exp[i(\mathbf{k} - \mathbf{k}_s)\cdot \mathbf{r}]dV \tag{7}$$

where now $|\mathbf{k}_s| = n|\mathbf{k}_0|$, and A represents constant factors.

At this point it is possible to gain some physical insight into the scattering process. To do this, we will consider that the fluctuations in the local dielectric constant can be represented mathematically by an infinite number of weighted sine and cosine functions, each pair having a different frequency (or wave vector) and phase argument. This is merely saying that the complicated fluctuations in the dielectric constant may be decomposed into an infinite number of well-defined periodic functions —*i.e.*, the fluctuations may be represented by a Fourier decomposition (*54*). Using complex notation, this concept may be written:

$$\delta\varepsilon(\mathbf{r}, t) = C_0 \int \delta\varepsilon(\mathbf{K}, t) \exp[i\mathbf{K} \cdot \mathbf{r}] d^3K \tag{8}$$

where C_0 is a constant and K is a dummy variable. Substituting Equation 8 into Equation 7 yields

$$\mathbf{E}_s(\mathbf{R}, t) = A\mathbf{k}_s\mathbf{x}(\mathbf{k}_s\mathbf{x}\mathbf{E}) \exp[i(\mathbf{k}_s \cdot \mathbf{R} - \omega_0 t)] \int \delta\varepsilon(\mathbf{K}, t) \left\{ \int_V \exp[i(\mathbf{k} - \mathbf{k}_s + \mathbf{K}) \cdot \mathbf{r}] dV \right\} d^3K \tag{9}$$

where now we need only integrate over the illuminated volume as all other space will give negligible contributions to the integral. The expression in braces is a three-dimensional Dirac delta function $\delta[\mathbf{K} - (\mathbf{k} - \mathbf{k}_s)]$ which has the property of being zero for all conditions except when $\mathbf{K} = \mathbf{k} - \mathbf{k}_s$. This in effect is telling us that the scattering is caused by one particular component of the dielectric constant fluctuation, namely that fluctuation whose wave vector $\mathbf{K}$ is equal to the difference between the incident and scattered wave vectors. Fixing the scattered angle θ and the wavelength of the incident light λ_0 will determine the particular scattering wave vector ($\mathbf{K}$) from which the scattering is being observed. From Figure 1 we see that

$$|\mathbf{K}| = 2k \sin(\theta/2) \tag{10}$$

where we have assumed $|\mathbf{k}| = |\mathbf{k}_s|$ which is justified since $\Delta K/K \sim 10^{-12}$. Interpreting this in terms of wavelength shows

$$\lambda = 2\Lambda \sin(\theta/2) \tag{11}$$

where $K = 2\pi/\Lambda$. Equation 11 is completely analogous to the Bragg equation for x-ray scattering from a crystal of lattice spacing Λ (*55*), and shows the similarity between the two processes. Equation 9 now becomes

$$\mathbf{E}_s(\mathbf{R}, t) = A\mathbf{k}_s\mathbf{x}(\mathbf{k}_s\mathbf{x}\mathbf{E}) \exp[i(\mathbf{k}_s \cdot \mathbf{R} - \omega_0 t)] \delta\varepsilon(\mathbf{K}, t) \tag{12}$$

which shows explicitly that information concerning the fluctuations of

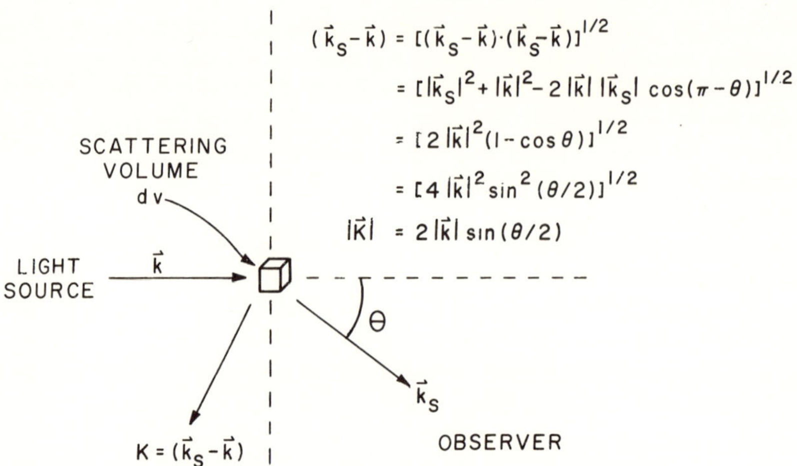

Figure 1. Derivation of the magnitude of the scattering vector **k** *in terms of the incident wave vector and the scattering angle*

the dielectric constant is contained in the scattered electric field. Equation 12 is essentially that developed by Einstein (*8*) who assumed nothing specific about the mechanism of the oscillating dielectric constant. Since $\delta\epsilon(\mathbf{K}, t)$ is a randomly varying quantity, it is necessary to consider the fluctuations of the scattered field with time.

In considering the statistical nature of the scattered light, it is useful to use the concept of the time autocorrelation function which is defined by

$$C_x(\tau) = \lim_{T\to\infty} \frac{1}{2T} \int_{-T}^{T} x(t)x^*(t + \tau)dt = \langle x(t)x^*(t + \tau)\rangle \quad (13)$$

where $x(t)$ is a parameter measured at time t and $x^*(t + \tau)$ is the complex conjugate of the same parameter measured τ units in time later. (For those unfamiliar with this function, Appendix A is provided which briefly describes some properties of $C_x(\tau)$.) The utility of this concept can be seen in applying the Wiener–Khinchine theorem (*56, 57*) which relates the autocorrelation function of a signal to the spectral power density of that signal by a Fourier transform

$$I_x(\omega) = \int C_x(\tau) \exp(i\omega\tau)d\tau \quad (14)$$

where $I(\omega)$ is the signal intensity at frequency ω. Applying this theorem to the scattered electric field should then yield the spectral line shape of the scattered light. Substitution of Equation 12 into Equation 13 gives

$$C_{E_s}(\omega) = \langle E_s(t)E_s^*(t + \tau)\rangle = \quad (15)$$

$$A^2 \exp[i\omega_0\tau] < \delta\varepsilon(\mathbf{K}, t)\delta\varepsilon^*(\mathbf{K}, t + \tau) >$$

where now we must evaluate the correlation function for the fluctuation of the dielectric constant. This calculation has been performed by Pecora (*16*) in the limit of infinitely dilute solutions where it can be safely assumed there is no molecule–molecule interaction or correlation between molecular movements.

Because of the length and complexity of Pecora's calculation, it will not be reproduced here. Instead, we will perform a somewhat less rigorous, but plausible, evaluation of $C_{\delta\epsilon}(\tau)$ to reach the same result. In doing this, it must be assumed that the time dependence of $C_{\delta\epsilon}(\tau)$ is the same as that observed if a large fluctuation in concentration were artificially imposed on the system at time $t = 0$ and allowed to decay according to Fick's second law of diffusion (*58*) where

$$\frac{\partial C_{\delta\varepsilon}(\mathbf{r}, \tau)}{\partial t} = D_t \nabla^2 C_{\delta\varepsilon}(\mathbf{r}, \tau) \tag{16}$$

with D_t being the translational diffusion coefficient. Using a spatial Fourier transformation to change from the variable $\mathbf{r}$ to $\mathbf{K}$, we find

$$C_{\delta\varepsilon}(\mathbf{K}, \tau) = C_{\delta\varepsilon}(0) \exp[-\Gamma t] \tag{17}$$

with $\Gamma = D_t K^2$ and $C_{\delta\epsilon}(0)$ being a constant of integration. Equation 15 now becomes

$$C_{E_s}(\tau) = \text{constant} \exp(i\omega_0\tau) \exp(-\Gamma t) \tag{18}$$

Taking the Fourier transform gives

$$I(\omega) = \frac{(\text{constant})\Gamma}{\Gamma^2 + (\omega - \omega_0)^2} \tag{19}$$

where we see that the spectral profile of the scattered light is Lorentzian in shape and centered around the incident frequency with half-width at half-height of

$$\Gamma = D_t K^2 = D_t \frac{4\pi n}{\lambda_0} \sin^2 (\theta/2) \tag{20}$$

This line shape was verified experimentally (*23, 24*) in 1967 for solutions of PSL spheres.

Detection of Signal and Experimental Apparatus

Equation 19 predicts that the spectrum of the scattered light contains frequencies that differ from the incident ($\sim 5 \times 10^{14}$ Hz) by only

$\sim 10^2$–10^3 Hz. Since conventional spectrometers (*e.g.*, a grating spectrometer or a Fabry-Perot interferometer) are not capable of resolving such small frequency shifts, it is necessary to rely on beating techniques to obtain the information contained in the scattered spectrum.

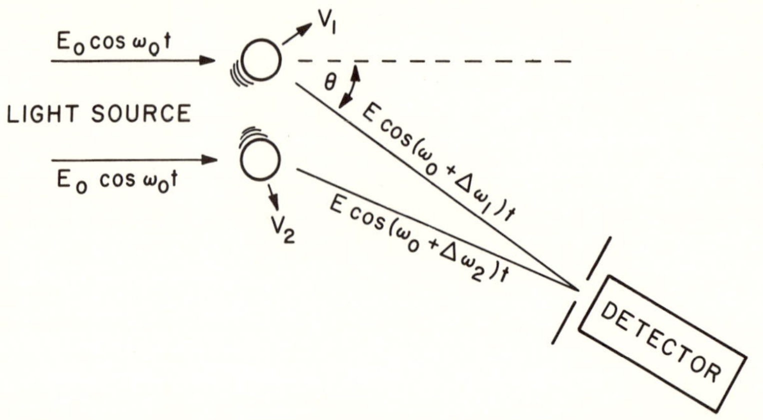

Figure 2. Light of incident angular frequency ω_0 is scattered from two particles undergoing random Brownian motion. Because the scattered light is Doppler shifted by the two particles respectively, the scattered signal contains two discrete frequencies.

In a standard beating experiment, a photomultiplier tube is used to detect the scattered light. Since the output photocurrent is proportional to the square of the electric field incident upon the tube, it is known as a square-law detector. Thus, if the scattered light contained only two frequencies, as would be the case if only two molecules were in the scattering volume (as in Figure 2), we find the magnitude of the photocurrent i to be

$$\begin{aligned} i \propto (E_1 + E_2)^2 &= E^2[\cos(\omega_0 + \Delta\omega_1)t + \cos(\omega_0 + \Delta\omega_2)t]^2 \\ &= E^2/2\,[2 + \cos^2(\omega_0 + \Delta\omega_1)t + \cos^2(\omega_0 + \Delta\omega_2)t] \\ &+ E^2[\cos(2\omega_0 + \Delta\omega_1 + \Delta\omega_2)t + \cos(\Delta\omega_1 - \Delta\omega_2)t] \end{aligned} \quad (21)$$

Since a photomultiplier tube responds to frequencies only up to about 10^9 Hz, the terms in Equation 21 containing optical frequencies in their argument are time averaged out to a dc component. Hence, the photocurrent output varies as

$$i \propto \mathrm{dc} + E^2 \cos(\Delta\omega_1 - \Delta\omega_2)t \quad (22)$$

where the low or beat frequency term in Equation 22 reflects the relative velocities of the two molecules. Real solutions, of course, contain many molecules in the scattering volume with a distribution of velocities, and the spectrum of the scattered light is the continuous spectrum of Equation 19 and not the discrete one discussed above. The photomultiplier tube has the added advantage, as Equation 22 shows, of shifting the center of the scattering profile from ω_0 to zero frequency where conventional wave analyzers may be used. In extracting the beat signal, two similar techniques may be employed, and it is useful to distinguish between them.

In the first technique, a signal is obtained from the incident light beam and allowed to combine with the scattered light to form beats. This signal is termed a local oscillator. Cummins *et al.* (*22*) first used this method and shifted the frequency of the local oscillator using an acoustical Bragg reflection modulator. More recently, it has been recognized that the shifting in frequency of the local oscillator is unnecessary so a local oscillator signal is produced by having the incident beam scatter from a stationary object placed in the scattering volume (usually a Teflon wedge). We shall call this technique homodyne spectroscopy. The second technique differs from the first in that the scattered light is allowed to interfere with itself on the photomultiplier tube to form beats. This is called self-beating and is a simpler experimental arrangement although the analysis of the results is more complicated because each part of the scattered spectrum beats against the entire spectrum. There is currently some ambiguity in the literature regarding the nomenclature regarding these two techniques. In keeping with common electrical engineering terminology, techniques involving a local oscillator derived from the same source as the signal are termed homodyne beating, independent of whether the local oscillator frequency is shifted from the incident frequency, while beating in the absence of a local oscillator is termed self-beating. This is the terminology we shall use here although in the literature the terms heterodyne and homodyne spectroscopy are sometimes used for what we call homodyne and self-beating spectroscopy, respectively. The spectral profile of the photocurrent is centered at zero frequency but is now given by the convolution of the original spectrum with itself.

$$P(\Omega) = \int_0^\infty I(\omega)I(\omega + \Omega)d\omega \tag{23}$$

or, in place of Equation 19

$$P(\Omega) = \text{constant}\,\frac{\Gamma}{\Omega^2 + (2\Gamma)^2} \tag{24}$$

where $P(\Omega)$ is the spectral power density of the photocurrent. The spectrum is still a Lorentzian, but now with a half-width at half-height of 2Γ. Non-Lorentzian power spectra will, in general, give more complicated results. The results presented here show then that measurements of the low, or beat, frequency power spectrum of the photomultiplier current may be interpreted in terms of the diffusion coefficients of the molecules responsible for the scattering.

As an alternative to measuring the power spectrum of the photocurrent, it is equally possible to measure its autocorrelation function to obtain the information contained in the scattered light. Recently, a number of real time correlation function computers have become available commercially, and these instruments can calculate approximations to $C(\tau)$, as defined in Equation 13, directly, from an electrical signal. These computers typically obtain $C(\tau)$ for 100 different values of τ simultaneously. In principle, they are capable of obtaining data 100 times as rapidly as a single-channel wave analyzer. As with the wave analyzer, the signal correlator may be used to process the scattered light from either a homodyne or self-beat experiment. If a signal correlator is used, however, a somewhat different development of the theory is needed.

We defined the autocorrelation for the scattered field as

$$C_E(\tau) = < E(\tau)E^*(T + \tau) > \tag{25}$$

However, in measuring the photocurrent's autocorrelation function we are interested in the quantity

$$C_i(\tau) \propto < I(t)I(t + \tau) > = < E(t)E^*(t)E(t + \tau)\, E^*(t + \tau) > \tag{26}$$

where the photocurrent is proportional to the incident intensity $I(t)$. $C_E(\tau)$ and $C_i(\tau)$ must now be related. This relationship may be calculated if it can be assumed that the scattered field is a Gaussian random variable (59). Stating the result, we have

$$C_i(\tau) = I_0^2[1 + C_E^2(\tau)] \tag{27}$$

where I_0 is the time-averaged scattered intensity.

Substituting $C_E(\tau)$ from Equation 18 into Equation 27 yields

$$C_i(\tau) = I_0^2 + a \exp(-2\Gamma t) \tag{28}$$

where a is a constant.

In discussing experimental correlation functions, it is convenient to define the normalized function

$$C_N(\tau) = \frac{C_i(\tau) - C_i(\infty)}{C_i(0) - C_i(\infty)} = \frac{C_E^2(\tau)}{C_E^2(0)} \tag{29}$$

where Equation 29 shows the direct connection between the measured correlation function and the quantities of interest

$$C_N(\tau) = \frac{< \delta\varepsilon(t)\delta\varepsilon(t + \tau) >^2}{< \delta\varepsilon(t)^2 >^2} \tag{30}$$

Up to now, we have given a general theoretical development of the self-beat technique. As a practical illustration of the experimental apparatus used to detect autocorrelation functions in scattering experiments, the equipment currently used in our laboratory will now be described. While our treatment of the autocorrelation function has been in terms of an analog signal, the computer that measures this function is actually a digital device. This is based on the fact that it is also valid to count the scattered photons in order to calculate $C_i(\tau)$ as the optical intensity signal is essentially determined by the number of photons that strike the photocathode per unit time. We have then

$$I(t) \propto \mathbf{E}(t)\mathbf{E}^*(t) \propto n(t) \tag{30a}$$
$$I(t + \tau) \propto \mathbf{E}(t + \tau)\mathbf{E}^*(t + \tau) \propto n(t + \tau)$$

where $n(t)$ and $n(t + \tau)$ are the number of photons striking the photocathode in a small interval of time at times t and $t + \tau$, respectively. Substituting Equations 30a into Equation 26 yields

$$C_i(\tau) \propto n(t)n(t + \tau) \tag{30b}$$

The photocurrent which is a series of discrete pulses gives $n(t)$ and $n(t + \tau)$ directly and can be used to calculate $C_i(\tau)$ digitally.

A block diagram of the experimental apparatus is presented in Figure 3. The light from a Spectra Physics He–Ne laser of ~50 mw power ($\lambda = 6328$ A) is focused to a point inside the scattering cell by lens L_1. The cell itself is housed inside a copper water jacket for temperature control. Light scattered at an angle θ is focused on a pinhole by lens L_2 and detected by a photomultiplier tube. For the purposes of measuring correlation functions, an ITT FW 130 photomultiplier should be used. Foord *et al.* (*60*) have shown that this tube gives the fewest spurious correlations. From the photomultiplier tube, the amplified signal is fed into a discriminator which generates pulses of uniform shape for each photomultiplier pulse of sufficient magnitude. The correlation computer itself was designed and built at this laboratory and uses digital logic to compute the correlation function of the photocurrent in the form

$$C_i(\tau) = \sum_j n(t_j)n(t_{j+k}) \tag{31}$$

where $k = \tau/\Delta T$ and $n(t_j)$ is the number of pulses during a time interval ΔT at time t_j. The computer can calculate $C_i(\tau)$ for 21 values of τ simultaneously. Since we expect exponential signals, data near $\tau = 0$ are of more interest than those at large values of τ. Consequently, the first 10 channels obtain $C(\tau)$ at intervals of ΔT, the next six at intervals $2\Delta T$, and the last five at intervals $5\Delta T$. The last channel thus gives $C_i(58\Delta T)$. The range of ΔT varies from 0.1 to 0.2×10^{-6} sec with the computer having a memory storage of 16 bits per channel.

The correlation function data are then transformed by a digital analog converter and displayed on an oscilloscope where the evolution of $C_i(\tau)$ may be conveniently watched. Alternatively, the digital output may be punched on paper tape for subsequent analysis. A typical solution will allow $C_i(\tau)$ to develop in ~5–15 sec with sufficient signal to noise ratio to obtain the time constant with an accuracy of 1% or better.

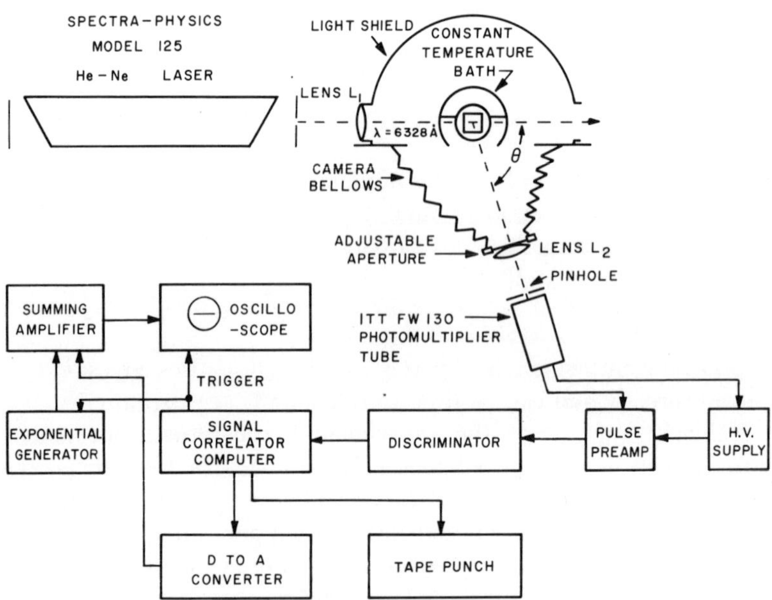

Figure 3. Block diagram of the experimental apparatus for an optical mixing measurement

Experimentally, the time constant of $C_i(\tau)$ may be evaluated in one of two ways. First, an internally generated exponential with variable time constant and magnitude may be subtracted from $C_i(\tau)$ until a null is detected on the oscilloscope screen. The time constant is then read off a calibrated dial; this procedure takes several minutes. In the second method, the coordinates of $C_i(\tau)$ are punched onto paper tape, and the data are then analyzed using a CDC 3600 computer. The program used for this computes a least-squares fit to $C_i(\tau)$ and then determines D_t. This latter procedure is somewhat longer than the first as it depends on the availability of the University computer etc., but it is much more precise and is routinely used. The technique of measuring $C_i(\tau)$ directly, as opposed to measuring the Lorentzian spectral profile, has the advantage of being quick, accurate, and easy to analyze. Also, long term drifts in the laser output are unimportant.

The optical arrangement as seen in Figure 3 is straightforward although it is important that it be designed with care as the final signal to noise ratio depends critically on this design. An extensive discussion of the criteria to be considered has been given by Dubin (*61*). We will present here only an introduction to the subject and a general statement of the results.

Since we wish to obtain full modulation of the signal at the photomultiplier, it is necessary to have the spherical wave fronts scattered by molecules at opposite extremes of the scattering volume essentially parallel over the full area of the detector. Otherwise, the two waves would interfere with each other constructively over part of the detector and destructively over the remainer leading to a relatively small change in the detector output as the molecules move with respect to each other. If, instead, the waves interfere either constructively or destructively over the entire detector surface, the signal will undergo excursions from zero up to twice the average signal. The area at the detector satisfying the above criteria is called the coherence area. Since the size of the coherence area is roughly that of the central bright spot that would be obtained if a pinhole the size of the sample were illuminated with parallel light, we find that a small sample area permits a large detector area.

The concept of the coherence area may also be described as follows. Suppose one of the sample dimensions in the plane perpendicular to the sample-detector line is a. Then, the angular range of light accepted in that direction $\Delta\alpha$ must be on the order of

$$\Delta\alpha \sim \lambda/a \tag{32}$$

in order to obtain a single coherence area. In terms of the apparatus described here, the sample size is determined by the pinhole and $\Delta\alpha$ by the aperture. A single coherence area will thus be obtained if

$$\frac{\lambda^2 r^2}{a^2} \sim 1 \tag{33}$$

where r is the distance from sample to detector. The total light intensity obtained in one coherence area is

$$I \propto \frac{I_0 R \lambda^2}{a} \tag{34}$$

where R is the familiar Rayleigh ratio. Equation 34 tells us that the signal may be increased by focusing the laser beam to a smaller diameter. A focal length of 10 cm for L_1 is a convenient compromise as a significantly shorter focal length will lead to several experimental difficulties including a broadening in the range of scattering angle accepted.

As in conventional light scattering, beating spectroscopy requires considerable care in eliminating dust and other debris from the solutions. In self-beating experiments, dust acts as a local oscillator source and thus introduces a small deviation in the measured time constant. It can be shown that in order to reduce this shift to less than 1%, the intensity of light entering the detector from undesirable sources (dust, stray light, etc.) must be kept below 0.5% of the sample light scattered intensity (*62*). The homodyne spectrometer, on the other hand, is relatively insensitive to dust as long as it is very much larger than the molecule of interest. Unfortunately, the homodyne spectrometer introduces other difficulties as any local oscillator noise from either laser power fluctuations or vibrations in the apparatus will overwhelm the true signal. Because large particles scatter so intensely, particular care must be taken in experiments where θ is small or dilute solutions are used. The techniques used to clean solutions and scattering cells are those established by conventional light-scattering experimenters (*13*).

The first molecular system studied with this type of spectroscopy was suspensions of PSL spheres (*22*) whose spherical shape allowed theoretical calculation of the diffusion coefficient according to the Stokes–Einstein equation (*58*)

$$D_t = \frac{k_B T}{f} = \frac{k_B T}{6\pi\eta r} \tag{35}$$

where k_B is the Boltzmann constant, T is the absolute temperature, η is the viscosity, r is the effective radius of the sphere, and f is the frictional coefficient. Agreement between calculated and measured values of D_t was good for PSL particles with r as large as 1830 A (*23*). When the spectrum is studied as a function of scattering angle, Γ is found to be linearly related to $\sin^2(\theta/2)$ as Equation 20 predicts. Verification of this has been shown for PSL spheres (*23*), and Figure 4 shows this relationship also holds for solutions of monodisperse polystyrene ($M = 1.8 \times 10^6$ daltons) in 2-butanone (*31*).

Application of Beating Spectroscopy to Molecular Weight Determinations

Introduction. Previously we have described the way in which beating spectroscopy may be used to obtain values of the diffusion coefficient

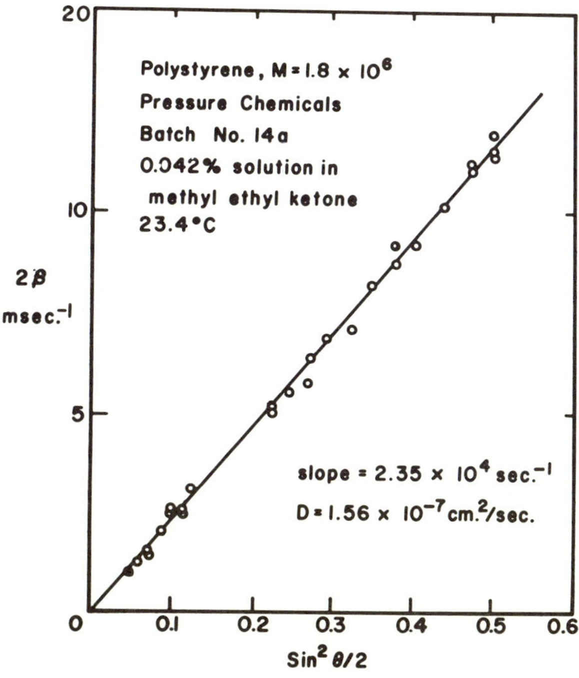

Figure 4. Experimental verification of Equation 20 for polystyrene in methyl ethyl ketone. Polystyrene is in the random-coil conformation (31).

for macromolecules in solution. In this section we will interpret these diffusion coefficients in terms of molecular weights. These ideas and interpretations are not necessarily new, and have been used before in describing conventionally measured diffusion coefficients. However, the fact that measurements of diffusion coefficients are now more readily available does give greater emphasis to these methods. Thus this section is essentially a review of the older techniques together with examples taken from recent beat spectroscopy experiments.

Diffusion coefficients can be related to molecular weight in three ways: first by application of the Stokes–Einstein equation, second by combination with sedimentation data, and third by consideration of homologous polymer solutions. In the first method, an equivalent spherical size of the molecules is calculated from D_t, and an approximate molecular weight is found by combining these data with the appropriate density. In the second method, diffusion measurements are coupled with those of sedimentation velocity to give molecular weights, and in the third method, molecular weights may be determined directly from measurements of diffusion coefficients alone once a calibration has been established. All three methods rely on fast accurate determination of diffusion coefficients for which optical mixing techniques are perfectly suited.

Stokes–Einstein Relationship. As was pointed out in the last section, diffusion coefficients may be related to the effective radius of a spherical particle through the translational frictional coefficient in the Stokes–Einstein equation. If the molecular density is also known, then a simple calculation will yield the molecular weight. Thus this method is in effect limited to "hard body" systems. This method has been extended for example by the work of Perrin (*63*) and Herzog, Illig, and Kudar (*64*) to include ellipsoids of revolution of semiaxes *a, b, b,* for prolate shapes and *a, a, b* for oblate shapes, where the frictional coefficient is expressed as a ratio with the frictional coefficient observed for a sphere of the same volume.

$$\frac{f_t}{f_0} = \frac{f_t}{6\pi\eta R_0} = \frac{[1 - b^2/a^2]^{1/2}}{(b/a)^{2/3} \ln \dfrac{1 + (1 - b^2/a^2)^{1/2}}{b/a}}; \; R_0{}^3 = ab^2 \tag{36}$$

$$\frac{f_t}{f_0} = \frac{[a^2/b^2 - 1]^{1/2}}{(a/b)^{2/3} \tan^{-1} [a^2/b^2 - 1]^{1/2}}; \; R_0{}^3 = a^2 b \tag{37}$$

where R_0 is the radius of a hydrodynamically equivalent sphere having a volume equal to that of the ellipsoid. Knowledge of the molecular dimensions and density will again yield an approximation to the molecular weight. A note of caution is needed here as the actual molecular dimensions may be quite different from those calculated from diffusion measurements. It is likely that a certain amount of solvent will adhere to the diffusing particle, and hence it must be considered as part of the macromolecule when the mass and volume of an equivalent hydrodynamic shape are computed. A change in diffusion coefficient then could be caused by an actual conformational change, a change in degree of solvation, or both, and it is impossible to determine the real case from a single diffusion experiment. Polydispersity in the sample will also lead to ambiguity in the interpretation of the molecular dimensions from diffusion data. More detailed discussion of these points is available (*58, 65, 66*).

Since it is possible to measure the rotational diffusion coefficient by optical mixing techniques, and since expressions equivalent to Equations 36 and 37 have also been calculated by Perrin (*67*) for rotational motion, we may solve the simultaneous equations

$$D_t = \frac{k_B T}{f_t} \text{ and } D_R = \frac{k_B T}{f_R} \tag{38}$$

and independently calculate the values of a and b subject to the ambiguities discussed above. A decision as to whether the molecules in question are best fitted by a prolate or an oblate shape cannot, however, be made without other evidence. Once this has been accomplished, an equivalent volume may be calculated and the molecular weight determined if the density is known. A procedure similar to this was used by Dubin *et al.* (*27*) who showed that the protein lysozyme can be best described by a prolate ellipsoid. Alternately, knowledge of D and M can lead to information on molecular dimensions.

Sedimentation Diffusion. If a molecule in equilibrium in a solution is suddenly subjected to a centrifugal field, it will tend to move to a new equilibrium position subject to the centrifugal force, the buoyant density of the fluid, the frictional restraining force, and the molecular conformation and weight. For a particular sedimenting force, the particle will approach this new equilibrium point with some terminal velocity (v_t) which it reaches after a short time. The ratio of this terminal velocity to that of the accelerating force (v_t/a, measured in 10^{-13} sec or Svedbergs) is termed the sedimentation coefficient, and is the measurement of interest in sedimentation velocity experiments. The sedimenting particles may be observed as a function of time with the aid of Schlieren optics which are sensitive to the difference in refractive index between solution and solvent.

An analytical expression relating the sedimentation coefficient to the molecular weight M is of the form

$$S \equiv \frac{v_t}{a} = \frac{v_t}{\omega^2 r} = \frac{M(1 - \bar{v}\rho)}{N_0 f} \tag{39}$$

with N_0 being Avogadro's number, ω the angular frequency at radius r, $\bar{v}$ the partial specific volume of the solute molecules, ρ the solvent density, and f the frictional coefficient. A comprehensive treatment of centrifugation techniques, both theoretical and practical, has been presented by Svedberg and Petersen (*68*), Schachman (*69*), and Tanford (*58*).

To minimize the effects of concentration dependence, the sedimentation coefficient is usually determined for a particular system at several concentrations and then extrapolated to infinite dilution (S^0). If, on the other hand, the diffusion coefficient is also measured at several concentrations and extrapolated to infinite dilution (D^0), then values of S^0 and D^0 may be related through the frictional coefficient to determine the molecular weight of the solute molecules

$$M = \frac{S^0}{D^0} \frac{RT}{(1 - \bar{v}\rho)} \tag{40}$$

where R is the gas constant. (It must be emphasized that the experimental conditions (solvent, ionic, strength, pH, temperature, etc.) must be as identical as possible in the two types of measurements.) This procedure is valid for any size or shape of molecule and is limited only by the experimental difficulties of obtaining S^0 and D^0. The solvent plays a very important role in sedimentation and diffusion experiments, and hence both S^0 and D^0 are usually reduced to some standard reference conditions (*e.g.*, H_2O at 20°C).

For large molecules, D^0 is difficult to obtain experimentally using conventional techniques because of the long times required for diffusion to take place and the necessity of preserving exacting experimental conditions over this period. [*See* Gosting (*70*) for a description of classical methods.] In this case, determination of D^0 by optical mixing techniques is clearly superior to the classical methods. For molecules of the coliphages T4, T5, and T7, Dubin *et al.* (*28*) measured D^0 by optical mixing light-scattering techniques and S^0 by conventional means and used previously determined values of $\bar{v}$ (*71*) to determine the molecular weights for these particles. A tabulation of their results is presented in Table I. Also, using the results of a chemical analysis (*71*) for the percentage of DNA in each phage, Dubin *et al.* were also able to evaluate the molecular weights of the respective DNA's. In this way, DNA molecular weight determinations were made without actually working with the DNA itself. Comparison of the molecular weights for the phages' DNA and T7 obtained by conventional means are also presented in Table I. In carrying out this work, Dubin *et al.* showed that the scattered spectral profile as a function of scattering angle was described by Equation 20 for T4, and it was assumed that this relationship held for the other phages.

Table I. Tabulation of Results for Molecular Weights of T4, T5, and T7 and Their Respective DNA's[a]

Particle	$D_{20,\omega}$, *10^{-7} cm²/sec*	$S^0_{20,\omega}$, *10^{-13} sec*	M *10^6 Daltons*	
T4	0.295 ± 0.003	890 ± 15	192.5 ± 6.6	
T5	0.397 ± 0.004	615 ± 10	109.2 ± 4.0	
T7	0.603 ± 0.006	453 ± 8	50.4 ± 1.8	49.4[b]
T4DNA			105.7 ± 3.8	104[c]
T5DNA			67.3 ± 3.1	65.7[c]
T7DNA			25.8 ± 1.0	23.2[c]

[a] Data from Dubin *et al.* (*28*). Molecular weights determined by other methods are included for comparison.
[b] Ref. *71*.
[c] Ref. *72*.

Homologous Polymers. As the size of a molecule increases, it is expected that its diffusion coefficient will decrease. Figure 5 shows a plot of D^0 *vs.* M for solutions of polystyrene in 2-butanone (methyl ethyl ketone) where the relationship between D^0 and M may be expressed by the semiempirical formula

$$D^0 = K_D M^{-b} \tag{41}$$

where K_D and b are constants. This equation is analogous to that used by Eigner and Doty (*73*) in relating sedimentation coefficients to molecular weight for solutions of DNA and also to the Mark–Houwink equation relating intrinsic viscosity to molecular weight ($[\eta] = KM^a$).

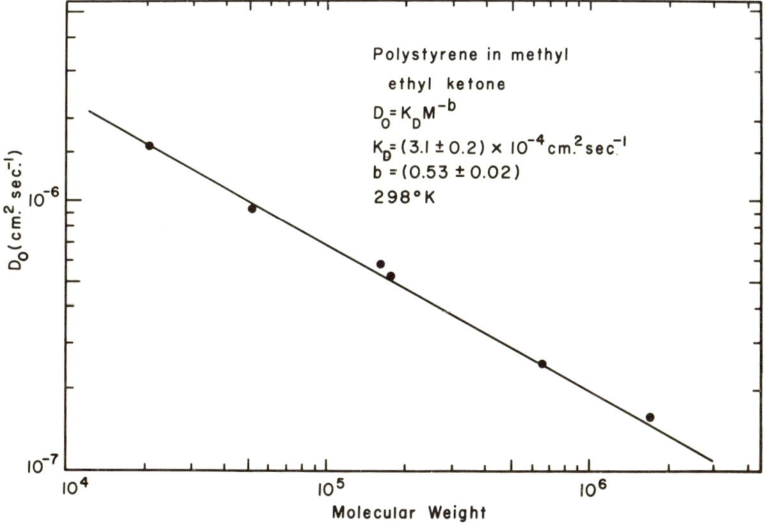

Discussions of the Faraday Society

Figure 5. Graph of the diffusion coefficient extrapolated to infinite dilution (D^0) vs. *molecular weight of polystyrene* (31).

In fact, for random-coil conformations, b in Equation 41 is explicitly related to the exponent in the Mark–Houwink equation by $3b = a + 1$. In this way values of M may be determined directly from measurements of D^0 provided K_D and b are known. Although both these parameters could in principle be calculated from a detailed knowledge of the geometry of the solute, it is usual to regard them as experimentally determinable parameters. A similar relationship has also been found to hold true for the polypeptide poly(γ-benzyl L-glutamate) (PBLG) dissolved both in 1,2-dichloroethane and in dichloroacetic acid. These results are shown in Figures 6 and 7, respectively.

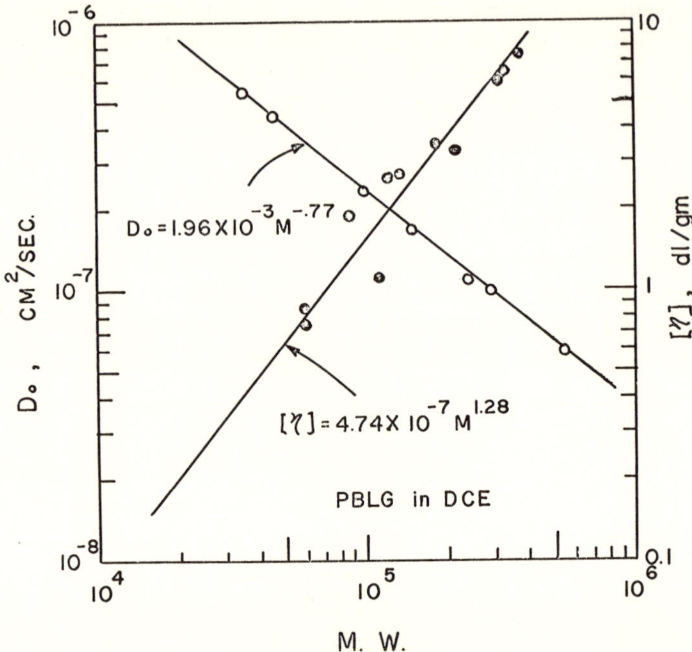

Figure 6. Graph of D⁰ vs. *molecular weight for poly(γ-benzyl-L-glutamate) (PBLG) in 1,2-dichloroethane. PBLG is in a helical conformation.*

Polydispersity. The determination of molecular weight is meaningful only if monodisperse solute molecules are used or, if used with polydisperse solutes, the type of molecular weight obtained (weight average, number average, etc.) is known. The case of a single component is the easiest of all to consider, but it is the least often encountered experimentally. Polydispersity is an almost universal phenomenon in synthetic polymer work, and in order to give a complete description of the self-beating technique, we will summarize an approach to this problem. A more general treatment has been given by Tagami and Pecora (*75, 76*). Other treatments are by Frederick, Reed, and Kramer (*71*) and Kopel (*62*).

In the detection of the autocorrelation functions in self-beat spectroscopy, solution polydispersity can lead to a non-exponential form. If we assume that there are no contributions to the autocorrelation function except those from translational diffusion for the different types of molecules, we can consider two simple cases: a continuous distribution of solute particle sizes and several distinct components in a solution. We shall approach the two cases by determining their effect on the observed correlation function.

In considering the first case, it is convenient to assume a Schulz distribution $W(M)$ for the molecular weight of the form

$$W(M) = \frac{1}{\Gamma_a(1+z)} \gamma^{z+1} M^z \exp[-\gamma M] \tag{42}$$

where $W(M)$ is the weight fraction of the molecules in the sample of molecular weight $M, \Gamma_a\ (1+z)$ is a gamma function of argument $(1 + z)$, and the parameters z and γ are related to the weight-average molecular weight by $M_w = (z+1)/\gamma$ (78). Using the definition of the normalized correlation function for the photocurrent, Equation 29, and remembering Equation 27, we find

$$C_N(\tau) = \frac{1}{M_w^2}\left[\int M f(M) e^{-\Gamma_D(M)\tau} dM\right]^2 \tag{43}$$

Assuming now $\Gamma = K^2 D_t = K^2 K_D M^{-b}$ where K_D and M are defined in Equation 41, Equation 43 may be evaluated to yield

$$C_N(t, z, b) = \left[\frac{(z+1)^{z+1}}{z!}\int x^{z+1} e^{-(z+1)x} e^{-T/x^b} dx\right]^2 \tag{44}$$

where we have used the change of variables $x = M/M_w$ and $T = K_D K^2 \tau / (M_w)^b$. Equation 44 is now a theoretical expression for the normalized correlation function observed for scattering from a solution containing a continuous distribution of molecular weights. Since Equation 44 cannot be evaluated explicitly, an approximate procedure will be described which can yield the degree of polydispersity (z) and the weight-average molecular weight (M_w).

Consider the two functions

$$Y_i = \frac{C_N(T_i) - C_N(T_{i+1})}{{}^1\!/_2[C_N(T_i) + C_N(T_{i+1})]\ \Delta T} \tag{45}$$

$$X_i = \frac{C_N(T_i) + C_N(T_{i+1})}{C_N^{(1)} + C_N^{(2)}}$$

where $C_N(T_i)$ and $C_N(T_{i+1})$ represent the magnitudes of the normalized autocorrelation functions at two successive values of T whose difference is represented by ΔT. Y_i is essentially a difference quotient which, in the limit, is related to the derivative dC_N/dT. X_i is basically a function of the magnitude of $C_N(T)$. $C_N(T, z, b)$ is only a function of T for particular values of b and z, and it may be calculated numerically. If C_N is a single exponential, as would be obtained for a monodisperse solution, then the plot of Y *vs.* X would be a horizontal straight line. If, however, the solution is polydisperse, a deviation from a horizontal line is expected. (A nontranslational diffusion contribution to the scattered signal could also cause a deviation.) Using numerical calculations and steps in T of $\Delta T = 0.08$, plots of Y *vs.* X may be obtained and are found to be very good approximations to straight lines for $0 \leqslant z \leqslant \infty$, $\frac{1}{3} \leqslant b \leqslant 1$, and $\frac{1}{3} \leqslant X \leqslant 1$. It should be noted that since the change of variables was made to T, it is unnecessary to know a specific value for the constant of proportionality between T and τ $(K_D K^2 / M_w{}^b)$, *i.e.*, it is not necessary to know M_w to construct a plot of Y *vs.* X. This plot may then be used to find the value of Y when $X = 0$ $[Y(0)]$ and $X = 1$ $[Y(1)]$, respectively, by extrapolation for particular values of z and b. The ratio of $Y(0)/Y(1)$ is then plotted *vs.* b for different values of z and is shown in Figure 8. Knowledge of b for the molecular system being studied and the value of $Y(1)/Y(0)$ determined from the experimentally measured correlation function will then give a value of z.

Making a straight line fit to Y, M_D may be calculated from $Y(1)$

$$M_D = \left[\frac{-2K_D K^2}{Y(1)} \right]^{1/b} \tag{46}$$

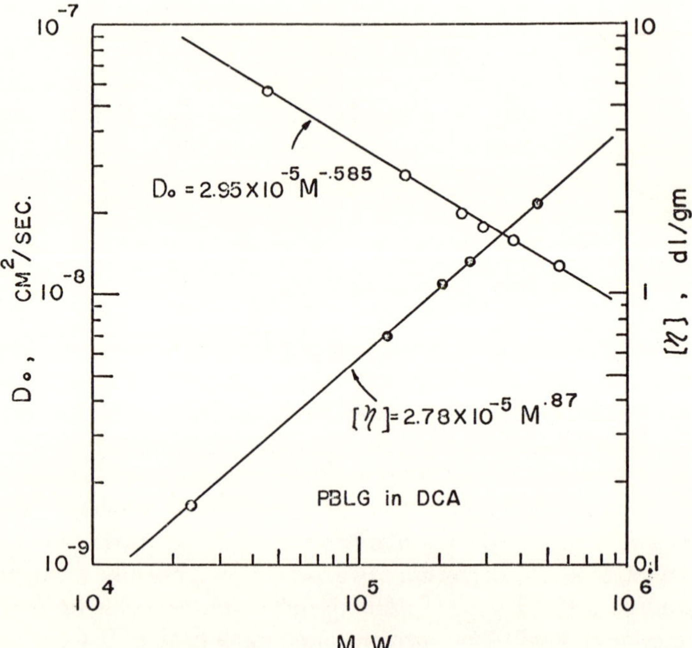

Figure 7. Graph of D^0 vs. *molecular weight for PBLG in dichloroacetic acid. PBLG is in a random-coil conformation.*

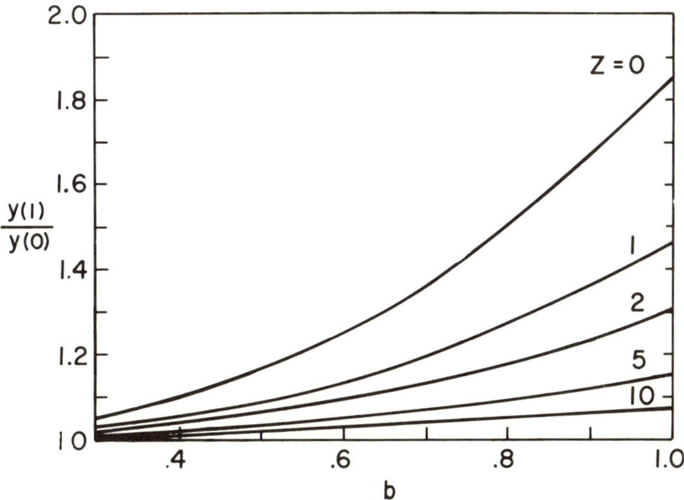

Figure 8. Graph of Y(1)/Y(0) vs. *the molecular parameter* b *for various values of the polydispersity parameter* z *of the Schulz distribution*

where M_D is the diffusion molecular weight obtained from Equation 44. The quantity $Y(1)$ may also be evaluated analytically from Equations 47 and 48 to yield

$$Y(1) = -2K_D K^2 \left(\frac{M_w}{z+1}\right)^{-b} \frac{\Gamma_a(z+2-b)}{\Gamma_a(z+2)} \tag{47}$$

Combining Equations 46 and 47 gives

$$\frac{M_w}{M_D} = (z+1)\left[\frac{\Gamma_a(z+2-b)}{\Gamma_a(z+2)}\right]^{1/b} \tag{48}$$

where now the weight-average molecular weight may be found. Equation 48 is shown as the dotted lines in Figure 9. Another way of arriving at Equation 48 can be made plausible by considering the following. The autocorrelation function derived in Equation 43 is essentially the square of a sum of weighted exponentials with each term in the sum representing a particular molecular weight present in the size distribution. Equation 48 can be derived if it is assumed that the autocorrelation function (Equation 43) can be approximated by a single exponential of form exp-$[-2\Gamma_D(M)\tau]$ and by imposing the condition that the first derivative of both forms of the autocorrelation function be equal at $\tau = 0$.

The results of Equations 47 and 48 require that $C_N(\tau)$ be measured in the limit $\Delta T \rightarrow 0$, a limit not easily achieved in practice because evalu-

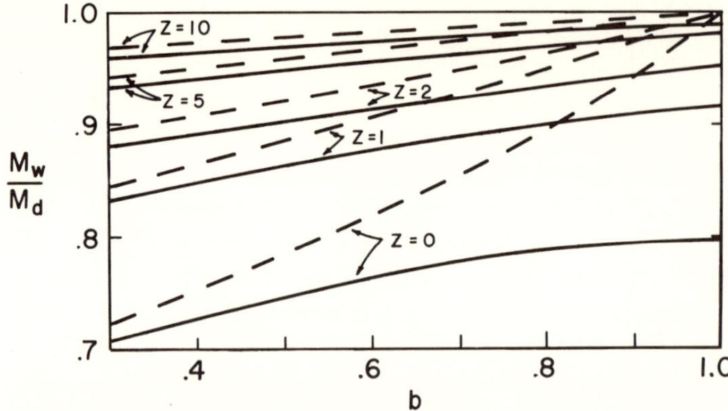

Figure 9. Dotted lines represent the theoretical Equation 48 relating M_w/M_D *to* b *for various values of* z*. Solid lines are determined from straight-line approximations of the lines in Figure 8.*

ation of Y also requires $C_i(\infty)$, from Equation 29. This problem can be seen in Figure 9 where the solid lines are determined from a straight-line fit to $Y(X)$ of the lines in Figure 8.

It is to be noticed that knowledge of K_D and b is required to interpret correlation functions in terms of M_w and z. The quantities K_D and b may be determined from separate experiments or possibly found in the literature. Thus a single measurement of the self-beat spectrum is adequate to give both the weight-average molecular weight (M_w) and the width of the weight distribution (z).

For the second case, that of a solution containing n distinct components, only the situation where $n = 2$ is sufficiently simple to consider. In this case, the photocurrent autocorrelation function is given by

$$C_i(\tau) = [Ae^{-\Gamma_1\tau} + Be^{-\Gamma_2\tau}]^2 \tag{49}$$

where A and B are constants and Γ_1 and Γ_2 are the time constants associated, respectively, with the two pure components. Assuming Γ_1 and Γ_2 are not too different, $C_i(\tau)$ is most easily analyzed by fitting to a single exponential. Setting $C_0(\tau) = A_0^2e^{-\Gamma_0\tau}$ and minimizing $[C_i(\tau) - C_0(\tau)]^2$ with respect to both A_0^2 and Γ_0, we may find the ratio A/B. Thus, if we know the individual scattering powers and diffusion constants of each component, it is possible to determine the ratio of the two components in a mixture. This analysis can therefore find use in cases of monomer–dimer equilibrium.

Summary

From the description presented here, it can be seen that self-beating spectroscopy as a technique for measuring the diffusion coefficient of molecular systems has several distinct advantages over conventional methods. While older methods rely on the production of artificial concentration gradients requiring precise experimental controls, the self-beating technique is performed on systems in thermal equilibrium thereby reducing the time necessary to make measurements to as little as a few seconds. The rapidity of the measurement, however, is not obtained at the expense of accuracy for accuracies on the order of 1% are routinely obtained. Values ~5% are typical for more conventional methods. Another advantage lies in the fact that measurements may be made on small quantities. This is especially important for hard to isolate biological molecules. As seen in the last section, measurements from self-beating spectroscopy can be related to molecular weights in several ways, depending on the type of molecule being studied, and can give information on the homogeneity of the sample. It is also possible to measure diffusion constants for a wide range of molecular shapes and sizes, as preliminary experiments have been performed on particles from as small as a few thousand daltons to such large molecules as rat liver mitochondria. With these advantages, self-beating spectroscopy should soon become a conventional method in the determination of molecular parameters.

Appendix A

The autocorrelation function is found by measuring the value of a signal at time t, and multiplying it by the magnitude of the same signal displaced in time τ units, and then averaging this product over all time. Equation 13 is essentially a mathematical expression of this concept where $x^*(t + \tau)$ indicates the complex conjugate of the signal. The autocorrelation function, unlike most average quantities, is not a number but is instead a function of the independent variable τ. Like most average quantities, it is a nonunique description of the signal. To illustrate some properties of this type of function, the autocorrelation function will be evaluated for a periodic signal of period T. Consider

$$y(t) = A \sin(\omega t + \theta) \tag{I}$$

where θ is some arbitrary phase angle and ω is the angular frequency $\omega = 2\pi/T$. Using Equation 13, we find

$$\begin{aligned} C_y(\tau) &= \langle A^2 \sin(\omega t + \theta) \sin(\omega t + \omega\tau + \theta) \rangle \\ &= \frac{A^2}{2} [\langle \cos(\omega\tau) \rangle - \langle \cos(2\omega t + \omega\tau + 2\theta) \rangle] \qquad \text{(II)} \\ &= \frac{A^2}{2} \cos(\omega\tau) \end{aligned}$$

The autocorrelation function of a periodic signal is itself periodic with a frequency equal to that of the original expression. It is to be noticed, however, that all information concerning the phase of the original signal is lost illustrating the nonunique description aspect.

By examining Equation II we can make the following observation

$$C_y(\tau) = C_y(\tau + T) \quad \text{(III)}$$

$$|C_y(0)| \geqq |C_y(\tau)| \quad \text{(IV)}$$

$$C_y(\tau) = C_y(-\tau) \quad \text{(V)}$$

Equation III is merely stating that a correlation function of a periodic signal is also periodic while Equation IV states that the maximum amount of correlation of a periodic function occurs when the same point of that signal is compared with itself and that the correlation is diminished as one compares two points that are farther and farther apart (up to a difference of T). Equation V states that the autocorrelation function is symmetric about $\tau = 0$.

In considering random fluctuations or signals, it is found from the definition of autocorrelation function that Equations IV and V also hold and further that

$$C_y(0) = \langle x^2 \rangle \quad \text{(VI)}$$

where $\langle x^2 \rangle$ is the mean square of the fluctuation. If a random-noise source of zero mean is added to an informational signal like

$$S(t) = x(t) + n(t) \quad \text{(VII)}$$

the $C_s(\tau)$ can be found to be of the form

$$C_s(\tau) = C_x(\tau) + C_n(\tau) \quad \text{(VIII)}$$

or in setting $\tau = 0$

$$\overline{S^2} = \overline{x^2} + \overline{n^2} \quad \text{(IX)}$$

which states that the powers of the two signals are additive. If a periodic signal is obscured by noise, then the calculation of autocorrelation functions may be employed to withdraw information concerning the periodic function. Autocorrelation functions can also be used, as in our case, to determine the spectral distribution of a purely random fluctuation by use of the Wiener–Khinchine theorem.

Literature Cited

1. Lord Rayleigh, *Phil. Mag.* (1871) **41,** 107, 447.
2. Oster, G., *Phys. Tech. Biol. Res.* (1955) **1,** 51.
3. Putzeys, P., Brosteaux, J., *Trans. Faraday Soc.* (1935) **31,** 1314.
4. Mie, G., *Ann. Phys.* (1908) **25,** 377.
5. van de Hulst, H. C., "Light Scattering by Small Particles," 1957, Wiley, New York.
6. Geiduschek, E. P., Holtzer, A., *Advan. Biol. Med. Phys.* (1958) **6,** 431.
7. Zimm, B., *J. Chem. Phys.* (1948) **16,** 1093.
8. Einstein, A., *Ann. Phys.* (1910) **33,** 1275.
9. Debye, P., *J. Appl. Phys.* (1944) **15,** 338.
10. Debye, P., *J. Phys. Colloid Chem.* (1947) **51,** 18.
11. Oster, G., *Chem. Rev.* (1948) **43,** 319.
12. Doty, P., Edsall, J. T., *Advan. Protein Chem.* (1951) **6,** 35.
13. Timasheff, S. N., Townend, R., "Physical Principles and Techniques of Protein Chemistry," Part B, 1970, Academic Press, New York.
14. Forrester, A. T., Parkins, W. E., Gerjouy, E., *Phys. Rev.* (1947) **72,** 728.
15. Forrester, A. T., Gudmundsen, R. A., Johnson, P. O., *Phys. Rev.* (1955) **99,** 1691.
16. Pecora, R., *J. Chem. Phys.* (1964) **40,** 1604.
17. Komarov, L. I., Fisher, I. Z., *Sov. Phys. JETP* (1963) **16,** 1358.
18. van Hove, L., *Phys. Rev.* (1954) **95,** 249.
19. Pecora, R., *J. Chem. Phys.* (1965) **43,** 1562.
20. Pecora, R., Steele, W. A., *J. Chem. Phys.* (1965) **42,** 1872.
21. Pecora, R., *J. Chem. Phys.* (1968) **48,** 4126.
22. Cummins, H. Z., Knable, N., Yeh, Y., *Phys. Rev. Lett.* (1964) **12,** 150.
23. Dubin, S. B., Lunacek, J. H., Benedek, G. B., *P.N.A.S.* (1967) **57,** 1164.
24. Arecchi, F. T., Giglio, M., Tartari, U., *Phys. Rev.* (1967) **163,** 186.
25. Cummins, H. Z., Carlson, F. D., Herbert, T. J., Woods, G., *Biophys. J.* (1969) **9,** 518.
26. Wada, A., Suda, N., Tsuda, T., Soda, K., *J. Chem. Phys.* (1969) **50,** 31.
27. Dubin, S. B., Clark, N. A., Benedek, G. B., *J. Chem. Phys.* (1971) **54,** 5158.
28. Dubin, S. B., Benedek, G. B., Bancroft, F. C., Freifelder, D., *J. Mol. Biol.* (1971) **54,** 547.
29. Foord, R., Jakeman, E., Oliver, C. J., Pike, E. R., Blagrove, R. J., Wood, E., Peacocke, A. R., *Nature (London)* (1970) **227,** 242.
30. Ford, N. C., Lee, W., Karasz, F. E., *J. Chem. Phys.* (1969) **50,** 3098.
31. Ford, N. C., Karasz, F. E., Owen, J. E. M., *Discuss. Faraday Soc.* (1970) **49,** 228.
32. Yeh, Y., *J. Chem. Phys.* (1970) **52,** 6218.
33. Rimai, L., Hickmott, J. T., Cole, T., Carew, E. B., *Biophys. J.* (1970) **10,** 20.
34. French, M. J., Angus, J. C., Walton, A. G., *Science* (1969) **163,** 345.
35. Pusey, P. N., Schaefer, D. W., Koppel, D. E., Camerini-Otero, R. D., Franklin, R. M., "Resumes des Communications—La Diffusion de la Lumiere par les Fluides," 1971, Paris.
36. Yeh, Y., Schuster, T. M., Yphantis, D. A., "Resumes des Communications—La Diffusion de la Lumiere par les Fluides," 1971, Paris.
37. Ware, B. R., Flygare, W. H., Abstracts, 162nd National Meeting of the American Chemical Society, Chicago, Ill., Sept. 1971.
38. Bloomfield, V. A., Dewan, R. K., Abstracts, 162nd National Meeting of the American Chemical Society, Chicago, Ill., Sept. 1971.
39. Carlson, F. D., Herbert, T., "Resumes des Communications—La Diffusion de la Lumiere par les Fluides," 1971, Paris.
40. Rimai, L., Cole, T., Gill, D., Abstracts, 162nd National Meeting of the American Chemical Society, Chicago, Ill., Sept. 1971.

41. Berge, P., Volochine, B., Billard, R., Hamelin, A., *C. R. Acad. Sci. Ser. D* (1967) **265**, 889.
42. Nossal, R., *Biophys. J.* (1971) **11**, 341.
43. Mann, J. A., Gulden, G., Summers, K., Gibbon, I. R., Abstracts, 162nd National Meeting of the American Chemical Society, Chicago, Ill., Sept. 1971.
44. Berne, B. J., Deutch, J. W., Hynes, J. T., Frisch, H. L., *J. Chem. Phys.* (1968) **49**, 2864.
45. Blum, L., Salsburg, Z. W., *J. Chem. Phys.* (1968) **48**, 2292.
46. Yeh, Y., Keeler, R. N., *Quart. Rev. Biophys.* (1969) **2**, 315.
47. Yeh, Y., Keeler, R. N., *J. Chem. Phys.* (1969) **51**, 1120.
48. Berne, B. J., Pecora, R., *J. Chem. Phys.* (1969) **50**, 783.
49. Fabelinskii, I. L., "Molecular Scattering of Light," 1965, Plenum Press, New York.
50. Chu, B., *Annu. Rev. Phys. Chem.* (1970) **21**, 145.
51. Cummins, H. Z., Swinney, H. L., *Progr. Opt.* (1969) **8**, 133.
52. (a) Benedek, G. B., "Polarization, Matter and Radiation," 1969, Presses Universitaire de France, Paris; (b) Ford, N. C., *Chem. Scripta* (1972) **2**, 193.
53. Landau, L. D., Lifshitz, E. M., "Classical Theory of Fields," 1962, Addison-Wesley, Reading, Mass.
54. Spiegel, M. R., "Advanced Calculus," 1963, Schaum, Chicago.
55. Halliday, D., Resnick, R., "Physics for Students of Science and Engineering," 1960, Wiley, New York.
56. Davenport, W. B., Root, W. L., "An Introduction to the Theory of Random Signals and Noise," 1958, McGraw-Hill, New York.
57. Carlson, A. B., "Communication Systems—An Introduction to Signals and Noise in Electrical Communications," 1968, McGraw-Hill, New York.
58. Tanford, C., "Physical Chemistry of Macromolecules," 1961, Wiley, New York.
59. Mandel, L., *Progr. Opt.* (1963) **2**, 181.
60. Foord, R., Jones, R., Oliver, C. J., Pike, E. R., *Appl. Opt.* (1969) **8**, 1975.
61. Dubin, S. B., Ph.D. Thesis, 1970, Massachusetts Institute of Technology.
62. Koppel, D. E., *J. Chem. Phys.* (1972) **57**, 4814.
63. Perrin, F., *J. Phys. Radium* (1936) **7**, 1.
64. Herzog, R. O., Illig, R., Kudar, H., *Z. Phys. Chem.,Abt. A* (1934) **167**, 329.
65. Yang, J. T., *Advan. Protein Chem.* (1961) **16**, 323.
66. Benoit, H., Freund, L., Spach, G., "Poly-α-Amino Acids," 1967, Marcel Dekker, New York.
67. Perrin, F., *J. Phys. Radium* (1934) **5**, 497.
68. Svedberg, T., Petersen, K. O., "The Ultracentrifuge," 1940, Oxford University Press, London.
69. Schachman, H. K., "Ultracentrifugation in Biochemistry," 1959, Academic Press, New York.
70. Gosting, L. J., *Advan. Protein Chem.* (1956) **11**, 429.
71. Bancroft, F. C., Freifelder, D., *J. Mol. Biol.* (1970) **54**, 537.
72. Schmid, C. W., Hearst, J. E., *J. Mol. Biol.* (1969) **44**, 143.
73. Eigner, J., Doty, P., *J. Mol. Biol.* (1965) **12**, 549.
74. Lee, W., Ph.D. Thesis, 1970, University of Massachusetts.
75. Tagami, Y., Pecora, R., *J. Chem. Phys.* (1969) **51**, 3293.
76. Pecora, R., Tagami, Y., *J. Chem. Phys.* (1969) **51**, 3298.
77. Frederick, J. E., Reed, T. F., Kramer, O., *Macromolecules* (1971) **4**, 242.

RECEIVED January 17, 1972. Work supported by research grants from the National Science Foundation (GB20491A1), the National Institutes of Health (GM 1494506), and the U.S. Air Force (AF 72-2170).

5

Thin-Layer Methods for Determining Molecular Weight Distribution

E. P. OTOCKA

Bell Laboratories, Murray Hill, N. J. 07974

Thin-layer chromatography (TLC) is a common laboratory technique for separating complex mixtures of solutes, usually by an adsorbtion mechanism. Several laboratories have applied the technique to the separation of polymer fractions and characterization of polymer molecular weight distributions. This work reviews the experimental results and theoretical approaches to the fractionation mechanisms.

Thin-layer chromatography (TLC) is a well-known technique for the separation of mixed solutes by adsorption. Compared with related techniques in column chromatography, TLC offers several advantages. The use of adsorbents having large surface area (particle size ~10 μ *vs.* 50 μ) results in excellent resolution. Separation and analysis time are reduced because many samples can be run simultaneously and material need not be eluted for quantitation. Several references provide a comprehensive background on TLC (*1–3*).

TLC has been a widely accepted technique in biochemistry and organic chemistry for a number of years. With modern quantitation TLC is being used today in such diverse applications as air pollution analysis and clinical medicine.

The initial applications of TLC to problems in polymer chemistry were directed to the separation of polymer blends, stereoisomers, and a variety of copolymers (*4–8*). From these investigations occasional molecular weight effects were noted (*8*). Recently efforts in three separate laboratories have been successful in determining the molecular weight distribution of a polymer sample by TLC techniques (*9–12*). An explanation of the facility and high resolving power observed has been sought through a number of continuing experiments (*11–16*). The purpose of this report is to review the results of these studies and to comment on the fractionation mechanisms postulated.

Three types of interactions occur during a TLC run which determine the results. The following is a discussion of these factors, with special emphasis on their relation to the chromatography of polymers.

The solvent–substrate interactions depend on the type of adsorption sites on the substrate particles and on the nature of the functional groups and dipole moment of the eluting solvent. The most common expression of the strength of the solvent–substrate interaction is the so-called eluotropic series. Common solvents are rated in order of polarity and therefore displacing power. A more quantitative rating is found using the solubility parameter δ or solvent strength parameter ϵ° (*3, 17*). In practical terms, the greater the solvent–substrate interaction, the more successful is the solvent in competition with the solute for the adsorption sites on the substrate, and the more completely will the solute be eluted. The interaction of polar solvents with substrates can result in significant changes in the composition of the mobile phase when a mixed solvent elution is being executed. The relative magnitude of this fractionation effect increases as the initial concentration of the more polar component is decreased. Indeed, the chromatographic purification of solvents is the most outstanding example of this phenomenon. These considerations remain largely unaffected by the nature of the solute (*i.e.*, polymeric *vs.* monomeric).

Another consideration unique to TLC is the variation in solvent concentration in the direction of elution (*18*). The phase ratio r is simply the weight of solvent per weight of adsorbent measured at various distances from the dip line. The change in r generates the solvent concentration profile. The decrease in phase ratio is an important consideration in subsequent discussions of the fractionation mechanism.

The solute–substrate interactions are responsible for the separation of low molecular weight solutes in conventional TLC. Solutes are partitioned between the mobile phase and substrate. Adsorption takes place at the "head" of the spot and desorption at the "tail." For solutes with varying affinities for the substrate, relatively more or less time is spent in the absorbed state resulting in different migration distances.

The adsorption and desorption behaviors of polymers have been the object of extensive study for a number of years (*19–21*). The low rates of polymer adsorption and desorption as determined by conventional methods would eliminate these processes from consideration as participating in TLC separations. A recent study has shown, however, that the removal of polystyrene from silica gel occurs rapidly in benzene, an eluting solvent, and is several orders of magnitude slower in CCl_4, a non-eluting solvent (*13*). Both CCl_4 and benzene are good solvents for polystyrene, but benzene is a stronger eluent. An important feature of polymer adsorption is the strong molecular weight dependence of the equilibrium surface coverage and total number of adsorbed segments per molecule (*19–21*). It is very difficult for significant amounts of polymer to adsorb onto any type of substrate from a good solvent.

Some mention should be made concerning the state of the polymer at the beginning of TLC experiment. Normally the sample is applied to the plate from solution in a relatively nonpolar solvent. This solvent is then evaporated, and the chromatographic plate is eluted with the chosen eluent. The drying step results in a polymer deposit which would be difficult to characterize; it is not simply a precipitate, and it probably is not a simple adsorbed (multi) layer. Redissolution and entry into the mobile phase under displacement conditions occur in a minute or less (*13*).

The effects of solute–solvent interactions play a greater role in the chromatography of polymers than in conventional TLC. Normally, solubility plays a very small role in the TLC behavior of monomeric compounds. For polymers, however, the effects of solute–solvent interaction are critical. Solvents with δ's which match that of the polymer are "good" thermodynamic solvents, and should displace the polymer during development.

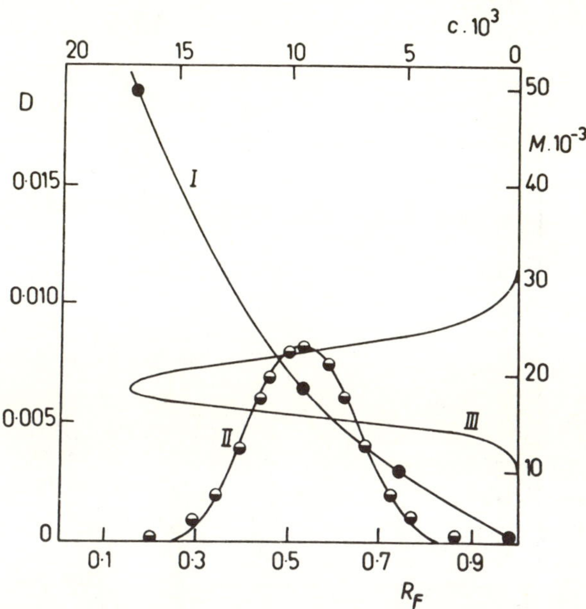

Figure 1. Determination of molecular weight distribution for narrow polystyrene standard. Curve I, calibration of R_f *vs.* M; *curve II, densitogram of two dimensionally developed sample; curve III, molecular weight distribution:* $\overline{M}_w/\overline{M}_n = 1.02$ (12).

Polymer Behavior in TLC

Figures 1, 2, and 3 show examples of molecular weight distributions (MWD) determined by TLC. For the analyses of Otocka and Inagaki, similar gradient elution methods were employed. Quantitation was achieved by densitometry *in situ* in one case and by densitometry of a photographic record of the plate in the other cases. Belenkii and Gankina employ a two-dimensional development method to overcome "chromatographic spreading" followed by photographic densitometry to determine the MWD of polystyrene fractions. These results are indicative of the broad range of polydispersities for which the technique is applicable. In order to advance in more generalized utility where other polymers can be studied, the mechanism of this high-resolution technique must be understood.

Several experiments by Inagaki lead to the postulation of a precipitation mechanism as the prime source of fraction in TLC. Samples of isotactic poly(methyl methacrylate) show migration with $0 < R_f < 1$ (R_f is defined as the solute migration from the starting line divided by the solvent front travel from the starting line) in mixed $CHCl_3$–CH_3OH in two separate eluent composition regions. In the first region (~80% $CHCl_3$), samples of different molecular weight show no difference in R_f. The sec-

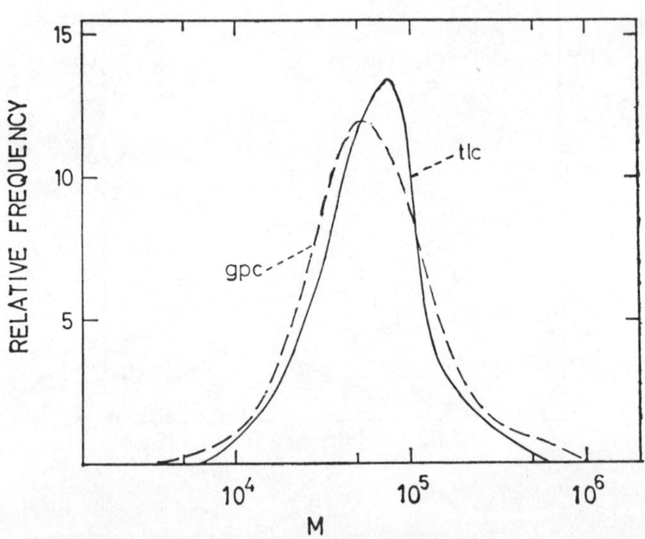

Figure 2. Comparison of GPC and TLC molecular weight distributions of polystyrene $\overline{M}_w = 76{,}600$, $\overline{M}_n = 45{,}400$ from TLC; $\overline{M}_w = 71{,}800$, $\overline{M}_n = 39{,}300$ from GPC (11).

ond region (~70% CH_3OH), where precipitation is imminent, shows a strong dependence of R_f on molecular weight (*14*).

Another indication that precipitation and not adsorption is operative depends on the observation that samples of the same molecular weight migrate identical distances from the dip line, independent of their starting positions on the TLC plate (*15*). This of course would not be the case if adsorption governed chromatographic mobility. To explain these phenomena, Inagaki has proposed that the solvent concentration profile results

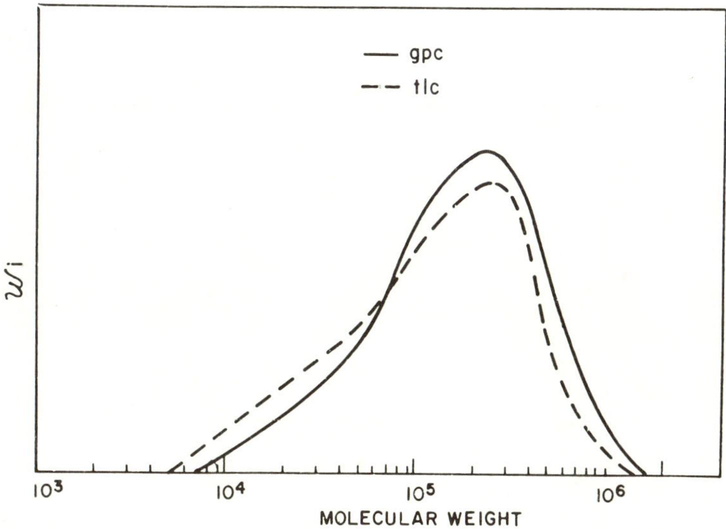

Figure 3. Comparison of GPC and TLC molecular weight distribution for a high polydispersity, high molecular weight polystyrene. $\overline{M}_w = 2.22 \times 10^5$, $\overline{M}_n = 8.16 \times 10^4$ *from GPC;* $\overline{M}_w = 2.14 \times 10^5$, $\overline{M}_n = 7.0 \times 10^4$ *from TLC* (10).

in precipitation. A typical solvent concentration profile is shown schematically in Figure 4. The exact contours and values of the solvent concentration profile depend on the method of development (ascending, descending, horizontal) and the nature of substrate (void volume) and solvent (polarity, vapor pressure). It was found that a measured solvent concentration profile and a series of R_f^* (distances from dip line) measurements indicated a polymer concentration of 0.01 volume fraction at precipitation. A separate measurement in the eluent gave ~0.03 as the concentration of polymer at precipitation for the same temperature (*15*).

More recently, Otocka measured the R_f^* values for several polystyrenes using dioxane–methanol and dioxane–2-propanol θ solvents (*16*). The results shown in Table I indicate that despite a great variety in the type of adsorbent nature, R_f* values are quite similar for similar development techniques. However, the agreement in calculated concentrations at precipitation is not very good between ascending, descending, and horizontal development. Based on Inagaki's solvent concentration profile and our dip line phase ratio, Otocka calculated concentrations at precipitation for ascending development between 0.003 and 0.006 gram of polymer per gram of solvent depending upon molecular weight. Light-scattering studies showed no turbidity at concentrations of polymer to 0.01 gram of polymer per gram of solvent (molecular weight = 160,000). Thus, while the precipitation theory of fractionation offers good qualitative explanation for the behavior of polymers in TLC, assessment of the quantitative aspects is not yet fully developed.

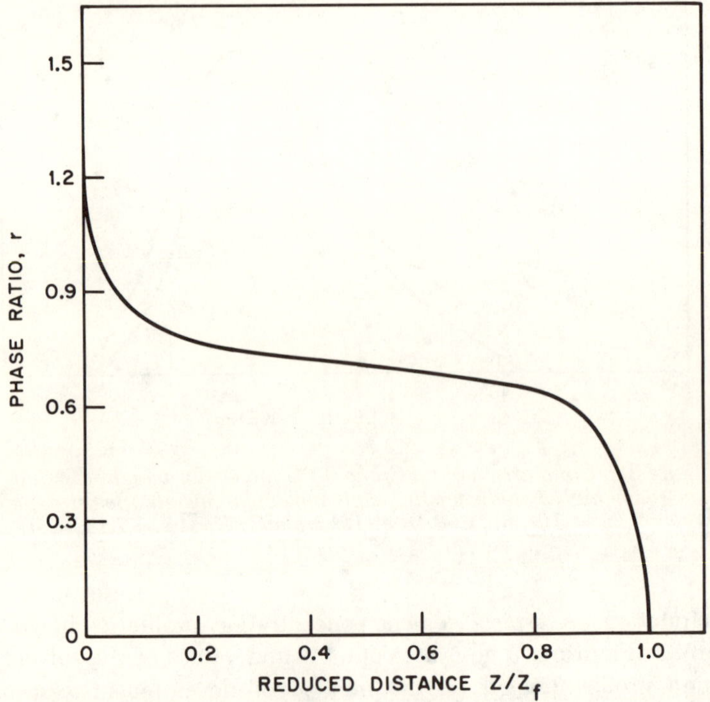

Figure 4. Schematic solvent concentration profiles. Phase ratio, r *(weight of solvent per weight of adsorbent) as a function of reduced distance* z/z_{final} *from the dip line.*

The adsorption concept of fractionation cannot account for the fact that samples reach a constant position from the dip line or that the nature of the substrate is unimportant in determining the retardation of sample migration. There are still several general observations that indicate adsorption effects can be detected in TLC. CCl_4 dissolves polystyrene of all molecular weights studied, yet it does not displace any. Acetone dissolves only low ($<\approx 20{,}000$) molecular weight polystyrene samples and displaces them quite readily. The addition of small amounts of acetone to CCl_4 permits displacement of molecular weights in excess of 10^5. Furthermore, it has been noted that when polystyrene samples are applied to plates from eluting solvents, the spot concentration profile is high at the edges and low in the center, while samples applied from CCl_4 show a near gaussian concentration profile. This pattern indicates differences in adsorption during sample application, depending on the displacing power of the carrier solvent.

Table I. Polystyrene Behavior for Various Development Techniques

	R^{f*} [a]					
	Plates [b]					
Sample	*S, A, a*	*S, D, a*	*C, a*	*S, A, h*	*S, D, h*	*S, A, d*
PS-2 (19,800)	0.75	0.77	0.75	0.66	0.72	0.70
PS-4 (160,000)	0.66	0.66	0.64	0.39	0.48	0.63
PS-6 (860,000)	0.63	0.61	0.63	0.30	0.30	0.60

[a] The eluent is dioxane (72%)–methanol (28%) θ solvent. Distances are measured from the dip line rather than the spotting line.
[b] S = silica, C = cellulose, A = activated, D = deactivated (fluorosilane), a = ascending, h = horizontal, and d = descending.

A number of experiments are possible which will add final clarification to questions concerning the details of the mechanism. Modern column chromatographic methods can be used to evaluate the role of the adsorbent. In a column, no solvent concentration profile exists, and there are at least two papers which indicate adsorption separations can be successful (*22, 23*). More extensive work is necessary to characterize the solvent concentration profile to improve the quantitative agreement between bulk precipitation concentrations and *in situ* precipitations in TLC.

Recent work has indicated that thin-layer gel permeation chromatography (TLGPC) is possible on wet plates (*16*). The resolution is much less than for the conventional TLC approach.

Conclusions

The precipitation mechanism proposed by Inagaki offers a good qualitative explanation for the fractionation of polymers by TLC. The quantitative agreement between TLC precipitation and bulk precipitation is lacking at the present time. Further experiments are needed to provide exact definition of the role of the adsorbent.

Literature Cited

1. Stahl, E., "Thin Layer Chromatography," 1962, Academic Press, New York.
2. Randerath, K., "Thin Layer Chromatography," 1963, Academic Press, New York.
3. Kirchner, J. G., "Techniques of Organic Chemistry, Vol. XII, Thin Layer Chromatography," 1967, Perry, E. S., Weissberger, A., Eds., Interscience, New York.
4. Langford, W. J., Vaughan, D. J., *J. Chromatogr.* (1950) **2**, 564.
5. Inagaki, H., Miyamoto, J., Kamiyama, F., *J. Polym. Sci., Part B* (1969) **7**, 329.
6. Inagaki, H., Matsuda, H., Kamiyama, F., *Macromolecules* (1968) **1**, 520.
7. Kamiyama, F., Matsuda, H., Inagaki, H., *Makromol. Chem.* (1969) **125**, 286.
8. Belenkii, V. G., Gankina, E. S., *Dokl. Akad. Nauk SSSR* (1969) **186**, 857.
9. Otocka, E. P., Hellman, M. Y., *Macromolecules* (1970) **3**, 362.
10. Otocka, E. P., *Macromolecules* (1970) **3**, 691.
11. Kamiyama, F., Matsuda, H., Inagaki, H., *Polym. J.* (1970) **1**, 518.
12. Belenkii, B. G., Gankina, E. S., *J. Chromatogr.* (1970) **53**, 3.
13. Otocka, E. P., Muglia, P. M., Frisch, H. L., *Macromolecules* (1971) **4**, 512.
14. Inagaki, H., Kamiyama, F., Yagi, T., *Macromolecules* (1971) **4**, 133.
15. Kamiyama, F., Inagaki, H., *Polym. J.*, in press.
16. Otocka, E. P., Hellman, M. Y., Muglia, P. M., *Macromolecules* (1972) **5**, 227.
17. Snyder, L. R., "Principles of Adsorption Chromatography," 1968, Marcel Dekker, New York.
18. Giddings, J. C., Stewart, G. H., Ruoff, A. L., *J. Chromatogr.* (1960) **3**, 239.
19. Rowland, F. W., Eirich, F. R., *J. Polym. Sci., Part A-1* (1966) **6**, 2033, 2401.
20. Koral, J., Ullman, R., Eirich, F. R., *J. Phys. Chem.* (1958) **62**, 541.
21. Stromberg, R. R., Grant, W. H., Passaglia, E., *J. Phys. Chem.* (1965) **69**, 3955.
22. Yeh, S. J., Frisch, H. L., *J. Polym. Sci.* (1958) **27**, 149.
23. Bannister, D. W., Phillips, C. S. G., Williams, R. J. P., *Anal. Chem.* (1954) **26**, 1451.

RECEIVED January 17, 1972.

6

The Use of Mass Chromatography to Measure Molecular Weights and to Identify Compounds Related to the Polymer Field

C. EUGENE BENNETT and DONALD G. PAUL

Chromalytics Corporation, Unionville, Pa. 19375

Mass chromatography is a new form of gas chromatography that uses two gas density detectors operated in parallel and provides (a) mass of components within 1–2% relative without determination of response factors, (b) molecular weight of components within 0.25–1% in the mass range 2–400, and (c) a powerful identification tool by the combined use of retention time and molecular weight data. The theoretical basis of the technique and its scope as a molecular weight analyzer, a qualitative identification tool, and a quantitative analyzer in the polymer field are discussed.

Mass chromatography is a new form of gas chromatography that maintains the retention data of GC while providing new information to calculate the molecular weight and absolute weight of each component. Although mass chromatography is a recent innovation of Paul (*1*, *2*), its basis was established by Liberti, Conti, and Crescenzi in 1956 (*3*). In the earlier work, the response of a compound was measured with a single gas density detector by making consecutive runs with a high and low molecular weight carrier gas. In Paul's work, simultaneous measurement occurs in a parallel chromatographic system including matched columns and two gas density detectors.

The predictable response of the gas density detector makes the overall method possible. The basic formula which describes its operation is

$$D_x = k \frac{MW_x}{MW_x - MW_{CG}} \qquad (1)$$

where D_x is the density of the unknown species x, k is a cell constant dependent upon operating conditions of the instrument, and MW_x and MW_{CG} are the molecular weights of the unknown x and carrier gas, respectively.

By definition, density is weight per unit volume. In case of the gas density cell, density is proportional to weight (wt_x) per chromatographic peak area (A_x). Thus it follows:

$$D_x \approx \frac{wt_x}{A_x} = k \frac{MW_x}{MW_x - MW_{CG}} \tag{2}$$

For quantitative analysis, the cell constant k is determined with a weighed internal standard and then used in calculating the amount of other compounds present. Starting with Equation 2, Paul and Umbreit (*1*) derived the following equation for molecular weight determination in a dual chromatographic system:

$$MW_x = \frac{K \frac{A_1}{A_2} MW_{CG_2} - MW_{CG_1}}{K \frac{A_1}{A_2} - 1} \tag{3}$$

where K is an instrument constant, A_1 and A_2 are the area responses for the unknown compounds in two detectors, and MW_{CG_1} and MW_{CG_2} are the the molecular weights of the two carrier gases.

With the Chromalytics Model MC-2 mass chromatograph, a sample is introduced into the unit, split into two portions, collected onto traps, and then analyzed simultaneously with two gas density detectors as shown in Figure 1. The peak height or area ratios from each detector are measured and the molecular weights calculated from Equation 3.

In actual operation, the molecular weight of an unknown can be determined by measuring the area or peak height ratio and using a K established for known compounds. Fortunately, it is not necessary to know the amount of sample introduced. For either molecular weight or

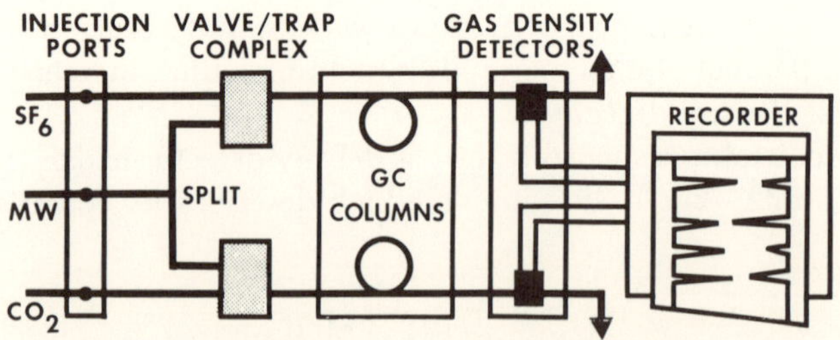

Figure 1. Schematic of the mass chromatograph showing flow, sampling, traps, columns, detector, and dual readout

quantitative analysis, it is only necessary for the split ratio to remain the same for standard and sample measurements, and internal standards can be used to avoid this requirement.

The main benefits of the mass chromatographic system can be summarized as follows. (1) Precise quantitative analysis can be performed without individual peak calibration. (2) Molecular weights are readily determined for compounds that can be gas chromatographed. (3) Peak identification is usually possible by the combined use of molecular weight and retention data (when such data are available). (4) The unique trap design and dual aspects of the instrument are ideally suited for evolved gas analysis from thermal analyzers, catalyst studies, etc. These benefits will be discussed throughout the paper with emphasis oriented to the polymer field.

Scope as a Molecular Weight Analyzer

The mass chromatographic method of measuring molecular weights can be applied to many thousands of compounds although it is limited to those materials which can be gas chromatographed or maintained in a vapor state. In many instances, volatile derivatives of nonvolatile or nonstable compounds can be utilized to extend the method. In general, organic compounds with molecular weights up to and over mass 400 can be analyzed. For polymers and other nonvolatile compounds, the sample can be thermally decomposed and the fragmentation products analyzed. The composition of these materials can often be established by this procedure.

The accuracy of the molecular weight measured on a theoretical basis can be determined from an analysis of Equation 3. The equation reveals that for any given pair of carrier gases, the error in a calculated molecular weight will be dependent upon the uncertainties of K and the peak ratio A_1 to A_2. Theoretical error curves for molecular weight can thus be obtained with pairs of carrier gases by varying the uncertainties of K and peak ratios. Such information has been reported by Swingle (*4*), and is reproduced in Figure 2. In Figure 2, CO_2 and SF_6 were the carrier gases used, and the molecular weight errors are shown as a function of molecular weight and different uncertainties of K. The data assume no error in peak ratio measurement. If the K could be measured to a standard deviation of $\pm 0.25\%$, then a standard deviation of ± 0.5 mass unit would hold up to molecular weight 250.

Swingle found, however, that with a variety of compounds, a $\pm 1\%$ standard deviation of K was more realistic, and a standard deviation of ± 1 mass unit at 200 and below would be achieved. Dutton (*5*) has found similar results in studies of hydrocarbons, aldehydes, and methyl esters of fatty acids. With careful work, however, our laboratory has determined *K* with a standard deviation of ±0.5%. In so doing, a standard deviation of ±1 mass unit at 250 is achievable.

The uncertainties of *K* and the response ratio depend upon the ability to measure detector response. Peak height ratios can be used with excellent results (generally better than areas) when the peaks are symmetrical and sharp. Peak hight ratios are also more useful than peak areas in overlapping peaks. For automated systems, peak areas are preferred since the ratios are readily measured, and the data are calculated with electronic integrators and computers.

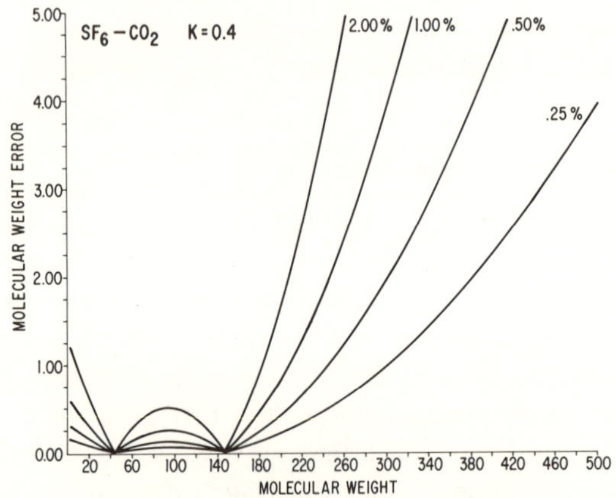

Figure 2. Theoretical molecular weight error curves as a function of K *from Equation 3*

The same precision as discussed above can be extended about 50 mass units by using N_2 (molecular weight 28) and perfluoropropane (molecular weight 188) compared with CO_2 and SF_6. For example, with a standard deviation in *K* of ±0.5, a mass error standard deviation of ±1 mass unit would be 300 instead of 250. Since the measurement of detector response is a function of the recorder (peak heights), integrator system (for areas), columns (absorption sites), electronics, temperature, etc., the overall precision of molecular weight measurement should be further improved in the future.

Mass chromatography has several advantages (2) over conventional methods of molecular weight measurement such as cryoscopy, ebulliometry, vapor density, and osmometry. For example, these methods require a pure sample and are dependent on ideal solution behavior or extrapolation to infinite dilution.

The main advantage of mass spectroscopy over mass chromatography is its precision of measurement and fragmentation patterns for structure determination. Mass chromatography, on the other hand, is a simple technique, and the determination of molecular weights represents a calculation rather than interpretation.

Scope as a Qualitative Identification Tool

As a qualitative tool, gas chromatography has been limited to retention data and a variety of ancillary techniques. Despite the limitation, retention data are widely used, and various tabulations of such information are available (2). Because mass chromatography provides both molecular weight and retention time for each peak, the technique represents a powerful means of identifying compounds as is illustrated in Figure 3 and Table I.

In Figure 3, a straight-line relationship between a retention index and molecular weight is illustrated for homologous series of *n*-alcohols, *n*-aldehydes, *n*-acetates, and *n*-hydrocarbons. The compound having molecular weight 130 and retention index 1180 corresponds to *n*-pentyl acetate. Since the relationship in Figure 3 holds for all types of related compounds, it is obvious that this technique could be widely used in identification of GC peaks.

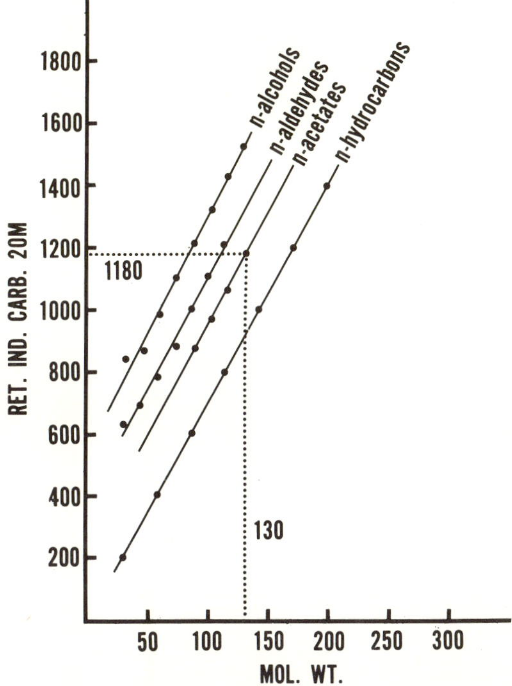

Figure 3. Relationship between molecular weight and retention index

The utility of the method is illustrated in Table I for the identification of an unknown mixture. The only information provided with the sample was that the compounds were from a list of 340 compounds in a book by McReynolds (*6*) listing retention data. The procedure used for identification was as follows. (1) The sample was chromatographed on an SE-30 column, and five separate peaks were observed. (2) Molecular weights and retention indices were determined, and the number of possible compounds from McReynolds having within ±2 mass units and ±50 retention index of the measured values were tabulated (Table I). Peak 2 was identified, and only two possibilities existed for Peak 3. (3) The sample was then chromatographed on a Carbowax 20M column, and an extra peak was separated making a total of six compounds. (4) Molecular weights and retention indices were determined. (5) A comparison of the molecular weights of six compounds along with retention indices on the two columns enabled positive identification as follows: peak 1, *n*-hexane or 2,3-dimethylbutane; peak 2, carbon tetrachloride; peak 3, chloroform; peak 4, toluene; peak 5, 2-, 3- or 4-heptanone; and peak 6, *n*-octyl alcohol. (6) Peaks 1 and 5 were identified as *n*-hexane and 2-heptanone by comparing retention data of standards.

Table I. Identification of Unknown Mixture Using Both Molecular Weight and Retention Index

GC Peak Identification

Peak No.	*Mol. Wt.*	*Retention Index*	*Possible Compds* [a]
SE-30			
1	100.2 [b]	600	6
2	154.2	650	1
3	91.9	760	2
4	113.9	870	11
5	138.5	1080	6
Carbowax 20M			
1	85.7 [b]	600	2 [c]
2	154.1	870	1 [c]
3	119.5 [b]	1010	1 [c]
4	92.3	1040	1 [c]
5	113.8	1175	3 [c]
6	129.9	1525	1 [c]

[a] From McReynolds' 340 Compounds, ±2 Mass Units, ±50 RI.
[b] Peak 1 (SE-30) = Peaks 1 and 3 (Carbowax (20M).
[c] Uses both SE-30 and Carbowax 20M data.

The mass chromatographic technique obviously has a wide scope of applications in the identification of volatile compounds. Its main limitation is the lack of retention data for all possible compounds. This is not a practical limitation where the worker deals in specialized areas such as monomers, plasticizers, additives, etc., or where data are available or can be readily prepared.

Scope as a Quantitative Analyzer

The gas density detector provides a predictable response as outlined in Equation 2 provided that the molecular weights of the various components in the sample are known. The detector has never achieved its full quantitative potential since molecular weight information is often not available.

Since the mass chromatograph provides molecular weight data, the instrument is ideally suited for quantitative gas chromatography. Considerable time and uncertainty are saved using the system compared with flame ionization and thermal conductivity detectors which require response factors for each and every compound.

In many laboratories, analysts do not determine response factors and hedge their results by reporting "area %" rather than "weight %." The justification for this procedure is that related compounds have similar response factors. This assumption is a dangerous one as is clearly demonstrated in Table II.

Table II. Quantitative Analysis of a Solvent Mixture

		Relative Error, %		
Compound	*Actual Wt. %*	*Gas Density* [a]	*Thermal Conductivity* [b]	*Flame Ionization* [b]
n-Octane	14.8	2.0	26.7	62.8
Tetrahydrofuran	18.6	1.6	14.4	1.1
Chloroform	30.9	0.3	23.6	84.5
n-Butyl alcohol	17.0	1.2	9.4	15.9
o-Xylene	18.5	1.6	4.9	75.7
Average Net Error, %		1.3	15.8	48.0

[a] Weight calculated using Equation 2.
[b] Weight calculated without determining individual response factors and assuming all compounds gave equal response by weight.

The large errors associated with the analysis using thermal conductivity (15.8%) and flame ionization (48.0%) detectors are understandable because of the diverse chemical nature of the solvents. Obvious good results could be achieved by individual calibration of response factors. The point being made is that the gas density detector gives excellent results without calibration even for unknown mixtures.

Other workers have found similar performance for the gas density detector. Walsh and Rosie (7) reported an average accuracy of ±1.6% for a series of ten compounds and ±1.1% for a different series of five compounds. Guillemin (8) even recommends the use of the gas density cell as a "detector calibrator" to determine the response factors for flame and thermal conductivity detectors.

One major aspect of quantitative analysis is sensitivity and dynamic range of linearity. Such data have been reviewed (2) for the gas density, thermal conductivity, and flame ionization detectors. Since response is a function of molecular weight in the gas density detector, it is difficult to make comparisons in a simple manner. In general, however, the sensitivity of the gas density cell is about twice that of comparable thermal conductivity cells and about one-tenth that of flame ionization detectors (when bleed of the column is limiting).

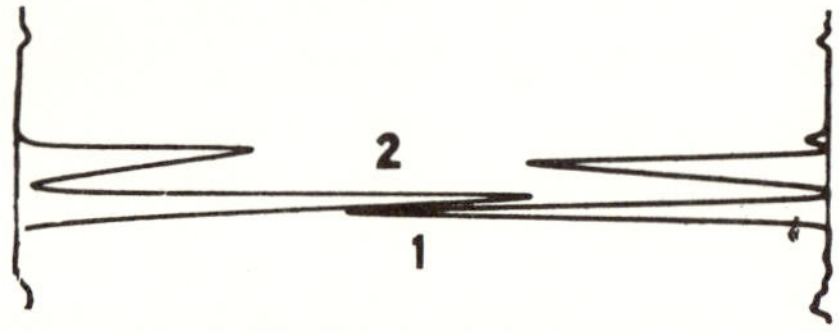

Col. 10% DC 410 – 12′x1/8″
Prog. 100-250°C @ 5°/min.

Figure 4. Mass chromatogram of reactor products 2,6-dibromohexane with sodium cyanide. Peak 1: 2,6-dibromohexane, molecular weight 244.0 (theory), 244.1 (MC-2). Peak 2: 2-cyano-6-bromohexane, molecular weight 190.1 (theory), 190.6 (MC-2).

Applications to the Polymer Field

Mass chromatography, by virtue of providing molecular weight, retention time, and quantitative analysis, is widely applicable to solving analytical problems. Some categories of application are summarized in the following paragraphs.

Identification of Reaction Products. Probably the biggest single application of mass chromatography is its use in analyzing reaction products. The mass chromatogram shown in Figure 4 answered an organic chemist's question as to whether peak 2 represented 2-cyano-6-bromohexane or 2,6-cyanohexane. He was developing a procedure to prepare 2-cyano-6-bromohexane by treating 2,6-dibromohexane with sodium cyanide. The retention time of peak 1 corresponded to that of the starting material, and neither product was available to compare retention times.

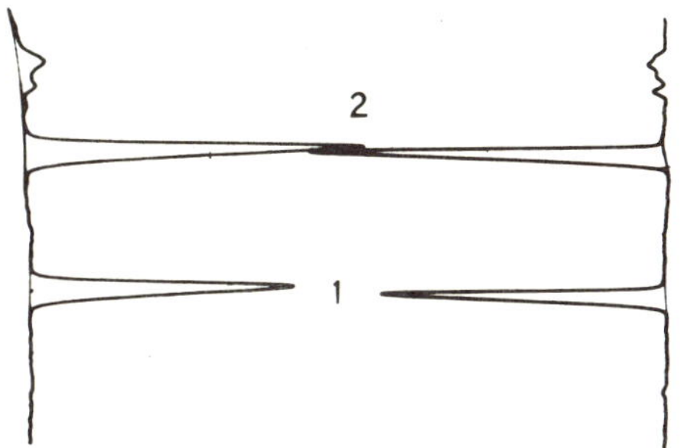

Col. 10% DC 410 – 12′x1/8″
Prog. 120-250°C @ 5°/min.

Figure 5. Mass chromatogram of photoinitiators in a polymer system. Peak 1: benzoin, molecular weight 212.24 (theory), 212.6 (MC-2). Peak 2: hydrobenzoin, molecular weight 214.26 (theory), 214.5 (MC-2).

A molecular weight determination clearly showed that the second peak corresponded to 2-cyano-6-bromohexane (molecular weight 190.1) and not 2,6-dicyanohexane (molecular weight 136.2). This procedure for identification offers clear advantages over collecting the peak for identification by infrared or other techniques.

Analysis of Volatiles in Polymers. A variety of volatile products appears in polymer products ranging from residual monomers, residual initiators, moisture, antioxidents, plasticizers, etc. The mass chromatograph is useful in providing both quantitative analysis and qualitative identification of such materials, especially in maintaining quality control and analyzing commercial products. An example of the latter application is shown in Figure 5 in the identification of residual photoinitiators used in a commercial polymer product. A comparison of molecular weights of common photoinitiators with those in the sample suggested that the compounds were benzoin and hydrobenzoin. Their retention times also corresponded to these compounds. Subsequent analytical work confirmed the results to be correct.

Study of Volatile Products from Thermal Analyzers. The MC-2 mass chromatograph is ideally suited for thermal analysis or pyrolysis studies for the obvious reasons of quantitative and qualitative analysis and also for its unique trapping assembly. With the traps, sample effluents can be collected and concentrated over extended periods of time prior to analysis.

On the other hand, the design of most gas chromatographic instruments requires that pyrolysis occur rapidly in order to prevent peak spreading. Thus, by necessity, pyrolyzers such as lasers, curie-point probes, and capacity discharge filaments have been designed to provide instantaneous high-temperature degradation. As a result, considerable fragmentation of products occurs, and a complex chromatogram results. Such pyrolyzers can be used with the mass chromatograph, but the more versatile technique of slow decomposition, trapping, and subsequent analysis is preferred.

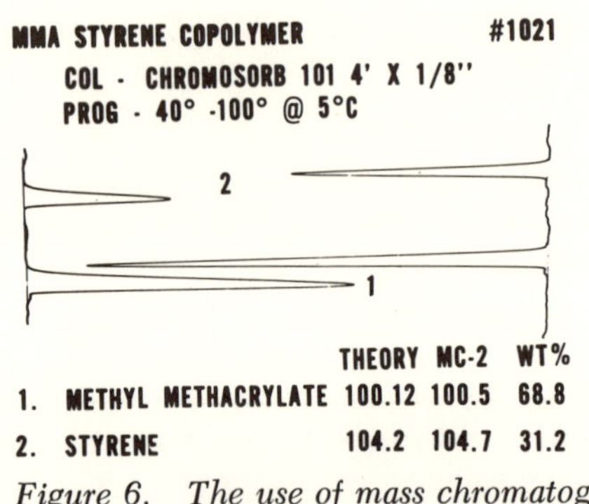

Figure 6. The use of mass chromatography to determine composition of a polymer sample

A simple but elegant example of the broad utility of mass chromatography is shown in Figure 6. In this case, a polymer sample was thermally decomposed at 400°C, and the products were analyzed. The molecular weight, retention time, and quantitative analyses of the major products indicated that the sample was a 69% methyl methacrylate–31% styrene copolymer.

Literature Cited

1. Paul, D. G., Umbreit, G. E., *Res. Develop.* (1970) **21**, 18.
2. Bennett, C. E., DiCave, L. W., Jr., Paul, D. G., Wegener, J. A., Levase, L. J., *Amer. Lab.* (1971) **3**, 67.
3. Liberti, A., Conti, L., Crescenzi, V., *Atti Dell. Accad. Naz. Dei Linc. Rend.* (1956) **20**, 623; *Nature (London)* (1956) **178**, 1067.
4. Swingle, R. S., *Ind. Res.* (Feb. 1972) 40.
5. Dutton, H. J., Lanser, A., Ernst, J. O., Kwolek, W. F., data submitted for publication.
6. McReynolds, W. O., "Gas Chromatographic Retention Data," Preston Technical Abstracts Co., Evanston, Ill., 1966.
7. Walsh, J. T., Rosie, D. M., *J. Gas Chromatogr.* (1967) **5**, 232.
8. Guillemin, C. L., Auricourt, F., Blaise, P., *J. Gas Cromatogr.* (1966) **4**, 338.

RECEIVED January 17, 1972.

7

Electrospray Mass Spectroscopy

MALCOLM DOLE, H. L. COX, Jr., and J. GIENIEC

Department of Chemistry, Baylor University, Waco, Tex. 76703

Electrospray mass spectroscopy, mass spectroscopy of ions created by an electrospray process, should be significant for the study of nonvolatile substances such as high polymers or thermally unstable lower molecular weight species. Intact gas-phase macroions of polystyrene, polyvinylpyrrolidone, and lysozyme were prepared by this technique, and low-resolution M/z *distributions were inferred from data obtained using a nozzle beam system with repeller-grid, Faraday-cage system as an analyzer–detector. Use of a time of flight mass spectrometer to improve resolution is not feasible, since the magnetic electron multipliers used to detect ions in TOF spectrometers have been shown to have negligible response to macroions. Applicability of the Plasma Chromatograph to determination of the charge states of gaseous macroions is also being investigated.*

Currently available techniques of determining molecular weights and molecular weight distributions permit only low-resolution analyses. Müller (*1*) has recently stated "If we could determine molecular weights [of macromolecules] to a tenth of a percent or better, a load could be taken from the microanalysts' shoulders." The primary (although not the only) problem which has heretofore prevented the development of a mass spectrometer for use with macromolecules is that of producing intact gas-phase macroions. Owing to the low vapor pressures of macromolecules, macroions cannot be produced in the gas phase by conventional techniques without extensive degradation and/or fragmentation. Electrospray mass spectroscopy provides a solution to this basic problem.

Electrospray mass spectroscopy (EMS) is the mass spectroscopy of gaseous ions produced by electrospraying into a suitable gas at atmospheric pressure a dilute solution containing as solute the macromolecules in question. (Although the use of other gases is possible, we have obtained our best results with nitrogen gas.) Although the technique is still in its infancy, it was first proposed (*2*) in 1966, and demonstrated in the case of polystyrene (*3, 4*) in 1968. Further studies (*5–7*) appeared in 1970 and 1971.

The Electrospray Technique

The prime requisite for any type of solution-spraying process suitable for use as an ion source in a mass spectrometer is that it produce droplets small enough so that on final evaporation of the solvent single macroion species will result. Various methods of aerosol production have been considered for this application (*8*). Mechanical and ultrasonic methods of aerosol production are generally limited to droplet diameters greater than 1 μ. In order to produce a reasonable proportion of single macromolecules, it is necessary that the solute concentration be such that an average of no more than one macromolecule be present in each of the final droplets. For a homogeneous droplet diameter of 1 μ, this requires a maximum solute concentration of the order of 3 pg ml^{-1} amu^{-1}. This represents a very low concentration. We have demonstrated experimentally (*5*) that such techniques do not produce individual intact macroions at reasonable solute concentrations. If, however, we can produce droplets of 0.1 μ diameter, solute concentrations up to 3 ng ml^{-1} amu^{-1} are permitted. This represents a reasonable concentration for macromolecules. It was demonstrated in 1952 (*9*) that the electrospray process can produce aerosols with rather homogeneous droplet diameters in the neighborhood of 0.1 μ or less.

The electrospray process consists of feeding a liquid through a metal capillary which is maintained at a high electrical potential with respect to some nearby surface. As the liquid reaches the capillary tip, the liquid is dispersed into fine electrified droplets by the action of the electric field at the capillary tip. If the liquid is volatile, as the liquid evaporates the droplets shrink in size, become electrically unstable, and break down into smaller size droplets. This process has been experimentally demonstrated by Doyle, Moffett, and Vonnegut (*10*) and by Abbas and Latham (*11*). If the liquid contains macromolecules, after the solvent has evaporated completely the macromolecules are left as electrically charged particles in the gas phase, that is, as gaseous macroions.

Although numerous attempts to provide theoretical explanations of the electrospray process have been made (*see,* for example, ref. *12–17*), a good quantitative theory of the phenomenon would require simultaneous solutions of the hydrodynamic and electrostatic differential equations, and, to our knowledge, no such theory has yet been proffered. However, experimental observations have provided some insight into the process.

Experimental observations (*19*) have indicated that liquids with molecular dipole moments less than 10^{-18} dyn$^{1/2}$ cm^2 or with specific conductivities outside the range 10^{-5}–10^{-13} ohm^{-1} cm^{-1} cannot be dispersed electrically. Schultze (*18*) investigated the electrospraying quality of various liquids, and observed that the liquids producing the finest and most stable sprays had specific conductivities in the range 2×10^{-8} to 5.6×10^{-6} ohm^{-1} cm^{-1}. He also observed that application of hydrostatic pressures to force the liquid out of the capillary resulted in a coarser spray with the coarseness of the spray increasing as the pressure increases. Gieniec (*7*) has obtained data indicating that the dielectric constant of the liquid is also an important factor. Liquids with low dielectric constants such as dioxane cannot be electrosprayed while liquids with high dielectric constants such as distilled water cannot be electrosprayed easily. However, a liquid having a high dielectric constant can be mixed with a liquid having a low dielectric constant to produce a liquid with an intermediate dielectric constant which can be electrosprayed successfully. An example is 70% dioxane (dielectric constant 2.209 at 25°C) and 30% distilled water (dielectric constant 80.37 at 25°C) by volume.

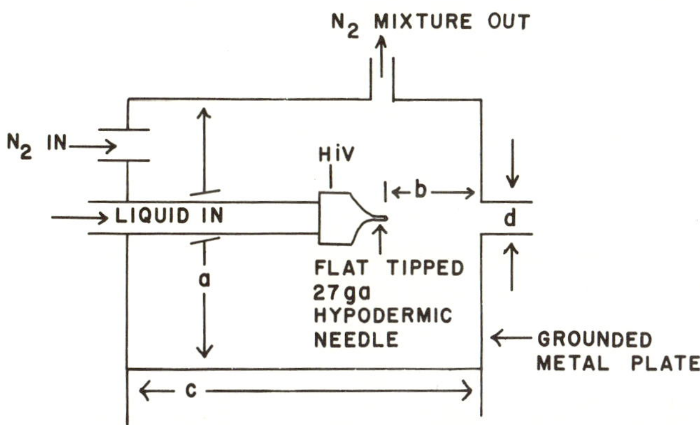

Figure 1. Schematic of a typical electrospray apparatus as used in electrospray mass spectroscopy (see *text for explanation of dimensions)*

It has also been observed that differences in the physical structure of the spraying system can influence the spray. A typical spraying system as used in EMS is illustrated schematically in Figure 1. The N_2 gas flow is required to dry the droplets and to help sweep the ions away from the needle. In the system employed by Dole *et al.* (*3–6*), the dimensions used were usually a = 13 inches, b = 100 mm, c = 24 inches, and d = 0.1 mm. The orifice of dimension d is used to sample the gaseous mixture and is not functional in the production of the spray. Using voltages on the needle of 4.5 kV and up they were able to obtain good sprays of both positive and negative polarity. However, best results were obtained with negative sprays. In a similar system used by Gieniec (*7*), the dimensions were $a = 44$ mm, $b = 5.5$ mm, $c = 25$ cm, and $d = 9$ mm. In this system the orifice of diameter d is a collimating orifice, and sampling takes place 9 cm downstream from this orifice. In addition, all of the gaseous mixture exits through this orifice rather than through a separate orifice. Again, using high voltages of 4.5 kV, good sprays are obtained with positive voltages, but negative sprays could not be formed. Using variations of these basic systems, Drozin (*19*) and Vonnegut and Neubauer (*9*) obtained both positive and negative sprays but observed that a spray with smaller droplets is obtained when the capillary is charged positively.

All of the work with macroions reported to date has been done using mixtures of either ethanol and water or acetone and benzene. Further investigations into the electrospray process should permit optimization of the system and production of a wide variety of solvent mixtures suitable for use in the electrospray process.

Formation of Macroion Beams

After the macroions are produced by the electrospray process, they are injected into a vacuum by use of a nozzle-beam system of the type first suggested by Kantrowitz and Grey (*20*) and later modified by Becker and Bier (*21*). Articles describing in detail the principles of operation of such a system have been published, among them being articles by Anderson, Andres, and Fenn (*22, 23*).

A typical nozzle-beam system with associated electrospray apparatus is shown schematically in Figure 2. Although nozzle-beam systems differ widely in the details of construction, the basic principles of operation are the same.

The gaseous mixture at the stagnation pressure (p_0) (atmospheric pressure in our systems) expands nearly isentropically through the nozzle orifice (100 μ diameter in our systems) into the region between the nozzle and skimmer, kept at pressure p_1 (normally 25 μ in our systems). As the expansion takes place, much of the energy of random thermal motion is converted into energy of forward-directed mass motion. Supersonic flow and aerodynamic shocks result. The shock region is barrel shaped and closed off at the downstream end. Inside the barrel shock, near molecular flow results, and, for points more than a few nozzle diameters downstream of the nozzle orifice and away from the shock boundaries, the flow approximates that of a radial source with its point of origin located a few nozzle diameters downstream of the nozzle orifice

(*24*). That is, the gas density varies as the inverse square of the distance downstream of this point.

Since the shock region is essentially a closed surface, the shock surfaces disrupt this flow. It has been shown [see, for example, Anderson *et al.* (*22, 23*)] that this disruption can be prevented, and molecular beams of rather uniform velocities and high intensities formed by immersing a conical "skimmer" into the shock region. The shocks become attached to the skimmer surface, and molecular flow through the skimmer orifice results. This also permits differential pumping of the system so that lower background pressures (approximately 5×10^{-5} mm in our systems) can be achieved in the analyzing and detecting region. Anderson *et al.* (*23*) among others have discussed the optimum design and placement of the skimmer.

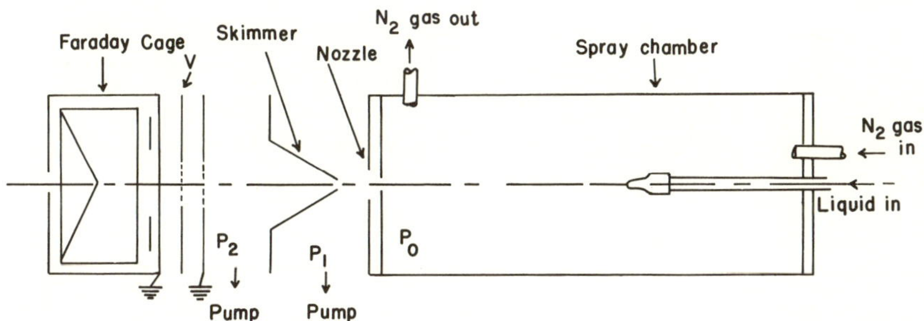

Figure 2. Electrospray chamber–nozzle beam–mass analyzer assembly [*after Dole* et al. (4)]

Since, during the expansion, thermal energy is converted into directed motion, the gas in the jet becomes very cold with local temperatures of tens of degrees Kelvin being possible. Because of this low temperature, quite narrow velocity distributions result (*24*). For a homogeneous gas with ratio of specific heats 1.4 (such as nitrogen) the final beam velocity is approximately $2.17v_0$ where v_0 is the speed of sound in the gas at pressure p_0. For nitrogen gas, the final beam velocity is approximately 750 meters/sec.

If the gas is a mixture of several pure gases of different molecular weights, with one of the gases representing only a minor constituent of the mixture, then the minor constituent takes on the velocity of the major constituent. This is the "seeded beam" technique as first verified experimentally by Becker *et al.* (*26, 27*). Since the macroions constitute only about 10^{-8} mole % of the final gaseous mixture, intermediate energy beams of macroions can be produced with the macroions having narrow velocity distributions (*2*).

The seeded-beam technique has a further advantage in that the final beam is enriched in the heavier constituent (*25, 26*). This is the so-called "Mach focussing factor."

Methods of Macromass Analysis

In principle, a conventional magnetic analyzer type of mass spectrometer could be used to measure the M/z (mass to charge) ratios of the macroions resulting from the electrospray process. However, for singly charged ions of mass 10^6 amu (atomic mass units), a magnetic field strength of about 450,000 G would be required to bend the ion path into a 10-cm radius circle if the initial ion velocity were that produced by an accelerating voltage of 1000 V. Thus, although use of magnetic analyzers with low mass macroions (15–50 kamu) is feasible, it is not practical for heavier macroions.

In the case of a time of flight mass spectrometer, the estimates are more reasonable. Inasmuch as the time of flight is proportional to the square root of the mass, one can calculate that for an accelerating voltage of 3000 V and a flight path of 100 cm the flight time for an ion of M/z equal to 10^6 would be 1.4 msec. Measurement of these flight times is entirely feasible. However, two difficulties remain. The first difficulty concerns itself with introducing the macroions into the mass spectrometer in such a way that the initial velocity of the macroions is as near zero as possible. In our work we have attempted to do this by slowing the molecular beam (*4*) of macroions down to thermal velocities and then by pulsing the ions at right angles into the flight tube of the mass spectrometer. To slow the ions down and to deflect them at right angles without loss of intensity represent two difficult problems. At the moment we are working on these two problems but without positive results as yet.

The second major difficulty is that of detecting the macroions since the magnetic electron multiplier (MEM) on which commercial time of flight mass spectrometers are based do not respond sufficiently to macroions. Our observations show no response to negative ions and a small response, amplification factor of about 20, to positive ions. A factor of about 10^6 is needed for operation of the time of flight mass spectrometer. The response of the MEM to positive ions is due to an Auger effect (*28*) which can exist when the ionization potential of the ion is more than twice the work function of the sensitive metal surface of the cathode of the MEM. In general the MEM detects ions because of either a kinetic or potential effect. Schram *et al.* (*29*) have obtained data on the rare gas ions which demonstrate that for operation of the MEM a critical velocity of about 10^6 cm sec^{-1}, independent of the mass, is required. If the extremely long extrapolation of these results is made to ions of macromolecular size, the conclusion can be reached that the kinetic effect would not function when using the MEM to detect macroions. In the case of negative ions an Auger or potential effect is not possible and, since in our experiments neither the positive nor the negative ions attained the threshold velocity, the kinetic effect could not contribute to the response of the MEM. An ion detector of the sensitivity of the MEM but applicable to macroions is critically needed.

The quadrupole mass filter (QMF) is a mass analyzer on whose operation use of an MEM is not necessarily dependent. The ion currents produced are of sufficient magnitude to be measured by means of a Faraday cage and a suitable amplifier such as a vibrating-reed electrometer. The QMF is a true M/z filter which requires no magnetic fields. Since first being proposed by Paul and Steinwedel (*30*), the QMF has been investigated extensively, and the principles and methods of operation are well known (*see*, for example, ref. *31*).

The QMF is comprised of four parallel rods excited by a combination of static and oscillating fields. Ions of a selected M/z range traverse the entire axial length of the mass filter while other ions are forced into unstable trajectories and are ejected from the filter. Usually the ratio of the oscillating and static fields is kept constant, and mass scanning is accomplished by varying this ratio. Frequency scanning could also be used. For stable oscillations of an ion passing through the quadrupole mass filter the values of the ratio $V/mr_0^2\omega^2$ where m is the mass of the ion, $2r_0$ is the distance between electrodes, ω is $2\pi f$ where f is the frequency, and V is either the dc or ac voltage (both are important) must be within certain limits as determined by equations of motion of the Mathieu type (*32*). Hence the stability of the ion path can be maintained for high-mass ions by increasing V or decreasing r_0 and/or ω. Of importance also are the initial beam velocity, the length of the field, and the mass resolution desired. As yet quadrupole mass filters are not commercially available for the macromass range since no need has heretofore existed. However, calculations indicate that production of such an instrument for use with ions with M/z up to 10^6 and resolutions of 0.1% should be relatively straightforward (*7*).

The Plasma Chromatograph (PC, registered trademark of the Franklin GNO Corp.) is also an instrument which does not depend on the use of magnetic fields (*33*). It should be well suited for the study of macroions because (a) the currents measured are well within the range of a Faraday cage–vibrating-reed electrometer detector system, and (b) the instrument operates at atmospheric pressure, thus making unnecessary the reduction in pressure from that of the electrospray chamber.

The fundamental property of the ions obtained from the PC is the electrical mobility of the macroions in nitrogen at atmospheric pressure. To a first approximation the mobility should be directly proportional to the charge on the ion, and probably inversely proportional to its mass to the two-thirds power. Thus the drift time in the PC is more dependent on the ion charge than on its mass. With the PC we hope to determine the charge distribution of the electrosprayed macroions. If the mobility is accurately a function of $M^{2/3}/z$, and if M/z of the same ions could be measured in a quadrupole or time of flight mass spectrometer, then M and z could be determined separately. There are, however, problems associated with the introduction of the macroions into the PC from the electrospray chamber. First of all, the velocity of the nitrogen gas in the spray chamber, which sweeps out the solvent vapor, must not be so high as to cause turbulence in the drift region of the PC, and second, since the input end of the PC is 2500 V above ground, the electrospray chamber has to be floated above ground to about 2500 V. Results, as yet, are inconclusive.

In the work to date, we have employed primarily a retarding grid system as an analyzer with a Faraday cage detector. The grid system is shown in Figure 2. This analyzing system is essentially an energy analyzer yielding an E/z spectrum. However, as mentioned above, in the

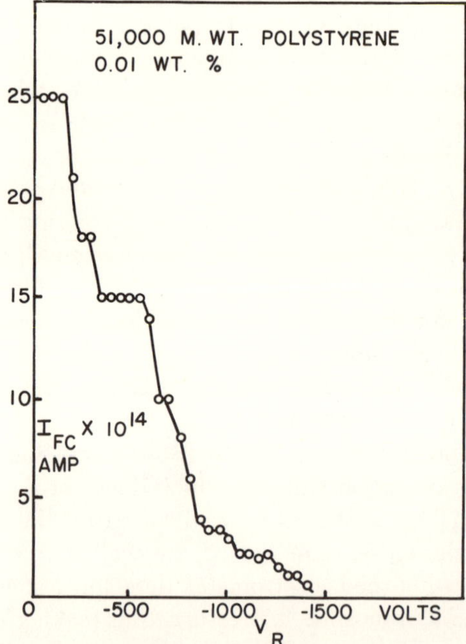

Figure 3. Repelling voltage curves for an electrosprayed 0.01 wt % solution of 51,000 amu polystyrene in two parts acetone to one part benzene by volume (4)

nozzle beam system the final velocity of the ions in the beam can be calculated, and the velocity distribution should be rather narrow (*4*). Therefore, from an E/z spectrum the M/z spectrum can be inferred.

Figure 3 shows a typical result obtained with a fractionated sample of polystyrene of average molecular weight 51,000 in a solvent of two parts acetone to one part benzene by volume. The repelling voltage required to stop a singly charged macroion of molecular weight 51,000 traveling at the calculated beam velocity of 743 meters/sec is 150 *V*. A large decrease in beam current is seen at this voltage. This result is to be

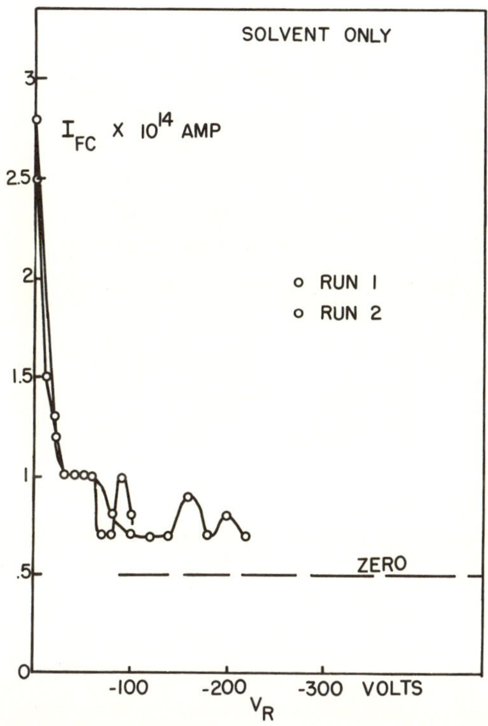

Figure 4. Repelling voltage curves for electrosprayed mixture of two parts acetone to one part benzene by volume (4)

contrasted to that shown in Figure 4 when only the pure solvent was electrosprayed.

Similar results obtained with lysozyme (a protein of molecular weight 14,600) in a 75% ethanol–25% water solution are shown in Figure 5. The vertical bars indicate the repelling voltage required to stop singly charged monomers, dimers, and trimers of lysozyme traveling at the calculated beam velocity of 754 meters/sec. Sharp dropoffs are seen to occur at the position of the first two bars. A small dropoff of curve 2 is seen to occur at the approximate position of the third bar.

These data provide strong evidence that singly charged macroions are being observed, and that the ions are traveling near the calculated beam velocity with a relatively narrow velocity distribution. Thus this technique can be used for low-resolution (approximately 5–10%) M/z analyses.

Degradation in the Electrospray

In some of the early work (*4*) data were obtained which were interpreted in terms of multicharged macroions; specifically, macroions of mass 411,000 amu seemed to be carrying two, three, or five unit charges per molecule. However, these results could equally well have been interpreted in terms of degradation into two, three, or five fragments each

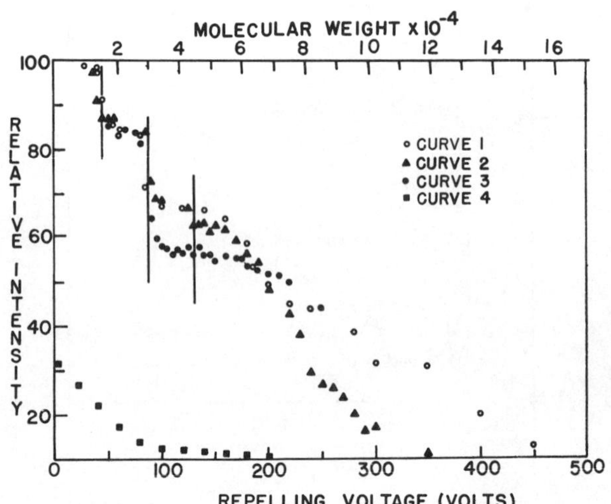

Figure 5. Repelling voltage curves for electrosprayed solution of lysozyme (molecular weight 14,600) in three parts ethanol to one part water by volume. Curve 4 is the repelling voltage curve for electrosprayed solvent only, drawn to the same scale (7).

one of which carried one unit charge. We have been studying (*34*) degradation in the electrospray, and the first results indicate that in the case of molecules of number-average (M_n) molecular weight of about 160,000 there was degradation sufficient to lower M_n to about 140,000 in one electrospraying operation. This is equivalent to about 14% degradation. With the Plasma Chromatograph we should be able to obtain a more definitive answer as to whether the low values of M/z previously observed (*4*) were due to degradation (low values of M) or to high z values. The degradation studies are continuing.

Acknowledgment

This investigation was supported by Public Health Service Research Grant No. 7 RO1 GM 17504 from the National Institute of General Medical Sciences, by Grant No. GP-27672 from the National Science Foundation, and by income from the chair in chemistry at Baylor University endowed by a gift from The Robert A. Welch Foundation. For all of this support we express our grateful appreciation. We are also indebted to Alfred Rheude for the construction of a cell which enabled us to compare the response of the MEM to that of a Faraday cage.

Literature Cited

1. Müller, R. H., *Anal. Chem.* (1968) **40**, 119A.
2. Dole, M., *Prepr., Int. Symp. Macromol. Chem.* (1966) **6**, 132.
3. Dole, M., Hines, R. L., Mack, L. L., Mobley, R. C., Ferguson, L. D., Alice, M. B., *Macromolecules* (1968) **1**, 96.
4. Dole, M., Hines, R. L., Mack, L. L., Mobley, R. C., Ferguson, L. D., Alice, M. B., *J. Chem. Phys.* (1968) **49**, 2240.
5. Mack, L. L., Kralik, P., Rheude, A., Dole, M., *J. Chem. Phys.* (1970) **52**, 4977.
6. Clegg, G. A., Dole, M., *Biopolymers* (1971) **10**, 821.
7. Gieniec, J., Ph.D. Dissertation, University of Wisconsin, 1971.
8. Neher, E., M.S. Thesis, University of Wisconsin, 1967.
9. Vonnegut, B., Neubauer, R. L., *J. Colloid Interface Sci.* (1952) **7**, 616.
10. Doyle, A., Moffett, D. R., Vonnegut, B., *J. Colloid Interface Sci.* (1964) **19**, 136.
11. Abbas, M. A., Latham, J., *J. Fluid Mech.* (1967) **30**, 663.
12. Ailam, G., Gallily, I., *Phys. Fluids* (1962) **5**, 575.
13. Pfeifer, R. J., Hendricks, C. D., *Phys. Fluids* (1967) **10**, 2149.
14. Ryce, S. A., Wyman, R. R., *Can. J. Phys.* (1964) **42**, 2185.
15. Ryce, S. A., Patriarche, D. A., *Can. J. Phys.* (1965) **43**, 2192.
16. Ryce, S. A., Wyman, R. R., *Can. J. Phys.* (1970) **48**, 2571.
17. Shorey, J. D., Michelson, D., *Nucl. Instrum. Methods* (1970) **82**, 295.
18. Schultze, V. K., *Z. Angew. Phys.* (1961) **13**, 11.
19. Drozin, V. G., *J. Colloid Interface Sci.* (1955) **10**, 158.
20. Kantrowitz, A., Grey, J., *Rev. Sci. Instrum.* (1951) **22**, 328.
21. Becker, E. W., Bier, K., *Z. Naturforsch. A* (1954) **9**, 957.
22. Anderson, J. B., Andres, R. P., Fenn, J. B., *Advan. Chem. Phys.* (1966) **10**, 275.
23. Anderson, J. B., Andres, R. P., Fenn, J. B., *Advan. At. Mol. Phys.* (1965) **1**, 345.
24. Anderson, J. B., Andres, R. P., Fenn, J. B., Maise, G., "Rarified Gas Dynamics," Supplement 3, Vol. II, p. 106, de Leeuw, J. H., Ed., Academic Press, New York, 1966.
25. Ashkenas, H., Sherman, F. S., "Rarified Gas Dynamics," Supplement 3, Vol. II, p. 84, de Leeuw, J. H., Ed., Academic Press, New York, 1966.
26. Becker, E. W., Bier, K., *Z. Naturforsch. A* (1954) **9**, 975.
27. Becker, E. W., Bier, K., Burghoff, H., *Z. Naturforsch. A* (1955) **10**, 565.

28. Hagstrum, H. D., *Phys. Rev.* (1954) **96**, 325; (1956) **104**, 672.
29. Schram, B. L., Boerboom, A. J. H., Kleine, W., Kistemaker, J., *Physica* (1968) **32**, 749.
30. Paul, W., Steinwedel, H., *Z. Naturforsch. A* (1953) **8**, 448.
31. Brubaker, W. M., "The Quadrupole Mass Filter," presented at the Congress International des Techniques et Applications du Vide, Paris, June 20–24, 1961.
32. Kiser, R. W., "Introduction to Mass Spectrometry and Its Applications," p. 86, Prentice-Hall, Englewood Cliffs, N. J., 1965.
33. Cohen, M. J., Karasek, F. W., *Res. Develop.* (1970) **8**, 330.
34. Teer, D., unpublished data.

Received January 17, 1972.

8

Recent Trends in the Determination of Molecular Weights and Branching in Elastomers

WILLIAM S. BAHARY[a]

Texas-U. S. Chemical Co., Parsippany, N. J. 07054, and Fairleigh Dickinson University, Teaneck, N. J. 07666

The methodology in determining number- and weight-average molecular weights of polybutadienes by osmometry and light scattering is described and applied to characterization of linear and branched samples. The ratio of intrinsic viscosities, $\bar{g}'$*, and 10% solution viscosities,* $\bar{g}''$*, of branched and linear polymers having the same weight-average molecular weight was used to measure the degree of long-chain branching. Results on 22 polybutadienes prepared by six different catalyst systems and ranging from* 2×10^5 *to* 2×10^6 *in molecular weight and 1.2 to 18 in polydispersity are presented. Whereas the intrinsic viscosity range was 2 to 4.5 dl/gram, the 10% solution viscosity for the same polymers ranged from* 5×10^2 *to* 5×10^4 *cps. The theoretical basis, advantages, and limitations of the method are discussed.*

This paper discusses the determination of molecular weights and molecular sizes of polybutadienes prepared by different catalyst systems. Molecular dimensions have been predominantly measured in dilute solution by light scattering (*1*, *2*), intrinsic viscosity (*3*), and more recently by gel permeation chromatography (*4*, *5*). Only a few studies exist on the effects of long-chain branching on molecular coil size in concentrated solutions. Onogi *et al.* (*6*) and Greassley and Prentice (*7*) have examined the effect of chain branching on the concentrated solution viscosities of poly(vinyl acetate) samples and report a reduction in viscosity with branching. This is in contrast to the enhancement of melt and bulk viscosities with branching observed by Long, Berry, and Hobbs (*8*) for poly(vinyl acetates) and by Kraus and Gruver (*9*) for polybutadienes.

[a] Present address: 291 N. Middletown Rd., Pearl River, N.Y. 10965.

This work examines the effect of long-chain branching on the low-shear concentrated solution viscosity of polybutadienes over a broad range of molecular weights and polydispersity. It will show that the reduction in molecular coil dimension arising from long-chain branching is more sensitively measured in concentrated than in dilute solutions for the polymers examined.

To this end, branching has been determined conventionally in dilute solution by the ratio of intrinsic viscosities of branched and linear polymers having the same molecular weight, defined by (*3*):

$$\frac{[\eta]_B}{[\eta]_L} = f(g, \lambda, z) \quad \text{at constant } M \tag{1}$$

where g is the ratio of the mean-square radii of gyration of the branched and linear polymers, λ is a branching structure parameter, and Z is an interaction parameter depending on solvent type. Since λ, Z, and polydispersity are unspecified, for the purpose of this work the experimentally determined ratio of intrinsic viscosities is represented by $\bar{g}'$ and called the average branching factor:

$$\frac{[\eta]_B}{[\eta]_L} \equiv \bar{g}' \quad \text{at constant } \overline{M}_w \tag{2}$$

where $\overline{M}_w$ is the weight-average molecular weight.

Branching factors were also determined by the ratio of concentrated solution viscosities of branched and linear polymers having the same weight-average molecular weight denoted by $\bar{g}''$:

$$\frac{\eta_{SB}}{\eta_{SL}} \equiv \bar{g}'' \quad \text{at constant } \overline{M}_w \tag{3}$$

where η_{SB} and η_{SL} are the low shear 10% solution viscosities of branched and linear polymers having weight-average molecular weights above the critical value (*10*). The powerfulness and usefulness of this approach are derived from the theoretical and experimental relation between $\bar{g}'$ and $\bar{g}''$ as shown below.

Experimental Procedure

Samples. Polybutadienes were chosen because samples prepared by different catalyst systems and containing broad ranges of molecular weights, polydispersities, and degrees of branching were available (*11*). The first group of samples consists of five linear polybutadienes having narrow molecular weight distributions, and these were prepared using a butyllithium catalyst system according to the method of Hsieh (*12*). The weight-average molecular weights of these samples ranged from 200,000 to 700,000. The second group, consisting of branched and polydisperse polybutadienes, was composed primarily of commercial samples although a few were especially prepared to cover the broadest practical range of branching and polydispersity. Sample numbers, catalyst systems, and sources are listed in Table I. Identification of catalyst systems in most cases is based on the transition metal used in conjunction with the metal alkyl halide cocatalyst.

Microstructure of polybutadienes refers to the percentage of cis, trans, and vinyl isomers. These were unavoidably variable because of the catalyst systems used (*13*): 2–98% cis, 1–76% trans, and 1–22% vinyl isomer. The vinyl contents of each sample are given in Table I for the purpose of light-scattering experiments.

Osmometry and Light Scattering. The osmometer employed was a Stabin-Shell high-speed automatic osmometer, Model M-100, obtained from Dohrman Co. The membrane was gel cellophane No. 600 obtained from J. Stabin, Brooklyn, N. Y. The performance of the apparatus was checked periodically with Standard Polystyrenes NBS 705 and 706 obtained from the National Bureau of Standards, and the results agreed to within a few percent. In the procedure used with polybutadienes, samples were dissolved and run without removal of the low molecular weight, nonrubber constituents. This procedure was followed after demonstrating that the addition of a few percent low molecular species has only a minor effect in lowering the osmotic number-average molecular weight. This phenomenon is explicable in terms of rapid diffusion of the smaller molecules or the Staverman effect (*14*).

The light-scattering apparatus and procedure have been formerly described (*15*). The only modification and precaution with polybutadienes were to use the appropriate refractive index differential for each polymer as this was observed to decrease with vinyl content; values used were 0.118, 0.112, and 0.106 dl/gram for 2, 10, and 20% vinyl polybutadienes in cyclohexane at 25°C and 546 mμ.

Viscometry. Intrinsic velocities were run in toluene at 30°C using Ubbelohde viscometers. Concentrated solution viscosities were determined with a Weissenberg rheogoniometer, Martin Sweets Co., and a Brookfield viscometer, Model LVT, Brookfield Engineering Co. The calibration and operation of both were checked periodically with standard viscosity silicone fluids obtained from Brookfield Engineering Co. The procedure followed was to dissolve 50.00 grams of the diced polybutadiene in 450.0 grams of toluene by slow shaking overnight. The viscosity of the 10% solution was determined using a Couette concentric cylinder attachment with the rheogoniometer. Evaporation was minimized by placing a Plexiglass compartment around the Couette and saturating the atmosphere with solvent with moist filter paper. The viscosity was meased over a 100-fold range of shear rates to specify the range in which the high molecular weight polydisperse samples exhibited Newtonian viscosities. The Model LVT was used routinely on 500-gram solutions after it was shown to eliminate the evaporation problem and to yield the low shear Newtonian viscosity simply and adequately within the experimental error of 5%.

Table I. Description of Branched

Sample No.	Identification No.	Catalyst System	% Vinyl Isomer
1	174	Butyllithium	10
2	181		10
3	55N		10
4	177		10
5	172		10
6	140		10
7	141		10
8	142		10
9	182	Nickel based	1
10	185	Titanium based	3–4
11	186		3–4
12	184		3–4
13	187		3–4
14	188	Cobalt based	1–2
15	189		1–2
16	36A		1–2
17	126	Emulsion (5°C)	18–20
18	126M		18–20
19	136		18–20
20	136M		18–20
21	190		18–20
22	191	Alfin	20–22

Results

The weight and viscosity data obtained for reference linear polybutadienes are given in Table II, where $\overline{M}_w$ is the weight-average molecular weight; $\overline{M}_n$, the number-average molecular weight; $\overline{M}_w/\overline{M}_n$, the polydispersity; $[\eta]_L$, the intrinsic viscosity of the linear polymer; and η_{SL}, the 10% solution viscosity of linear polymer. The relationship between intrinsic viscosity and weight-average molecular weight for these samples is shown in Figure 1 together with the corresponding equations of Cooper *et al.* (*16*) (Equation 4) and Kraus and Gruver (*9*) (Equation 5). The relationship of this work is expressed by Equation 6 which is intermediate between the other two.

$$[\eta]_L = (1.45 \times 10^{-4})\overline{M}_w^{0.76} \qquad \text{Benzene, 30°C} \qquad (4)$$

$$[\eta]_L = (2.17 \times 10^{-4})\overline{M}_w^{0.75} \qquad \text{Toluene, 25°C} \qquad (5)$$

$$[\eta]_L = (1.67 \times 10^{-4})\overline{M}_w^{0.76} \qquad \text{Toluene, 30°C} \qquad (6)$$

The difference between Equations 4 and 6 is due in part (*ca.* 8%) to the solvent effect (*17*) and in part to the different refractive index differ-

Polybutadiene Samples

Source	*Ref.*
Texas-U. S. Chemical Co.	*12*
Diene 35, Firestone Tire & Rubber Co.	*39*
Diene 55, Firestone Tire & Rubber Co.	
Texas-U. S. Chemical Co.	
” ” ” ”	
Diene 35, Firestone Tire & Rubber Co.	
Diene 35NF, Firestone Tire & Rubber Co.	
Diene 55, Firestone Tire & Rubber Co.	
BROI, Japan Synthetic Rubber	
BROI, Japan Synthetic Rubber	*40*
Cis BR 1203, Phillips Petroleum Co.	*41*
Duragen, General Tire & Rubber Co.	
Cisdene 100, American Synthetic	
Taklene 1220, Polymer Corp.	
Ameripol CB–220, Goodrich-Gulf	*11*
Ameripol CB–220, Goodrich-Gulf	
Texas-U. S. Chemical Co.	*42,43*
U. S. Industrial Chemicals	*44*

ential employed. The difference between Equations 5 and 6 may be attributed to small variations in branching among butyllithium polybutadienes discussed below since solvents and refractive index differentials employed were identical (*18*). Variations resulting from light-scattering instrument calibration cannot be ruled out, however. Although Equation 6 may not be superior to Equations 4 and 5, it was chosen since η_{SL} values were available only for samples of Equation 6, and their use eliminated artifacts arising from differences in reference linear samples used in calculating $\bar{g}'$ and $\bar{g}''$. Use of Equation 5, which presumably represents more linear reference polybutadienes, does not materially alter the conclusions drawn as shown in Figure 5.

In Figure 2 the weight-average molecular weight and 10% solution viscosities for the linear samples are plotted. Linear polybutadienes having higher molecular weights were not obtained, so the relationship depicted in Figure 2 had to be extrapolated to higher values. This was felt justifiable since the samples at the high end fell in line with the slope of 3.5 in good agreement with published results (*10*).

Molecular data for the branched polybutadienes are given in Table III. The values for the average branching factor $\bar{g}'$ were calculated from

Table II. Molecular Weight and Viscosity Data for Linear Polybutadienes

Sample	$\overline{M}_w \times 10^{-5}$	$\overline{M}_w/\overline{M}_n$	$[\eta]$	$\bar{g}'$	$\bar{g}'_{corr}$	η_{SL}
1	2.15	1.18	2.02	1.07[a]	1.07	1,440
2	2.6	2.58	1.96	0.91	0.94	1,250
3	3.05	1.89	2.44	0.99	0.99	2.960
4	4.07	1.31	3.05	1.00	1.00	8,870
5	6.67	1.26	4.45	1.00	1.00	49,000

[a]Calculated from Equation 6.

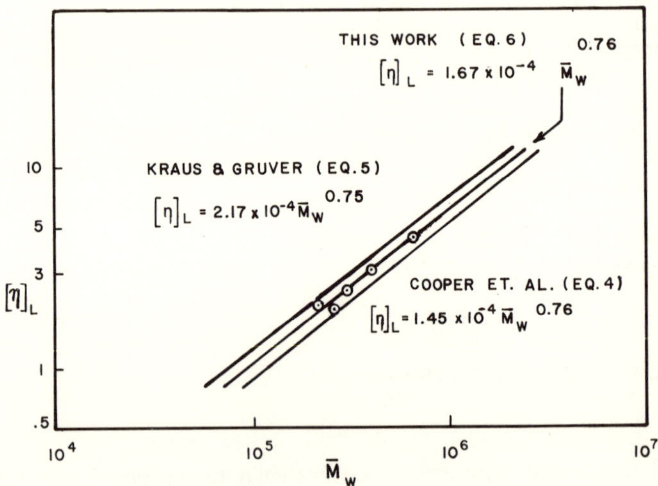

Figure 1. $[\eta]_L$–$\overline{M}_W$ *relations for linear polybutadienes*

the weight-average molecular weight of the polymer in question and Equation 6 to obtain $[\eta]_L$, the intrinsic viscosity of the linear polymer having the same weight-average molecular weight.

Values for the average branching factors obtained from the concentrated solution viscosity $\bar{g}''$ were obtained similarly by using the relationship shown in Figure 2. The log–log plot of $\bar{g}'$ and $\bar{g}''$ is given in Figure 3, and values for $\bar{g}''$ are given in Table IV.

Since intrinsic viscosity but not concentrated solution viscosity is known to be sensitive to polydispersity (*10, 19*), a correction has been applied to $\bar{g}'$ according to the calculations of Berger and Shultz (*20*). A value of zero was taken for b, exponent a was set equal to 0.75, and Q_{Br}/Q_{LIN} was related to $\overline{M}_w/\overline{M}_n$ using Equation 21 of Shultz (*19*). The branching factor corrected for polydispersity, $\bar{g}'_{corr}$, is given by

$$\bar{g}'_{corr} = \bar{g}' \Big/ \left(\frac{Q_{Br}}{Q_{LIN}}\right) \quad \begin{matrix} b = 0 \\ a = 0.75 \end{matrix} \qquad (7)$$

Values for $\bar{g}'_{corr}$ are included in column 7 of Table III. In essence, $\bar{g}'_{corr}$ is the ratio of the intrinsic viscosity of the branched polymer to the intrinsic viscosity of the linear polymer having a similar polydispersity.

The log–log plot of $\bar{g}'_{corr}$ and $\bar{g}''$ is shown in Figure 4, together with the small spread in $\bar{g}'_{corr}$ for the linear samples.

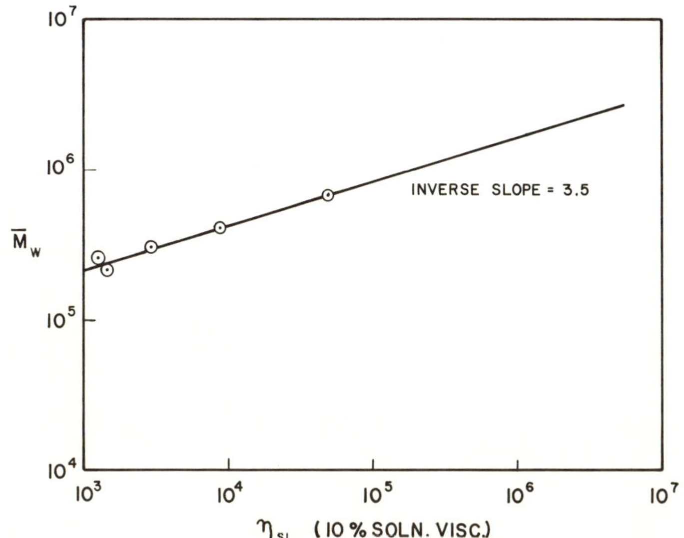

Figure 2. η_{SL}–$\bar{M}_W$ *relation for linear butyllithium polybutadienes*

Table III. Molecular Weight and Conventional Branching Factors for Polybutadienes

Sample No.	$\bar{M}_w \times 10^{-5}$	$\bar{M}_w/\bar{M}_n$	$[\eta]_B$, dl/gram	$\bar{g}'$	$\bar{g}'_{corr}$
6	2.7	2.68	1.90	0.812	0.900
7	3.5	3.50	1.91	0.721	0.764
8	3.4	2.23	2.33	0.896	0.908
9	4.8	4.66	2.99	0.879	0.965
10	6.4	4.32	2.41	0.571	0.620
11	7.7	7.26	2.53	0.522	0.608
12	6.0	4.20	2.44	0.607	0.658
13	7.9	5.72	2.41	0.482	0.544
14	4.1	4.77	1.96	0.653	0.718
15	6.7	6.98	2.03	0.467	0.541
16	4.8	6.23	2.37	0.697	0.796
17	17.9	17.9	2.17	0.231	0.301
18	9.5	9.9	2.08	0.362	0.438
19	19.0	17.3	2.67	0.272	0.352
20	9.5	9.0	2.59	0.447	0.533
21	11.1	15.4	2.17	0.334	0.378
22	9.1	13.6	3.06	0.552	0.692

Table IV. Branching Factors $\bar{g}''$ for Polybutadienes from 10% Solution Viscosity

Sample	η_{SB}, *cP*	η_{SL}, *cP*	$\bar{g}''$
6	869	2,220	0.391
7	916	5.400	0.170
8	2,437	5,000	0.487
9	9,930	15,900	0.625
10	2,529	42,000	0.0602
11	2,969	78,000	0.0381
12	2,178	38,000	0.0573
13	1,297	84,000	0.0154
14	712	9,400	0.0757
15	425	49,000	0.00867
16	1,547	15,800	0.0979
17	1,160	1,400,000	0.000829
18	1,060	160,000	0.00663
19	4,330	1,700,000	0.00255
20	3,460	160,000	0.0216
21	1,640	275,000	0.00596
22	17,810	135,000	0.132

Discussion of Results

The data of this work are analyzed in the light of published results before discussing their theoretical implications.

From the results of Tables II and III, the polybutadiene samples identified by different catalyst systems can be arranged in order of increasing polydispersity: butyllithium, nickel based, titanium based, cobalt based, alfin, emulsion. Considering variations in polydispersity from sample to sample, the agreement of this order with the published results of Alliger, Johnson, and Foreman (*21*) and Hulme and McLeod (*22*) is excellent.

Likewise, the samples can be arranged in the order of decreasing coil size and increasing branching, as determined by $\bar{g}'_{corr}$ and $\bar{g}''$, again using the catalyst systems to identify the samples. The most linear polymers are the reference butyllithium samples followed by the nickel-based polymer, butyllithium, alfin, cobalt based, titanium based, and emulsion. The correction to the branching factor for polydispersity makes the nickel-based and alfin polybutadienes less branched with respect to the other polymers examined.

The variation in branching between the two sets of butyllithium polybutadienes is attributable to polymerization impurities such as 1-butyne and vinylacetylene as shown by Adams and his co-workers (*23*). This factor might be responsible for the small difference between Equations 5 and 6. The agreement of these branching factors with published data for polymers prepared with similar catalyst systems (*9, 22, 24*) lends further credence to the experimental results of this work.

Two basic theoretical models exist to describe the degree of long-chain branching in dilute solution. According to the nondraining model depicted by Zimm and Kilb (*25*) and Kilb (*26*), the ratio of intrinsic viscosities of branched and linear polymers is given by:

$$\frac{[\eta]_B}{[\eta]_L} = g^{1/2} \quad \text{at constant } M \tag{8}$$

where g is equal to the ratio of mean-square radii of gyration of branched and linear polymer coils having the same molecular weight:

$$g = \bar{R}^2{}_B/\bar{R}^2{}_L \tag{9}$$

In this model, derived originally for star-shaped branched molecules, polymer molecules are represented by beads connected by identical Hookean springs, and the decrease in viscosity with branching is expressed by the $g^{1/2}$ rule.

The second is the hydrodynamic model of Flory and Fox (*27*) which represents the polymer molecule by an equivalent nondraining hydrodynamic sphere. Assuming the Flory constant to be the same for linear and branched polymers, the degree of branching is given by the $g^{3/2}$ rule:

$$\frac{[\eta]_B}{[\eta]_L} = g^{3/2} \quad \text{at constant } M \tag{10}$$

where g has the same significance as before.

For concentrated solutions of amorphous polymers, Bueche's mathematical model shows the ratio of zero shear viscosities of branched and linear polymer above the critical molecular weight in the entanglement region to be (*28*):

$$\frac{\eta_{SB}}{\eta_{SL}} = g^{7/2} \text{ for } M > M_C \tag{11}$$

where M_C is the critical molecular weight.

The Bueche model is based on random-flight statistics of freely draining polymer molecules. Accordingly, two possible relations exist between $\bar{g}'$ and $\bar{g}''$. (1) For the Zimm–Kilb model, combining Equations 2, 3, 8, and 11 the relation obtained is given by:

$$\bar{g}'' = (\bar{g}')^7 \tag{12}$$

(2) For the Flory–Fox model:

$$\bar{g}'' = (\bar{g}')^{7/3} \tag{13}$$

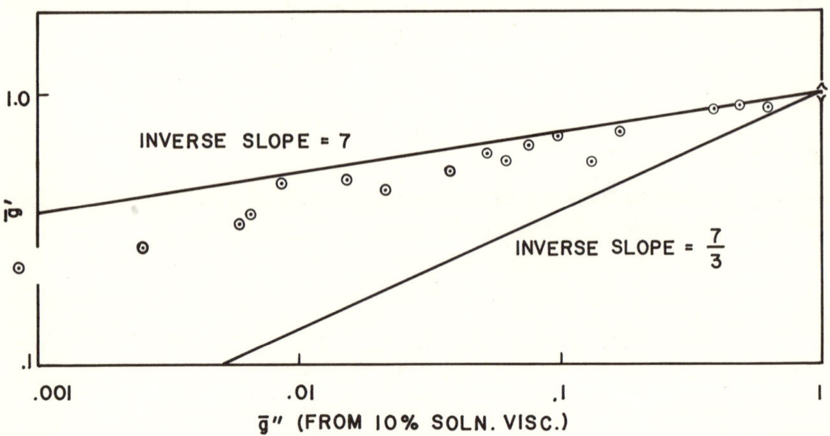

Figure 3. Relations between branching factors ḡ' and ḡ" (ḡ' from Equation 6)

Both of these relations are shown by the solid lines in Figure 3 for the ratio of intrinsic viscosities uncorrected for polydispersity and in Figure 4 for $\bar{g}'_{corr}$.

It is observed that Bueche's equation in combination with the $g^{1/2}$ rule explains the results of this work whereas in combination with $g^{3/2}$ it does not. The correction to $\bar{g}'$ for polydispersity brings closer the agreement between the data and the seventh power relation. The difference of a few percent between the expected and observed slopes of 7 and 6.4 may be attributed to an undercorrection for polydispersity; in this regard according to Graessley's findings current theories do not sufficiently account for the reduction in viscosity with polydispersity, whether the Beasley or the Stockmayer molecular weight distribution is employed (*29*).

Using Equation 5 instead of Equation 6 to calculate $\bar{g}'_{corr}$ as shown in Figure 5 leads to the same conclusion regarding the $g^{1/2}$ rule, thus bearing out the previous contention.

It has been hypothesized by Graessley (*30*) and others (*3, 31*) that the $g^{1/2}$ rule is applicable to star-shaped and slightly branched polymers whereas the $g^{3/2}$ rule describes the behavior of highly branched "comb-shaped" polymers for which $[\eta]_B/[\eta]_L < 0.5$. The results of this study indicate that the $g^{1/2}$ rule in combination with Equation 11 adequately represents the random trifunctionally and tetrafunctionally branched polybutadienes even though the ratio of intrinsic viscosities is as low as 0.3.

The foregoing discussion assumed that using a good solvent in place of a Θ solvent has no significant effect on the ratio of viscosities. With some exception (*32, 33*), the evidence in the case of $\bar{g}'$ shows the validity to be unimpaired for polydisperse as well as monodisperse systems (*29, 34–36*). It is expected that polymer expansion in good solvents would be in the same direction for branched and linear polymers in dilute and concentrated media so that any errors would be compensating.

Furthermore, microstructural variations of cis, trans, and vinyl contents have no discernible effect on the relation between $\bar{g}'$ and $\bar{g}''$ of Figures 3–5. Since the same linear butyllithium polymers containing 10% vinyl isomer were used as reference in calculating $\bar{g}'$ and $\bar{g}''$, it is likely that microstructural differences have been cancelled.

No enhancement of concentrated solution viscosity was observed in high molecular weight polybutadienes with branching as might have been expected from the bulk viscosities measured by Kraus and Gruver (*9*).

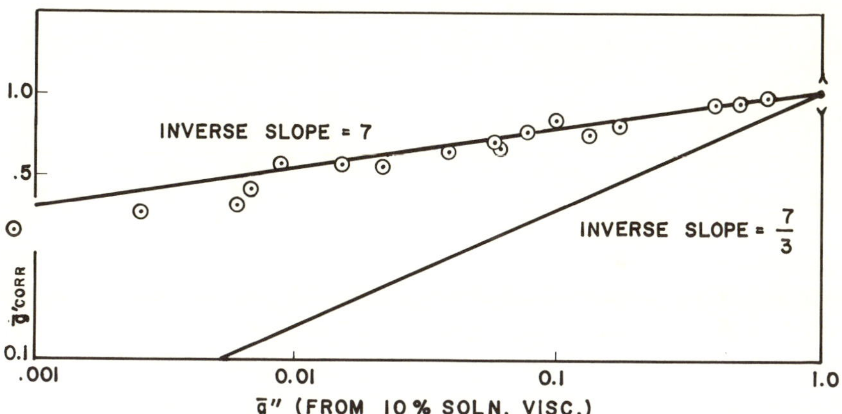

Figure 4. Relations between branching factors $\bar{g}'_{corr}$ and $\bar{g}''$ ($\bar{g}'_{corr}$ from Equation 6)

In analogy with the crossover molecular weight of 100,000 reported for bulk viscosities of tetrachain branched polybutadienes (*9*), it is possible to estimate a crossover molecular weight for 10% solutions. Using a constant value, $(MV)_C$, for the critical crossover composition, where V is the volume fraction of polymer (*37, 38*), the crossover molecular weight is estimated at 1.1 million for 10% solutions. Since no enhancement in concentrated solution viscosity was observed up to 1.9 million, it appears that the concept of critical entanglement composition is not applicable here. One might speculate that the long-chain branches that are entangled in bulk "disentangle" rapidly with dilution as was observed by Graessley

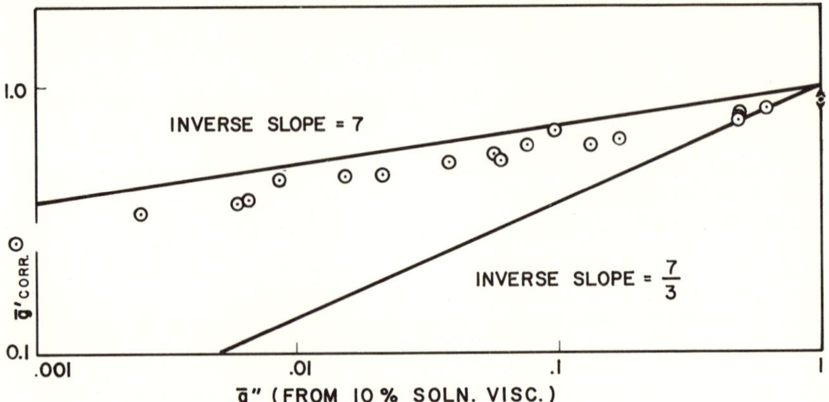

Figure 5. Relations between branching factors $\bar{g}'_{corr}$ and $\bar{g}''$ ($\bar{g}'_{corr}$ from Equation 5)

and Prentice (7). Viscosity measurements of branched and linear polymers at higher concentrations might clarify this point.

Practical considerations make the concentrated solution viscosity method for determining branching factors from molecular dimensions a powerful technique. First, it is considerably more sensitive than the intrinsic viscosity method. Although the range of intrinsic viscosities was 1.9–4.5 dl/gram the range of solution viscosities was 425–49,000 cP, a factor of 2.4 *vs.* 100. Secondly, the method is simpler when applied to polydisperse samples since the concentrated solution viscosity is relatively insensitive to polydispersity, and no correction is necessary. It should be added that the method is limited to concentrations and molecular weights above the entanglement composition and below the crossover composition. For the range of molecular weights examined, 10% was a valid choice.

In conclusion, evidence is provided in support of a sensitive procedure for the determination of the degree of branching in high polymers based on 10% solution viscosity. The method is more sensitive by a power of seven. The data cover the range of molecular weight 200,000–2 million and polydispersities of 1.2 to 18. The results of this work are explicable by the combination of the models of Zimm and Kilb and of Bueche. This is a fertile field for further research.

Literature Cited

1. Stacey, K. A., "Light Scattering in Physical Chemistry," Butterworths, London, 1956.
2. Berry, G. C., *Prepr. IUPAC Int. Symp. Macromol. Chem.* (1966) 144.
3. For a recent review, see for example, Berry, G. C., Casassa, E. F., *J. Polym. Sci., Part D* (1970) **4**, 1.

4. Moore, J. C., *J. Polym. Sci., Part A* (1964) **2,** 835.
5. Cazes, J., *J. Chem. Educ.* (1966) **43,** A567, A625.
6. Onogi, S., Kimura, S., Kato, T., Masuda, T., Miyanaga, N., *J. Polym. Sci., Part C* (1966) **15,** 381.
7. Graessley, W. W., Prentice, J. S., *J. Polym. Sci., Part A-2* (1968) **6,** 1887.
8. Long, V. C., Berry, G. C., Hobbs, L. M., *Polymer* (1964) **5,** 517.
9. Kraus, G., Gruver, J. T., *J. Polym. Sci., Part A* (1965) **3,** 105.
10. Bueche, F., "Physical Properties of Polymers," Wiley-Interscience, New York, 1962.
11. See, for example, Cooper, W., Eaves, D., Vaughan, G., Degler, G., Hank, R., "Elastomer Stereospecific Polymerization," *Advan. Chem. Ser.* (1966) **52.**
12. Hsieh, H. L., *J. Polym. Sci., Part A* (1965) **3,** 153.
13. For a complete list relating microstructure to catalyst systems, *see* Bahary, W., Sapper, D. I., Lane, J. H., *Rubber Chem. Technol.* (1967) **40,** 1529.
14. Bruss, D. B., Stross, F. H., *J. Polym. Sci.* (1961) **55,** 381.
15. Bahary, W. S., Bsharah, L., *J. Polym. Sci., Part A-1* (1968) **6,** 2819.
16. Cooper, W., Eaves, D. E., Vaughan, G., *J. Polym. Sci.* (1962) **59,** 241.
17. Cleland, R. L., *J. Polym. Sci.* (1958) **27,** 349.
18. Stacey, C. A., private communication.
19. Shultz, A. R., *J. Polym. Sci., Part A* (1965) **3,** 4211.
20. Berger, H. L., Shultz, A. R., *J. Polym. Sci., Part A* (1965) **3,** 4227.
21. Alliger, G., Johnson, B. L., Forman, L. E., *Kautschuk Gummi* (1961) **14,** WT 248.
22. Hulme, J. M., McLeod, L. A., *Polymer* (1962) **3,** 153.
23. Adams, H. E., Bebb, R. L., Eberly, K. C., Johnson, B. L., Kays, E. L., *Kautschuk Gummi* (1965) **18,** WT 709.
24. For a review of branching and polydispersity among polybutadienes see, for example, ref. 13.
25. Zimm, B. H., Kilb, R. W., *J. Polym. Sci.* (1959) **37,** 19.
26. Kilb, R. W., *J. Polym. Sci.* (1959) **38,** 403.
27. Flory, P. F., "Principles of Polymer Chemistry," Chapter 14, Cornell University Press, Ithaca, N. Y., 1953.
28. Bueche, F., *J. Chem. Phys.* (1964) **40,** 484.
29. Graessley, W. W., Mittelhauser, H. M., *J. Polym. Sci., Part A-2* (1967) **5,** 431.
30. Graessley, W. W., In "Characterization of Macromolecular Structure," National Academy of Science, Washington, D. C., 1968, Publication 1573.
31. Kurata, M., Fukatsu, M., *J. Chem. Phys.* (1964) **41,** 2934.
32. Berry, G. C., Hobbs, L. M., Long, V. C., *Polymer* (1964) **5,** 31.
33. Spiro, J. G., Goring, D. A. I., Winkler, C. A., *J. Phys. Chem.* (1964) **68,** 323.
34. Thurmond, C. D., Zimm, B. H., *J. Polym. Sci.* (1952) **8,** 477.
35. Morton, M., Helminiak, T. E., Gadgary, S. D., Bueche, F., *J. Polym. Sci., Part A* (1965) **3,** 4131.
36. Orofino, T. A., Wenger, F., *J. Phys. Chem.* (1963) **67,** 566.
37. Porter, R. S., Johnson, J. F., *Chem. Rev.* (1966) **66,** 1.
38. Berry, G. C., Fox, T. G., *Advan. Polym. Sci.* (1968) **5,** 261.
39. Ward, W. A., Willis, J. M., *Rubber Age* (1960) **87,** 815.
40. Jenkins, D. K., Timms, D. G., Duck, E. W., *Polymer* (1966) **7,** 419.
41. Grouch, W. W., *Rubber Plast. Age* (1961) **42,** 276.
42. McCall, C. A., Nudenberg, W., Goldstein, H. J., *Rubber World* (1963) **150,** 31.
43. Sarbach, D. V., Sturrock, A. T., *Rubber Age* (1961) **90,** 423.
44. Hensley, V. L., Greenberg, H., *Rubber Chem. Technol.* (1965) **38,** 103.

RECEIVED January 17, 1972.

9

Fractionation of Linear Polyethylene with Gel Permeation Chromatography.

V. IUPAC Samples

NOBUYUKI NAKAJIMA[1]

Plastics Division, Allied Chemical Corp., Morristown, N. J. 07960

The number and weight average molecular weights were determined for two samples of linear polyethylenes distributed by the Macromolecular Division of IUPAC. The methods used were GPC, osmotic pressure, infrared analysis, melt viscosity and intrinsic viscosity. Data interpretations are discussed for each method. By comparing the results the average molecular weights were obtained; for one sample, $\overline{M}_N$ *= 10,500 to 11,000 and* $\overline{M}_W$ *= 150,000 to 165,000: for another sample,* $\overline{M}_N$ *= 13,600 to 18,500, and* $\overline{M}_W$ *= 40,000 to 48,000.*

In the previous papers of this series (*1, 2, 3, 4*) calibration and reproducibility of gel permeation chromatography (GPC) have been extensively examined. This paper describes the application of GPC to two selected samples of linear polyethylenes, one having a narrow molecular weight distribution (NMWD) and another a broad molecular weight distribution (BMWD). These samples were distributed by the Macromolecular Division of IUPAC (*5*) for the "molecular characterization of commercial polymers." The average molecular weights by GPC are compared with the data obtained from infrared spectroscopy, osmotic pressure, melt viscosity, and intrinsic viscosity. Problems associated with data interpretation are discussed.

[1] Present address: B. F. Goodrich Co., Development Center, P.O. Box 122, Avon Lake, Ohio 44012.

Experimental

Operating Condition of GPC. The operating conditions were the same as before (*2*) except that the sample concentration was 0.5%. Solvent was 1,2,4-trichlorobenzene, temperature was 137°C, injection time was 120 sec, and the flow rate was 1 cc per minute. Four columns having nominal capacities of 7×10^6, 3×10^6, 10^5, and 10^3 were used.

Calibration. Figure 1 shows the calibration curve based on seven narrow distribution polystyrene standards. This is calibration No. 6 in the previous paper (*4*). At high molecular weights, where there was no calibration standard, a linear extrapolation was used in the semilogarithmic plot. This extrapolation is arbitrarily chosen. As long as the extrapolation is used, there is some degree of uncertainty (*3*). Some improvement resulted by using a broad distribution polyethylene as a supplementary standard—*i.e.*, a control sample (Figure 2). The molecular weight distribution of this standard was predetermined by the following

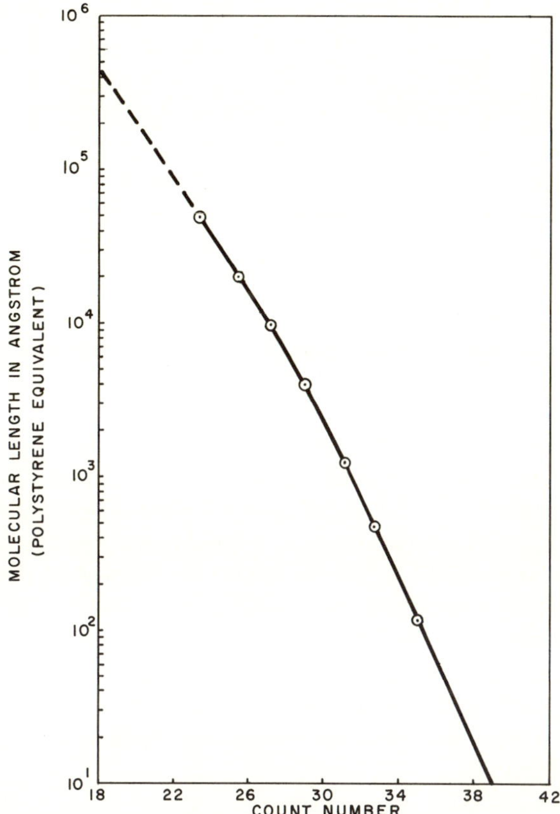

Figure 1. Calibration curve No. 6 with polystyrene standards

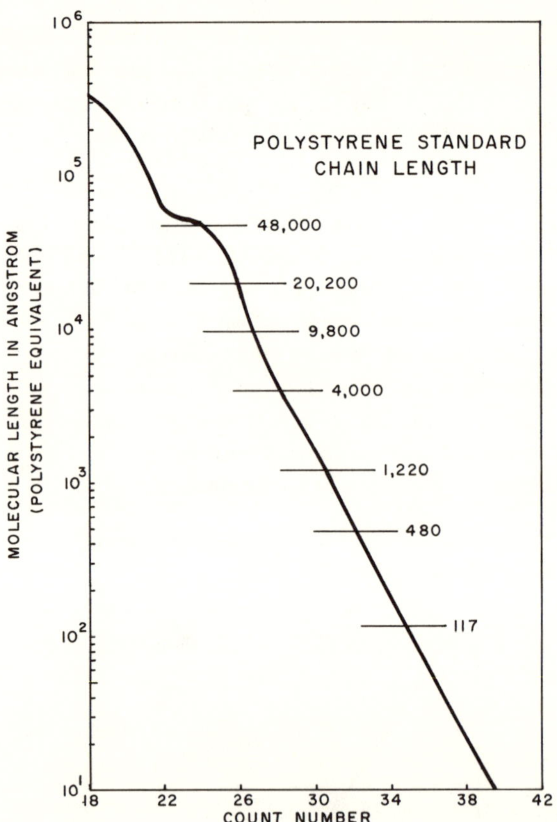

Figure 2. Calibration curve No. 6 corrected by using broad distribution polyethylene as a reference

sequence. First, a GPC fractionation curve was obtained. Then, the number average, $\bar{A}_N$, and the weight average molecular length, $\bar{A}_W$, were calculated by using a polystyrene calibration curve (calibration No. 3 of the previous paper) (*4*). Then, $\bar{A}_N$ and $\bar{A}_W$ were calibrated to the number average, $\bar{M}_N$, and weight average molecular weight, $\bar{M}_W$, respectively by a constant factor of 17.5. The $\bar{M}_N$ value was based on the end group analysis by infrared spectroscopy. The $\bar{M}_W$ had been calculated from the low shear Newtonian viscosity of the melt. For the control sample, polyethylene, two sets of values were obtained for $\bar{A}_N$ and $\bar{A}_W$. One set was based on calibration No. 3 and another on No. 6. These sets did not have the same values. Therefore, correction factors, f_N and f_W, were derived as follows:

$$f_N = \frac{(\bar{A}_N)_3}{(\bar{A}_N)_6} = 0.774 \qquad f_W = \frac{(\bar{A}_W)_3}{(\bar{A}_W)_6} = 0.840 \tag{1}$$

where the subscript numbers refer to the particular calibration. This enables us to convert $(\overline{A}_N)_6$ and $(\overline{A}_W)_6$ to $\overline{M}_N$ and $\overline{M}_W$, respectively.

$$17.5 f_N\ (\overline{A}_N)_6 = 17.5\ (\overline{A}_N)_3 = \overline{M}_N \tag{2}$$

$$17.5\ f_W\ (\overline{A}_W)_6 = 17.5\ (\overline{A}_W)_3 = \overline{M}_W \tag{3}$$

The correction factors, f_N and f_W, together with the conversion factor of 17.5 are applied to the samples in the present study.

Instead of correcting the average molecular length, each molecular length, A_i, may be corrected against the reference:

$$f_i = \frac{(A_i)_3}{(A_i)_6} \tag{4}$$

This was done in the following way. First, two cumulative distribution curves were prepared for the control sample; one is based on the calibration No. 3 and the other on No. 6.

$$\left(\int_0^i dW_i\right)_3 = \left(\int_0^i \frac{dW_i}{dA_i}\, dA_i\right)_3 \quad vs.\ (\log A_i)_3 \tag{5}$$

$$\left(\int_0^i dW_i\right)_6 = \left(\int_0^i \frac{dW_i}{dA_i}\, dA_i\right)_6 \quad vs.\ (\log A_i)_6 \tag{6}$$

Then, $(A_i)_3$ and $(A_i)_6$ were read from the curves at the same cumulative fraction,

$$\left(\int_0^i dW_i\right)_3 = \left(\int_0^i dW_i\right)_6 \tag{7}$$

The values of f_i were calculated by Equation 4 for all ranges of the cumulative fraction. The results of the calculation enabled us to compare the characteristics of GPC performances at two different times. In this treatment any difference in the GPC performance—*e.g.*, resolution—was expressed in terms of the correction factor, f_i, to the chain length, A_i.

The calibration curve of Figure 1 is a plot of $(\log A_i)_6$ *vs.* count number, N_6. The curve was corrected according to the procedure described above; the new calibration curve is a plot of $(\log A_i)_3 = \log f_i + (\log A_i)_6$ against N_6. A use of this calibration curve is discussed later with the samples of the present study. Noteworthy is a significantly different performance of two GPC runs at the molecular lengths from 20,000 to 100,000. With the polystyrene standards alone such detailed difference could not be detected. To illustrate this point the molecular lengths of the polystyrene standards are indicated on the graph.

Results

GPC runs of BMWD and NMWD samples were made within two days after calibration No. 6 was done, and the GPC traces are shown in Figures 3 and 4. The baseline was stable and reproducible before and

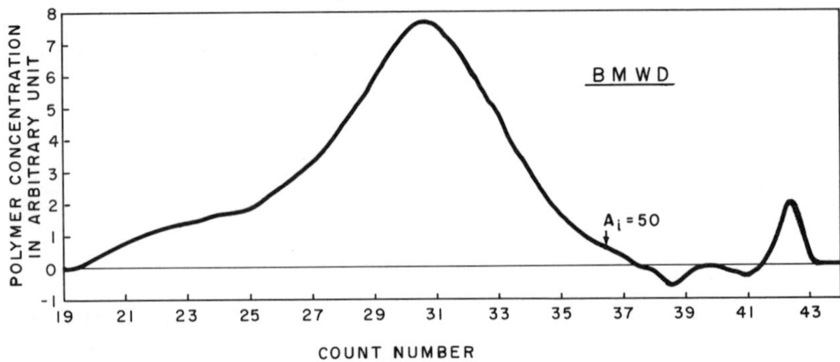

Figure 3. GPC trace of BMWD sample

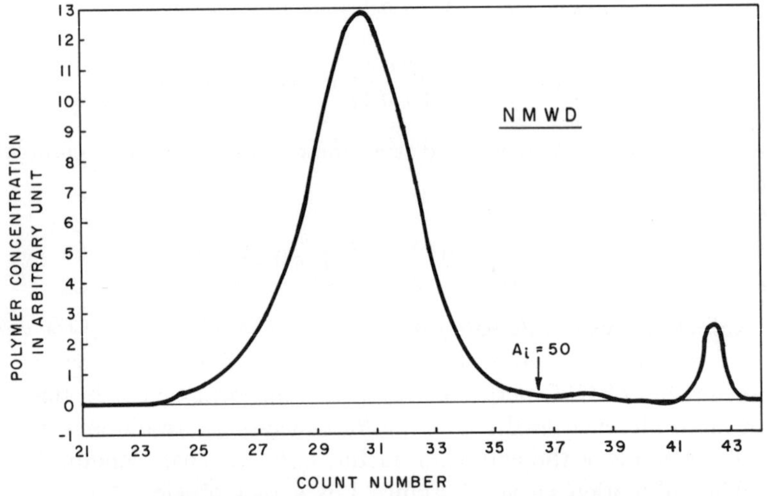

Figure 4. GPC trace of NMWD sample

after elution of the samples. However, the separation from the baseline at the low molecular weight region is subject to interpretation because of interference by the additives in the polymers and/or the impurities in the system. Treatment of the data in this region is discussed below. The height reading at every half count was fed into a computer program (*6*) together with the calibration data of Figure 1 or 2. Table I is a summary of the molecular weight data obtained by GPC and other methods. The experimental techniques and the data treatment are discussed individually.

GPC Data Treatment. As discussed above, treatment of the data at the low molecular weight region presented a considerable problem. Interpretation 1 treated data as if there were no molecules smaller than chain length $A_i = 50$; it gave an arbitrary cutoff at the low molecular weight. Interpretation 2 faithfully treated the data as it appeared in Figures 3 and 4. This meant that for BMWD sample there was no interference from the negative peak. For NMWD sample the chain length of as short as 10 A was included. For both interpretations the calibration was done according to the curve in Figure 1. These two interpretations gave a large difference in $\overline{M}_N$ values of NMWD sample obviously because of a low molecular weight peak influencing the difference.

The average molecular weights by interpretation 1 were corrected by the method described previously (*4*)—that is, the correction factors, f_N and f_W, were applied. The results are shown as interpretation 3.

The results by interpretation 4 were based on the calibration curve of Figure 2. For NMWD sample the low molecular weight peak at the count number larger than 37 was ignored, assuming that the peak resulted from an impurity. For all four interpretations the conversion factor, 17.5 was used to calculate the molecular weights from the chain lengths.

Osmometry. Measurements were made in tetrahydronaphthalene at 130°C by using a membrane osmometer (Mechrolab model 502). The plots of osmotic pressures at different concentrations are shown in Figure 5. Comparing the osmotic pressure results with those of GPC (interpretation 1) it appears that molecules larger than $A_i = 50$ may have permeated the membrane. If this is the case, the $\overline{M}_N$ by osmometry is too high.

Infrared Spectroscopy. Spectra were obtained on a Perkin-Elmer model 521 grating infrared spectrophotometer. The amount of branching in these materials is very small and appears to be of the methyl type, which is indicated by the small absorption near 1140 cm^{-1}. Ethyl branching was also sought but not detected at 770 cm^{-1}. The results are summarized in Table II.

Calculation of the methyl branch is based on the assumption that one double bond exists per molecule. The methyl absorbance at 1378 cm^{-1}, in excess of that required for methyls terminating the backbone, was interpreted as methyl branches. The value may be incorrect if the assumption is not valid. Also, a part or all of the methyl may correspond to long branches rather than methyl branch. The number average molecular weight by interpretation 1 is based on the assumption that there is one double bond per molecule. This assumes the presence of branches.

Table I. Molecular Weights of

Technique	*Interpretation*	*BMWD* $\overline{M}_N$	$\overline{M}_W$	$\overline{M}_W/\overline{M}_N$
GPC	1	13,500	193,000	14.3
	2	11,000	150,000	13.6
	3	10,500	162,000	15.4
	4	10,600	165,000	15.6
Osmometry		(17,700)	—	—
Infrared	1	10,500	—	—
	2	(8,600)	—	—
Melt viscosity				
150°C		—	(112,000)	—
170°C		—	(115,000)	—
190°C		—	—	—
Intrinsic viscosity	1	—	200,000	—
	2	—	(100,000)	—

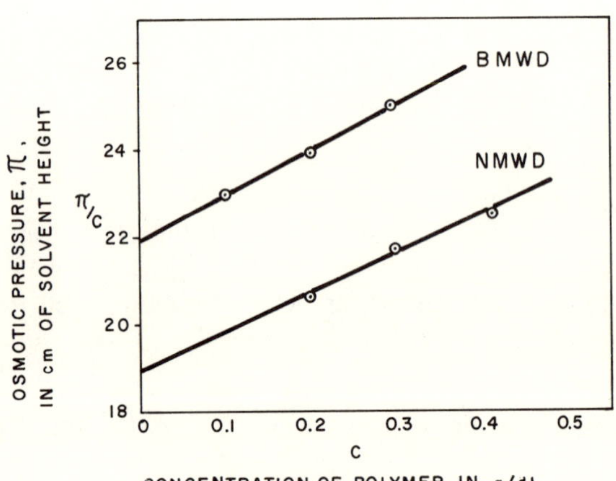

Figure 5. Osmotic pressure measurements with BMWD and NMWD samples

Table II. Infrared Results

	BMWD	*NMWD*
Methyl branch/1000 carbons	0.6	0.6
Unsaturation, wt %		
vinyl	0.248	0.080
trans	trace	0.003
vinylidene	0.008	0.006

Two Commercial Polyethylenes

NMWD		
$\overline{M}_N$	$\overline{M}_W$	$\overline{M}_W/\overline{M}_N$
17,500	56,100	3.20
10,700	55,600	5.20
13,600	47,200	3.47
13,700	45,800	3.35
20,400	—	—
(30,200)	—	—
18,500	—	—
—	39,800	—
—	41,700	—
—	44,700	—
—	(87,100)	—
—	47,900	—

The $\overline{M}_N$ value for NMWD sample appears to be too large. On the other hand, if we do not assume the presence of the branching at all, the $\overline{M}_N$ values are smaller, as given in interpretation 2.

Melt Viscosity. Low shear melt viscosities were measured by Kepe's cone-plate viscometer (*7*) at 150°, 170°, and 190°C. No stabilizer was added to the sample. The flow curves are shown in Figure 6. The viscosities of NMWD are in good agreement with those observed by others (*8*); the viscosities of BMWD by our measurements are somewhat lower. The Newtonian viscosities, η_0, were observed with NMWD sample. With BMWD sample, η_0 was estimated by extrapolation shown in the figure. The extrapolated values are uncertain; they may have been underestimated. The Newtonian viscosities are listed in Table III.

The $\overline{M}_W$ values were calculated by an equation from Tung (*9*):

$$\log \eta_0 = 3.4 \log \overline{M}_W + (1.64 \times 10^3/T) - 15.5$$

Intrinsic Viscosity. Intrinsic viscosity was determined in tetrahydronaphthalene at 130°C. BMWD = 1.80 dl/gram; NMWD = 0.98 dl/gram. These values were converted to the corresponding values in decahydronaphthalene at 135°C using a conversion factor of 1.16 given by Tung (*10*). Then, an equation by Billmeyer (*11*) for whole polymer was used to calculate $\overline{M}_W$ (interpretation 1). For NMWD sample the equation for the fractions (*11*) may be more appropriate; the calculated $\overline{M}_W$ values are given as interpretation 2.

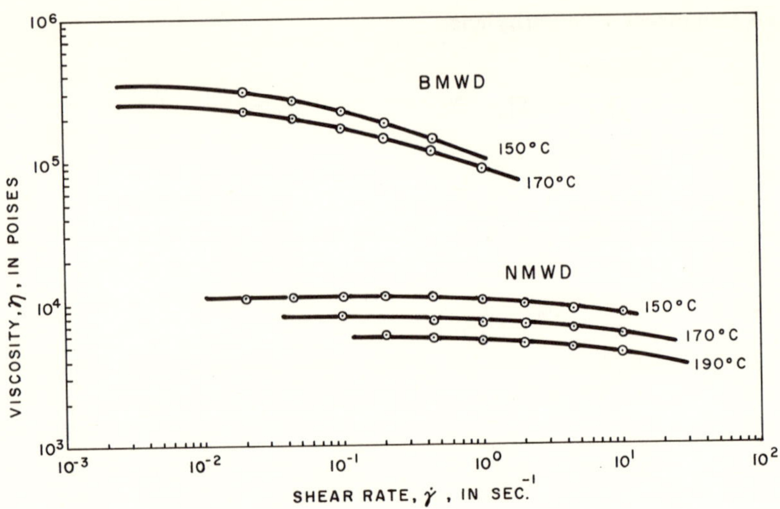

Figure 6. Melt viscosities of BMWD and NMWD samples

Table III. Newtonian Viscosities of Melt Flow

Temperature, °C	*BMWD*[a] $\eta_0 \times 10^{-5}$ *poise*	*NMWD* $\eta_0 \times 10^{-4}$ *poise*
150	(3.5)	1.10
170	(2.5)	0.80
190	—	0.58

[a] Value estimated

Discussion

BMWD Sample. The $\overline{M}_N$ of this sample appears to be about 10,500–11,000. The values obtained by GPC with interpretations 2, 3, and 4 are in good agreement. They also agree with the result from the infrared analysis with one-double-bond-per-molecule assumption. This automatically assumes the presence of small amounts of branches in the sample. If we assume the unbranched chain, $\overline{M}_N$ is 8600, which is somewhat smaller than other values. The GPC $\overline{M}_N$ by interpretation 1 is larger because of the artificial cut-off at the molecular length larger than 50 A. The osmotic pressure measurements gave a larger value, which is probably the result of the diffusion of low molecular weight species through the membrane.

The $\overline{M}_W$ is more difficult to estimate precisely. The melt viscosity results provide only an order of magnitude which is about 100,000. Since Newtonian viscosity was not observed, no refinement can be made on the data. The intrinsic viscosity gives an estimate of $\overline{M}_W$ of 200,000 according to an equation for whole polymer. If we use the equation for fractions, $\overline{M}_W$ is estimated as 110,000. Although the value agrees well with that from the melt viscosity, it must be in error because this polymer has a broad molecular weight distribution.

The $\overline{M}_W$ values by GPC are between the above two extremes. Perhaps the correct value lies between 150,000 and 165,000 because these are obtained by better data treatment interpretations 2, 3, and 4. Taking the best estimates of $\overline{M}_N$ and $\overline{M}_W$, the ratio is calculated as $\overline{M}_W/\overline{M}_N =$ 13.5 to 15.5.

NMWD Sample. The $\overline{M}_N$ values of this sample are in the range 10,000 to 30,000. However, 30,200 obtained from the infrared analysis appears to be too high. The value is based on the one-double-bond-per-molecule assumption. It automatically assumes the presence of branches. If no branches are present, $\overline{M}_N$ is calculated as 18,500, which is more in line with other data. The GPC $\overline{M}_N$ values are between 10,700 and 17,500. The lower value is the result of including the low molecular weight peak between 10 and 35 A. Probably this peak is not of polyethylene fraction because the lowest possible $\overline{M}_N$ estimated from the infrared is 18,500. $\overline{M}_N$ by osmometry may be too large for the same reason stated before. The correct $\overline{M}_N$ for this sample is probably 13,600–18,500.

The $\overline{M}_W$ values are in better agreements than $\overline{M}_N$'s. They are in the range 40,000–48,000. The $\overline{M}_W$ calculated from intrinsic viscosity favors equation for fractions rather than that for whole polymers. The best estimates of $\overline{M}_W/\overline{M}_N$ ratios are 2.0–3.5.

Acknowledgments

The author is indebted to R. T. Guliana for the osmotic pressure data and intrinsic viscosities, to G. A. Tirpak for the infrared measurements, and to R. D. Hoffman for the melt viscosities.

Literature Cited

1. Nakajima, N., *J. Polymer Sci., Pt A-2* (1966) **5**, 101.
2. Nakajima, N., *J. Polymer Sci.* (1968) **C-21**, 153.
3. Nakajima, N., *Separation Sci.* (1971) **6** (2), 275.
4. Nakajima, N., *J. Appl. Polymer Sci.* (1971) **15**, 3089.
5. Benoit, H., Macromolecular Division of IUPAC, Centre de Recherches sur les Macromolecules, Strasbourg, France.
6. Pickett, H. E., Cantow, M. J. R., Johnson, J. F., *J. Appl. Polymer Sci.* (1966) **10**, 917.
7. Kepe, A., *J. Polymer Sci.* (1956) **22**, 409.
8. Wales, J. L. S., *Pure Appl. Chem.* (1969) **20**, 331.
9. Tung, L. H., *J. Polymer Sci.* (1960) **46**, 409.
10. Tung, L. H., *J. Polymer Sci.* (1959) **36**, 287.
11. De La Cuesta, M. O., Billmeyer, F. W., Jr., *J. Polymer Sci.* (1963) **A-1**, 1721.

RECEIVED January 17, 1972.

10

Reproducibility of Molecular Weight Measurements by GPC with Infrared Detectors

J. H. ROSS, Jr. and R. L. SHANK

Research and Development Department, Chemicals and Plastics, Union Carbide Corp., South Charleston, W. Va. 25303

The variability of tests for physical properties and structural features of polymers usually exceeds ±5% at the 95% confidence interval. Molecular weight measurements of polymers are generally accepted to have large variances, in particular, the interlaboratory round robins. With the advent of gel-permeation chromatography (GPC) and with the increased attention given measurements of this kind, significantly lower levels of variability may be expected. This paper shows that the capability of the GPC technique for measurement of molecular-weight parameters, $\overline{\mathrm{M}}\mathrm{w}$ *and* $\overline{\mathrm{M}}\mathrm{n}$*, for polyethylenes is better than ±2.5% in repetitive measurements over periods as long as a month. The excellent performance of this GPC system is attributed to the high photometric precision and sensitivity of the infrared detection device, to the data acquisition and computational procedures, and to the stabilizer used.*

Industry has rapidly accepted the GPC technique and exploited it for a variety of uses including quality control, guidance of product blending and polymer syntheses, and establishment of physical and structural property relationships. In each of these areas, requirements for precision have increased as more information was obtained.

Operation of the Waters Associates model 100 instrument at elevated temperatures, necessitated by dissolution of polyethylene, requires several modifications in order to improve baseline stability and sensitivity (*1*). Usually the first modification is to reduce the unacceptably high temperature gradients in the oven which may be as much as 20°C from top-to-bottom when the oven is operated at 130°C. The refractometer requires thermostating to better than ±0.01°C, and the original refractometer is generally replaced with the Waters Associates R-4 refractometer. With these and even other modifications, it has not been possible to perform the analysis with the precision required by many of the applications for the data. Several of the Waters Associates units have been relegated to operation at relatively low temperatures only, including the one in the author's laboratory. Other workers have accepted the poor precision and the difficult task of operation of the chromatograph at high temperatures and have attempted to draw conclusions from these data even though the reproducibility is not as high as desired (*2*).

Statistical studies of the precision of measurements of molecular weight parameters by GPC are not common (*3*). Interlaboratory round robins of GPC analyses conducted by ASTM and other GPC-oriented groups have shown high variance. Gamble *et al.* indicate a precision of about ±5% for the replicate measurement of butyl rubber analyses of $\overline{Mw}$ and $\overline{Mn}$ (*4*). The test apparatus described here evolved as an effort to maximize precision and accuracy of the GPC technique. We chose the National Bureau of Standards (NBS) Standard Reference Material No. 1475 to indicate both the precision and accuracy capabilities of the system. It was assumed that the NBS sample is homogeneous and that the values of molecular weight parameters have high accuracy, although NBS considers it impossible to provide a statement as to the absolute accuracy at this time (*5*).

The purpose of this paper is to demonstrate that the precision and accuracy capabilities of the GPC technique when operating at high temperatures for long periods can be quite high. It is not an attempt to make comparative systematic appraisal of the factors affecting precision and accuracy of this system with that of the Waters Associates chromatograph, since in the author's laboratory even after quite vigorous effort the precision and accuracy of the latter instrument were so completely unacceptable at the high temperatures required for polyethylenes as to make such a comparison impossible. With the high precision and accuracy of the system discussed herein, many more applications for the data are revealed.

Experimental

Apparatus. A GPC system using infrared spectrophotometric detection of the column effluent was described earlier (*6*). Because only branched polyethylenes were examined initially, certain modifications were necessary for the separation of linear polyethylenes. Certain improved components have been added although essentially the same apparatus as that already described was employed. For example, the oven containing the columns was designed to have variations in temperature less than ±0.01°C from end to end. Porous glass-bead column packing was replaced with the cross-linked polystyrene gels (Styragel) as improved columns became available. The four columns contained gels having nominal porosities of 10^4, 10^5, 10^6, and 10^7 A and had plate counts of better than 700 plates/foot as determined with nonane.

A Hughes Series II micropump was used to pump degassed perchloroethylene at a reproducible rate of 1 ± 0.003 ml/minute. This pump produces 300 overlapping pulsations per minute and was used without a pulsation damper because pressure variations were less than ±1 psi. A Waters Associates automatic injection unit was used to add 2 ml polymer solution containing approximately 1 to 2 mg polymer to the stream for each determination. The concentration of sample is selected so that the height of the curve is about 7 inches. The concentration may range between 0.04 and 0.075%. This injection unit when loaded is capable of automatically injecting seven samples within a 24-hour period. Degradation was avoided by adding 100 ppm 3,5-ditertiary butyl catechol to the solution.

A schematic of the detection recording and data acquisition system is shown in Figure 1. The detector (Perkin-Elmer model 112) had a double-pass optical system and a calcium fluoride prism set at 3.4μ (2940 cm^{-1}). For the lowest signal-to-noise ratio, it was necessary to replace the globar with a Sylvania FTC tungsten-iodine, 375 W lamp operated at 100 W. The stability of this lamp was found to exceed that of other infrared sources by a factor of at least ten. Life time of the tungsten-iodine bulb operating at reduced wattage is at least three

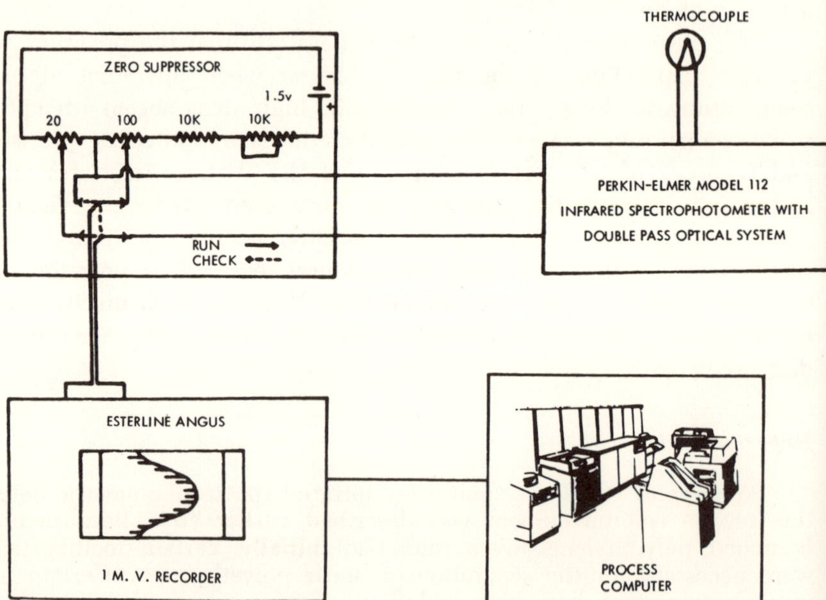

Figure 1. GPC infrared detection, recording, and data acquisition system

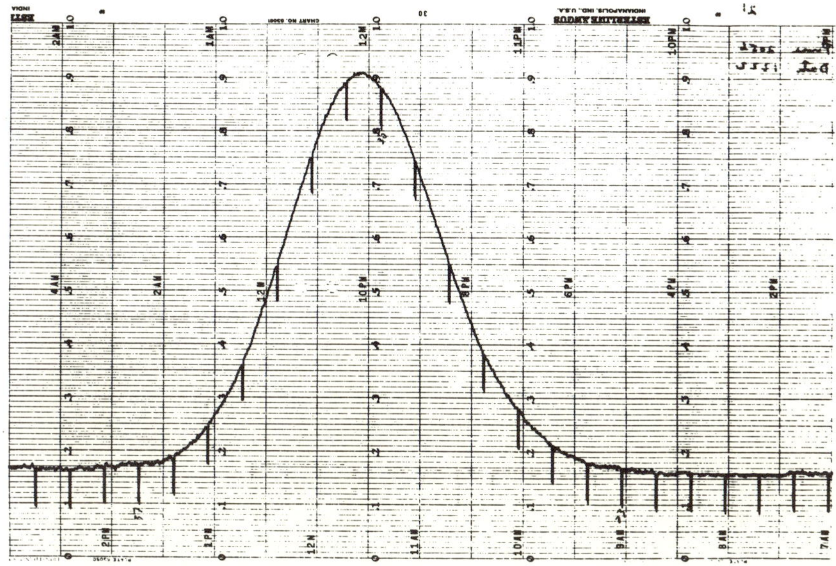

Figure 2. Recorder output on a 10-inch chart for 2 mg of NBS SRM No. 1475

months and some have exceeded a year. Baseline stability is better than ±1% in a 24-hour period.

A special zero suppressor was connected to the output of the preamplifier so that the transmission from 80 to 100% of the incident energy could be recorded. The relationship of concentration to signal output was found to be essentially linear in this range. An example of the recorder output on a 10 inch chart is shown in Figure 2.

Data Acquisition. A Control Data Corp. (CDC) 1700 process control computer was used for on-line data acquisition and storage without manual intervention, accelerating the collection of experimental data by eliminating the burden of handling the large quantity of data involved. Data reduction, which included smoothing, baseline correction, interpretation, and average molecular-weight calculations, was performed at a convenient time with simple operator initiation, producing tabular and graphical displays of the molecular weight distribution. The computer usage improved the accuracy of the results by employing a statistical analysis of the data instead of depending upon personal interpretation.

The recorder drive gear was connected to a potentiometer containing an applied voltage. The potentiometer output was directly proportional to the infrared absorption. The output signal was sent directly to a process computer where an analog-to-digital converter produced a digital fixed-point quantity. This value was stored in a current value table, in main storage, containing 15 locations.

The converter output signal was monitored continuously. When no sample was in the system the signal was zero and the scan frequency was 60 second. When a sample was injected, the signal increased to a previously selected baseline value, the scan interval changed to 5 seconds, and the contents of the current value table, when filled, were stored on the computer's disk files. At the end of the run the converter output signal was automatically reduced to zero, and the file for that particular set of data was closed. It was possible to inject another sample immediately and store the data in another file. The sample injection and data acquisition were entirely automatic, requiring no manual supervision. Data reduction required manual initiation and was done at a convenient time. This may also be automated, but at the present time software problems have not been completely resolved.

The experimental data were smoothed by fitting short segments to a first-degree polynomial. The volume markers were located during the smoothing process. All calculations were performed on the smoothed data. Number-, weight-, and Z-average molecular weights were calculated using standard methods described in most texts on polymers.

Besides the calculation of average molecular weights, several other means of characterizing the distribution were produced. These include tables of percentile fractions *vs.* molecular weights, standard deviation, skewness, and kurtosis. The data for the tables were obtained on punched cards as well as printed output. The punched cards were used as input to a CAL COMP plotter to obtain a curve as shown in Figure 3. This plot is normalized with respect to area. No corrections were made for axial dispersion.

Calibration. The accepted method of calibrating a GPC system was used. Narrow molecular-weight distribution high-density polyethylene polymers were characterized by light scattering, osmometry, and sedi-

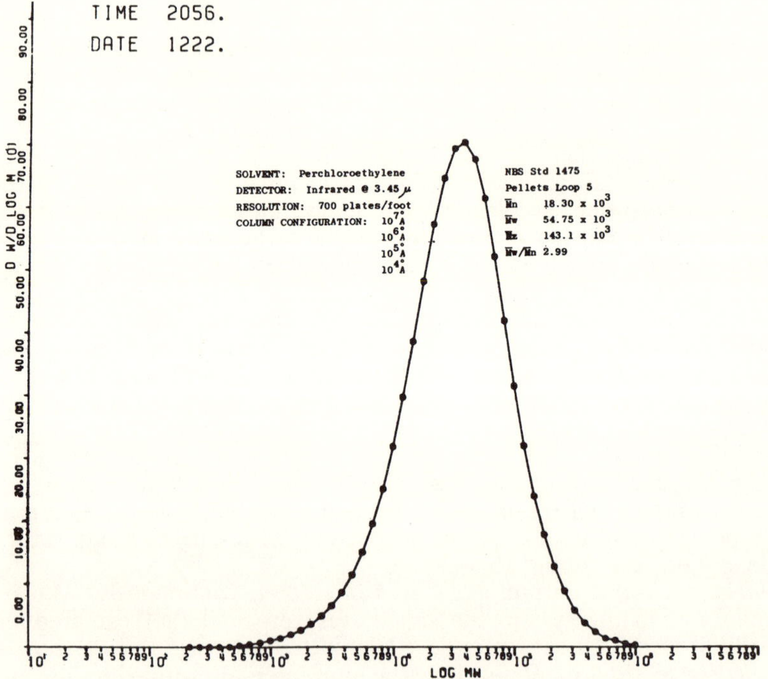

Figure 3. CAL COMP plot of NBS SRM No. 1475

mentation techniques. These data were used to obtain the elution volume–molecular weight relationship.

A universal calibration curve was established by plotting the product of the limiting viscosity numbers and molecular weight, $\overline{M}_w[\eta]$, *vs.* the elution volume, *EV*, for a variety of characterized polymers. The major usefulness of the universal calibration curve was to validate individual molecular-weight values and to provide extended molecular-weight calibration at the ends of the calibration curve where fractions of narrow dispersion of the polymer being analyzed are not available. The calibration curve was monitored daily with polystyrene fractions certified by Pressure Chemicals. The relationship between the polyethylene fractions and polystyrene fractions was determined using the universal calibration curve.

The calibration was represented in the computer program by a fifth-degree polynomial. The conventional method of least-squares was followed to determine the coefficients of the polynomial. The sensitivity of the normal equations made round-off error a significant factor in the calculations. The effect of round-off error was greatly reduced when the calculations were performed with double-precision arithmetic. The molecular weights corresponding to selected count numbers were calculated from the coefficients. The coefficients were input information for the data-reduction program.

Discussion

Precision and Accuracy. Table I shows the results of tests on the linear polyethylene NBS Standard Reference Material No. 1475. This reference material has a pellet-to-pellet coefficient of variation of 3% in the limiting viscosity number according to the NBS Certificate. At least 50 pellets are recommended for a representative sample on which limiting viscosity number is obtained. Nine analyses were performed over a period of about one month on the pellets, using approximately three pellets per determination. The analysis of these data is shown in Table I along with a comparison with the NBS data and their estimate of precision.

Subsequently, 3 grams (at least 60 pellets) of the pellets were homogenized by dissolution and precipitation as suggested by H. L. Wagner of NBS. Table II shows the results of 13 analyses on the homogenized fine powder performed over a period of about 10 days.

These data in Tables I and II show that the variation of the $\overline{M}_n$, $\overline{M}_w$, and $\overline{M}_w/\overline{M}_n$ values at the 95% confidence interval is less than ±2.5% for this particular polyethylene. Variation of $\overline{M}_z$ appears to be about twice this level. Comparison with the NBS molecular-weight parameters indicates excellent absolute accuracy except for $\overline{M}_z$. The tables also show that pellet-to-pellet variations in molecular weight could not be detected.

Table I. GPC Analysis of NBS

Time	Date	Loop	$\overline{M}_n/1000$
336	1203	1	17.06
1744	1218	3	17.59
2109	1218	4	17.71
34	1219	5	18.70
400	1219	6	17.88
1404	1222	3	18.29
2056	1222	5	18.30
1506	104	5	17.17
438	105	3	17.90
$\bar{x}$			17.84
Variability of mol wt parameters by GPC (South Charleston)			17.84 ± 0.437 ± 2.4%
Variability of mol wt parameters (NBS)			18.3 ± 1.2

[a] Each analysis performed on approximately three pellets per determination or 0.075 gram.

Table II. GPC Analysis of NBS Standard No. 1475

Time	Date	Loop	$\overline{M}_n/1000$
206	1208	1	18.15
857	1208	3	17.74
106	1209	1	18.79
439	1210	4	16.88
1618	1210	1	16.80
234	1211	4	16.77
1513	1211	1	17.07
127	1212	4	17.09
1902	1214	4	18.66
1450	1215	4	17.49
105	1216	1	17.98
135	1217	1	17.40
42	1218	6	18.10
$\bar{x}$			17.61
Variability of mol wt parameters by GPC (South Charleston)			17.61 ± 0.384 ± 2.2%
Variability of mol wt parameters (NBS)			18.3 ± 1.2

[a] Three grams of pellets were dissolved in perchloroethylene, slowly precipitated in cold methyl alcohol, and dried, producing a fine powder. Each analysis was performed on 0.075 gram of the homogenized powder.

The excellent performance of the instrumentation results from a combination of contributions from several components, *i.e.*, the high sensitivity and stability of the infrared detection system, the data acquisition and computational procedures, and the stabilizer used.

tandard No. 1475 Pellets[a]

$\overline{M}_w/1000$	$\overline{M}_z/1000$	$\overline{M}_w/\overline{M}_n$
54.68	129.5	3.20
53.45	123.0	3.04
55.09	131.3	3.11
57.04	135.0	3.05
54.83	128.8	3.07
54.34	141.6	2.97
54.75	143.1	2.99
53.14	120.7	3.09
53.25	121.4	2.97
54.51	130.49	3.05
54.51 ± 0.992	130.49 ± 6.72	3.05 ± 0.061
± 1.8%	± 5.1%	± 2.0%
54.2 ± 2.0	149 ± 13	2.96

Dissolved and Precipitated in Methyl Alcohol[a]

$\overline{M}_w/1000$	$\overline{M}_z/1000$	$\overline{M}_w/\overline{M}_n$
56.06	132.9	3.09
56.48	133.8	3.18
57.72	158.5	3.07
54.60	135.1	3.23
53.35	125.8	3.18
53.53	128.4	3.19
52.28	119.7	3.06
54.32	131.5	3.18
58.30	139.9	3.12
56.11	136.7	3.21
52.70	114.7	2.93
54.77	129.4	3.15
54.95	128.4	3.04
55.01	131.91	3.12
55.01 ± 1.16	131.91 ± 6.57	3.13 ± 0.053
± 2.1%	± 5.0%	± 1.7%
54.2 ± 2.0	149 ± 13	2.96

Acknowledgment

We acknowledge the contributions of the computer section at this laboratory, in particular J. S. Bodenschatz, Linda Jarrett, Richard Settle, and T. J. McGovern. We gratefully acknowledge the contributions of

F. Rodriquez for the concept of the zero suppressor and for his technica advice. In addition, the assistance of M. E. Casto and W. E. Coiner ar recognized. We thank T. P. Wilson for his interest in this work anc Union Carbide Corp. for permission to publish these results.

Literature Cited

1. Williamson, G. R., Cervenka, A., *Eur. Polym. J.* (1972) **8,** 1009.
2. Nakajima, J., private communication, and ADVAN. CHEM. SER. (1973) **125** 98.
3. Johnson, J. F., Porter, R. S., *Prog. Polym. Sci.* (1970) **2,** 201.
4. Gamble, L. W., Westerman, L., Knipp, E., *Amer. Rubber Chem. Technol* (1965) **38,** 823.
5. Ross, G., Frolen, L., *J. Res. Natl. Bur. Std.* (1972) **76A,** No. 2, 163.
6. Ross, J. H., Casto, M. E., *J. Polym. Sci.* (1968) Part C, **21,** 153.

RECEIVED January 17, 1972.

11

Gel Permeation Chromatography Calibration

II. Preparative GPC Fractionation and Characterization of Poly(methyl methacrylate) for Calibration in 2, 2, 2-Trifluoroethanol

THEODORE PROVDER

Glidden-Durkee Division of SCM Corp., Strongsville, Ohio 44136

JAMES C. WOODBREY and JAMES H. CLARK

Monsanto Co., St. Louis, Mo. 63166

ESMOND E. DROTT

Monsanto Co., Texas City, Tex. 77590

Because of the insolubility of polystyrene standards in 2,2,2-trifluoroethanol (TFE), poly(methyl methacrylate) (PMMA) is suggested as the standard for gel permeation chromatography (GPC) in TFE. PMMA standards used here were whole polymers made by free-radical polymerization and fractions from them. The primary molecular weight calibration curve is compared with the indirect molecular weight calibration curve generated with polystyrene standards in tetrahydrofuran (1). Differences among molecular weight averages and intrinsic viscosities calculated from the direct PMMA molecular weight calibration curve and corresponding experimental values are attributed to experimental errors and to an apparent molecular weight dependence of the specific refractive index increment in TFE. An error in an earlier paper (1) is corrected, and methods for obtaining secondary molecular weight calibration curves from hydrodynamic volume–calibration curves are reviewed.

The solvent 2,2,2-trifluoroethanol (TFE) is excellent for gel permeation chromatography (GPC) characterization of polyamides and polyacrylates and has many more desirable properties than *m*-cresol, the solvent commonly used. The advantages of using TFE rather than *m*-cresol for polyamide characterization have been cataloged elsewhere (*1*). The one main disadvantage of using TFE as a GPC solvent is the insolubility of the readily available characterized polystyrene standards. This insolubility prevents the generation of a hydrodynamic volume (HDV) calibration curve and the subsequent generation of secondary molecular weight calibration curves in TFE (*1, 2*).

Theory

Methods for Generating Molecular Weight Calibration Curves from Hydrodynamic Volume Calibration Curves. Hydrodynamic volume calibration curves can be constructed simply by plotting the product of the intrinsic viscosity $[\eta]$ and the weight-average molecular weight ($\overline{M}_w$) for a narrow molecular weight distribution polymer standard against the peak retention volume (PRV) of the standard for a given column set in a specific solvent at a given temperature. Once an HDV curve is obtained, secondary molecular weight calibration curves for polymers of interest could be obtained, provided the Mark-Houwink parameters ϵ_x and K_x are known (denoted by subscript x) from the relation where Z_s is the effec-

$$Z = [\eta]M = KM^{\epsilon+1} \tag{1}$$

$$M_x = (Z_s/K_x)^{1/(\epsilon_x+1)} \tag{2}$$

tive HDV obtained from the polymer standards and M_x is the molecular weight for the polymer of interest.

If the Mark-Houwink parameters are known or can be established for a polymer standard that is soluble in TFE as well as for the polymer of interest, the molecular weight calibration curve for the polymer of interest,

$$\log_{10}M_x = f_x(v), \tag{3}$$

can be expressed in terms of the molecular weight calibration curve for the polymer standard in TFE,

$$\log_{10}M_s = f_s(v) \tag{4}$$

Two polymer species eluting at the same retention volume have the same hydrodynamic volume. In terms of the polymer standard and the polymer of interest in TFE, this equality of hydrodynamic volume can be expressed as

$$([\eta]M)_x = ([\eta]M)_s \tag{5}$$

Substitution of Equations 1, 3, and 4 into Equation 5 leads to an expression first derived by Coll and Prusinowski (3)

$$\log_{10} M_x = \left(\frac{1}{1+\varepsilon_x}\right)\log_{10}(K_s/K_x) + \left(\frac{1+\varepsilon_s}{1+\varepsilon_x}\right) f_s(v) \tag{6}$$

If the Mark-Houwink parameters are unknown and there is insufficient data available for their direct generation, molecular weight calibration curves can be generated by (a) an empirical technique based upon the determination of the intrinsic viscosity of each polymer fraction obtained by the GPC syphon counter or (b) using at least two out of three experimental observables, number- and weight-average molecular weights $\overline{M}_n$, $\overline{M}_w$, and $[\eta]$ to fit mathematically for effective values of ϵ and K.

Meyerhoff (4) and Goedhart and Opschoor (5) have measured the viscosity of each eluting GPC fraction by coupling an automatic capillary tube viscometer with the GPC syphon. The low polymer concentration in each fraction necessitated an extremely accurate efflux time measurement to ±0.01 second since the flow time of each fraction containing polymer has flow times, t_i, greater than that of pure solvent, t_o, by at most 2.00 seconds. The specific viscosity $_{\text{sp}_i}$ of the ith polymer fraction is calculated from the flow times of the pure solvent and the polymer fraction.

$$\eta_{sp_i} = 1 - \eta_{ri} = 1 - (t_i/t_o) \tag{7}$$

The concentration C_i of the polymer fraction is given by

$$C_i = (A_i/A)\,(w/\Delta v) \tag{8}$$

where w is the total amount of injected polymer in grams, Δv is the volume of the syphon in deciliters, A is the total peak area, and A_i is the area corresponding to the ith polymer fraction. Since the concentration of each fraction is in the concentration range approaching infinite dilution ($C_i \leqslant 0.02$ wt-vol %), the intrinsic viscosity of each polymer fraction i can be taken as η_{sp_i}/C_i. A more accurate value of the intrinsic viscosity $[\eta]_i$ can be obtained from the following equation:

$$[\eta]_i = \left(\frac{\eta_{sp}/C_i - \ln \eta_{r_i}/C_i}{0.5C_i}\right)^{1/2} \tag{9}$$

where η_{r_i} is the relative viscosity of the ith polymer fraction. This equation is derived from a power series expansion of viscosity *vs.* concentration (6) and requires that the Huggins (7) constant $k_{\text{H}} = 0.333$ for the polymer–solvent system of interest. In this manner the intrinsic viscosity of all the experimentally measurable polymer fractions comprising the GPC trace can be determined. Then the average molecular weight of each fraction, M_i, can be determined from the HDV calibration curve constructed from polymer standards and from the equation

$$M_i = Z_i/[\eta]_i \tag{10}$$

where Z_i is the effective HDV of fraction i at the same retention volume as $[\eta]_i$. In this manner the secondary molecular weight calibration curve for the polymer of interest can be determined. The necessary information for the generation of the Mark-Houwink intrinsic viscosity–molecular weight relationship is now available. An important advantage of this

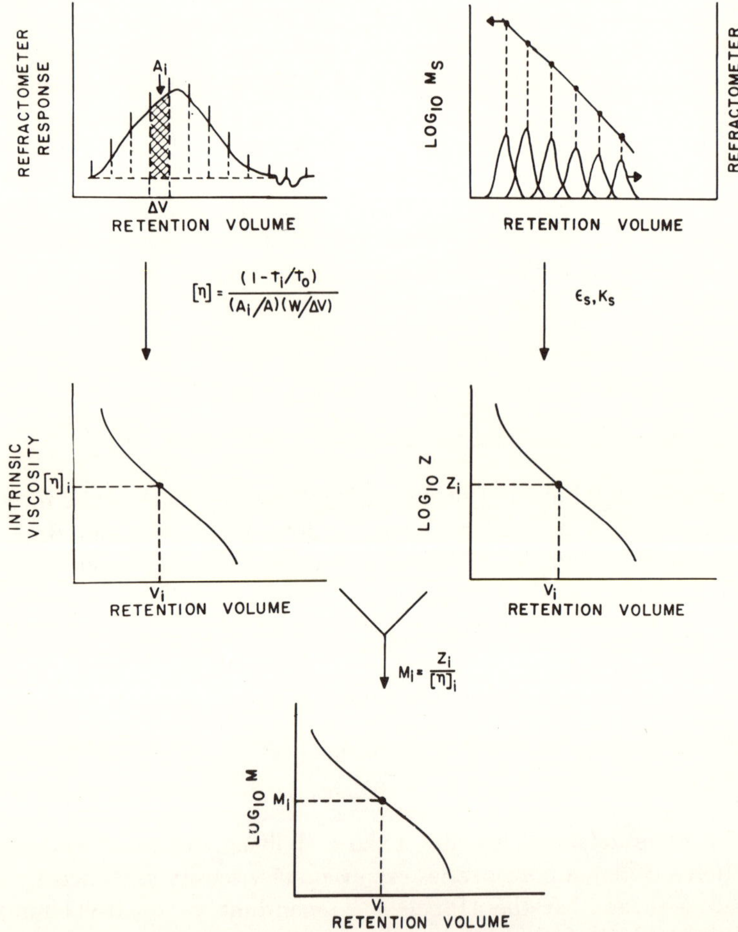

Figure 1. Illustrative method for constructing a secondary molecular weight calibration curve from an HDV calibration curve and intrinsic viscosity values of GPC polymer fractions

technique is that the polymer of interest can be highly branched and need not follow a linear $\log_{10} [\eta]$ *vs.* $\log_{10} M$ relationship over the retention volume range of the sample. The construction of the secondary molecular weight calibration curve is schematically illustrated in Figure 1. Goedhart and Opschoor (*5*) have successfully applied this method to polystyrene, poly(vinyl chloride), and poly(vinyl acetate) polymers in tetrahydrofuran. They give equations for determining the retention volume corresponding to the $\overline{M}_w$ of the fraction, for correcting for partial mixing of the preceding fraction in the tubing between the refractometer and the syphon, and for mixing in the syphon with the previous fraction as a result of incomplete emptying of the syphon.

If the determination of the intrinsic viscosity of each polymer fraction obtained from the GPC syphon counter is infeasible, use can be made of $[\eta]$, $\overline{M}_n$, $\overline{M}_w$, and the GPC trace of the whole polymer sample for the determination of the Mark-Houwink parameters. Provder and co-workers (*1, 2*) have derived expressions for $\overline{M}_n$, $\overline{M}_w$, and $[\eta]$ in terms of ϵ, K, and Z which are summarized below:

$$\overline{M}_n = \left[\int_{Z_L}^{Z_H} (Z/K)^{-1/(\epsilon+1)} (da/dZ)dZ\right]^{-1} \tag{11}$$

$$\overline{M}_w = \int_{Z_L}^{Z_H} (Z/K)^{1/(\epsilon+1)} (da/dZ)dZ \tag{12}$$

$$[\eta] = K\int_{Z_L}^{Z_H} (Z/K)^{\epsilon/(\epsilon+1)} (da/dZ)dZ \tag{13}$$

$$\frac{da}{dZ} = F(v_Z)\cdot[1/(df_H/dv)_{v_Z}]\cdot[\log_{10}e/Z] \tag{14}$$

where (da/dZ) is the weight-differential distribution of HDV, $(df_H/dv)_{v_Z}$ is the slope of the HDV calibration curve, and $F(v_Z)$ is the area-normalized, baseline-adjusted chromatogram height at retention volume v_Z. The limits of integration Z_L and Z_H correspond to the lowest and highest HDV species, respectively in the sample. Upon fitting to one of the parameter sets $\{\overline{M}_n,\overline{M}_w\}$ or $\{\overline{M}_n,[\eta]\}$ in a least-squares sense (*1, 2*) by using gradient search techniques such as that of Marquardt (*8*) or by using nongradient direct search techniques such as that of Nelder and Mead (*9*), effective values of ϵ and K can be obtained. Then the secondary molecular weight calibration curve for the polymer of interest can be obtained from Equation 2. These Mark-Houwink parameters are termed effective values since they reflect instrument spreading effects and experimental errors in the determination of $[\eta]$, $\overline{M}_n$, and $\overline{M}_w$, as well as chromatogram baseline errors. This technique can be applied to polymers of interest which are linear and branched provided the $\log_{10} [\eta]$ *vs.* $\log_{10} M$ relationship is linear over the retention volume range of the sample. These techniques have been applied successfully to PMMA and polystyrene in tetrahydrofuran, and PMMA, poly(vinyl acetate), poly(trimethylene oxide), and linear and branched polyamides in TFE (*1*).

Method for Generating Molecular Weight and HDV Calibration Curves in TFE Using Polystyrene Standards in Tetrahydrofuran. The indirect procedure used for constructing HDV calibration curves in TFE at 50°C using polystyrene standards in tetrahydrofuran at 25°C described previously (*1*) deserves some clarification because of an error generated in the use of this method. The correct procedure is summarized below:

(A) Generate retention volume chromatograms in tetrahydrofuran and TFE from "test" polymer samples, which are not necessarily narrow in molecular weight distribution and need not be characterized but are soluble in TFE and cover the retention volume range of interest in both solvents. The test polymers used were PMMA samples.

(B) Construct integral distribution curves of retention volume from the raw chromatograms.

(C) Construct a retention volume calibration curve by making a one-to-one correspondence between the retention volume in tetrahydrofuran, v_{THF}, and the retention volume in TFE, v_{TFE}, at points of equal weight per cent polymer on the integral distribution of retention volume curves of the test polymer samples. The construction of this curve assumes that the polymer molecules eluting at v_{THF} are identical to the polymer molecules eluting at v_{TFE}. Thus the molecular weight of the polymer molecules in tetrahydrofuran at v_{THF} is the same as the weight of the polymer molecules in TFE at v_{TFE}.

Thus far, the procedure is the same as that illustrated by the first three rows of plots in Figure 7 in (*1*). The retention volume calibration curve obtained from the PMMA test polymer samples is shown in Figure 9 in (*1*).

(D) Construct the HDV calibration curve in TFE from the polystyrene-generated HDV calibration curve in tetrahydrofuran in conjunction with the v_{THF} *vs.* v_{TFE} retention volume calibration curve by one of two possible routes described below and schematically illustrated in Figure 2.

Both of these routes involve transforming the retention volume axis v_{THF} to v_{TFE} *via* the retention volume calibration curve and transforming the HDV axis for PMMA in tetrahydrofuran to that for PMMA in TFE by use of the Mark-Houwink parameters for PMMA in tetrahydrofuran and in TFE.

Route 1: (a) Using the Mark-Houwink parameters of the PMMA test polymer in tetrahydrofuran, ϵ_{THF}^{PMMA} and K_{THF}^{PMMA}, construct the PMMA molecular weight calibration curve in tetrahydrofuran from the polystyrene HDV calibration curve by the use of Equation 2 where x is PMMA and s is polystyrene. This step is illustrated by the pictographs A to B in Figure 2.

(b) Using the retention volume calibration curve in conjunction with the PMMA molecular weight calibration curve in tetrahydrofuran, construct the PMMA molecular weight calibration curve in TFE. This step is illustrated by the pictographs B to C to D in Figure 2.

(c) Using the PMMA molecular weight calibration curve in TFE and the Mark-Houwink parameters of PMMA in TFE, ϵ_{TFE}^{PMMA} and K_{TFE}^{PMMA} construct the HDV calibration curve in TFE by the use of Equation 1. This step is illustrated by the pictographs D to E in Figure 2.

Route 2: (a) Using the retention volume calibration curve in conjunction with the polystyrene HDV calibration curve, construct an ap-

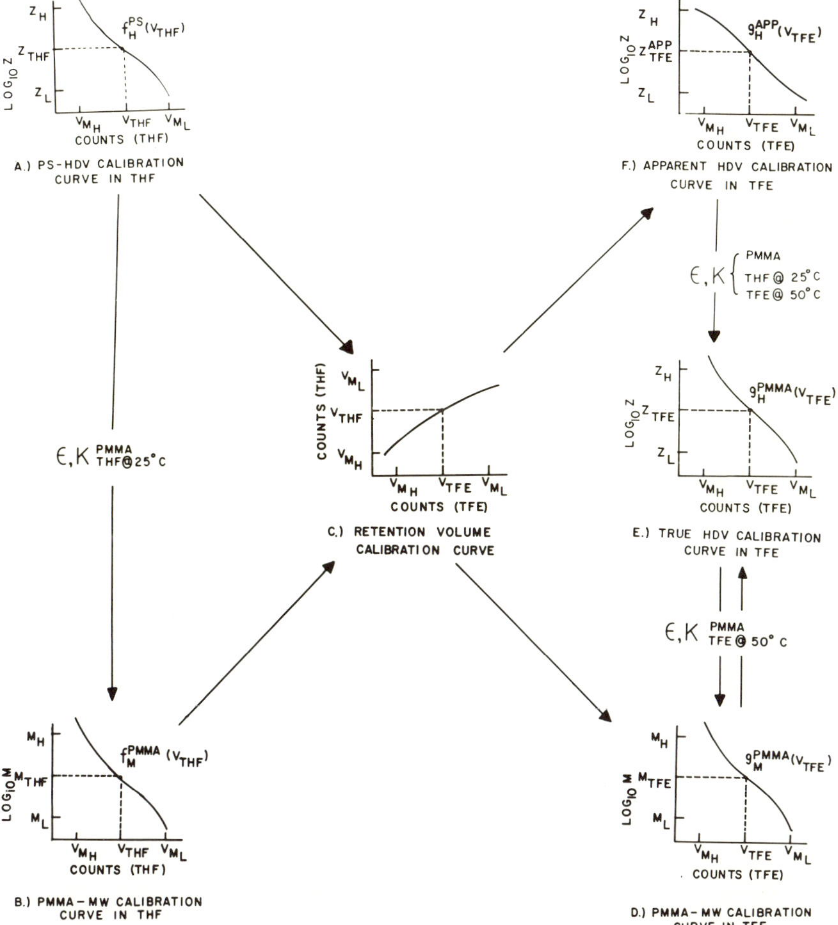

Figure 2. Illustrative method for generating PMMA molecular weight and HDV calibration curves in TFE using a polystyrene–HDV calibration curve in tetrahydrofuran

parent HDV calibration curve in TFE. In this step the retention volume axis is transformed from v_{THF} to v_{TFE}. However the HDV axis still corresponds to Z^{PS}_{THF} and therefore is denoted as Z^{APP}_{TFE}, the apparent HDV in TFE. This step is illustrated by the pictographs A to C to F in Figure 2.

(b) Using the Mark-Houwink parameters of PMMA in tetrahydrofuran and in TFE, convert the apparent HDV calibration curve. This step is illustrated by the pictographs F to E in Figure 2.

An equation can be derived relating Z^{PMMA}_{TFE} (v_{TFE}) to Z^{PS}_{THF} (v_{THF}). As was pointed out in step C, the retention volume calibration curve relating v_{THF} to v_{TFE} was constructed by relating v_{TFE} to v_{THF} at points of equal weight percent polymer on the integral distribution of retention volume curves in tetrahydrofuran and in TFE. At these points the molecular weight of the polymer species in tetrahydrofuran is the same as the molecular weight of the polymer species in TFE.

$$M^{PMMA}_{THF} = M^{PMMA}_{TFE} \tag{15}$$

Use of Equation 2 leads to the expressions

$$M^{PMMA}_{THF} = \left(\frac{Z^{PMMA}_{THF}}{K_{THF}}\right)^{\left(\frac{1}{\varepsilon_{THF}+1}\right)} \tag{16}$$

$$M^{PMMA}_{TFE} = \left(\frac{Z^{PMMA}_{TFE}}{K_{TFE}}\right)^{\left(\frac{1}{\varepsilon_{TFE}+1}\right)} \tag{17}$$

Noting that

$$Z^{APP}_{TFE} = Z^{PMMA}_{THF} = Z^{PS}_{THF} \tag{18}$$

and combining Equations 16, 17, and 18 with Equation 15 leads to the expression

$$\log_{10} Z^{PMMA}_{TFE} = \left(\frac{1}{\varepsilon_{THF}+1}\right) \log_{10}\left\{\frac{(K)_{TFE}^{\varepsilon_{THF}+1}}{(K)_{THF}^{\varepsilon_{TFE}+1}}\right\} + \left(\frac{\varepsilon_{TFE}+1}{\varepsilon_{THF}+1}\right) \log_{10} Z^{PS}_{THF} \tag{19}$$

Using the Mark-Houwink parameters for PMMA in tetrahydrofuran and in TFE leads to the following expressions relating Z^{PMMA}_{TFE} (v_{TFE}) to Z^{PS}_{THF} (v_{THF}).

$$\log_{10} Z^{PMMA}_{TFE} = 0.0480 + 1.0428 \log_{10} Z^{PS}_{THF}; \ \overline{M}_v < 31{,}000 \tag{20}$$

$$\log_{10} Z^{PMMA}_{TFE} = 0.02610 + 1.0554 \log_{10} Z^{PS}_{THF}; \ \overline{M}_v \geq 31{,}000 \tag{21}$$

Route 1 was used in Ref. *1* to construct a HDV calibration curve shown as the curve designated by + in Figure 10, Ref. *1*. Route 2 was used

Table I. Comparison of HDV Values in TFE

Retention Volume, counts	$Z_{TFE}(app.)$	$Z_{TFE}(corr.)$	$Z_{TFE}(true)$
32	1.06×10^6	2.28×10^6	1.94×10^6
36	1.07×10^5	2.03×10^5	2.20×10^5
40	0.804×10^4	1.25×10^4	1.25×10^4

as far as step 3, A to C to F, to construct an apparent HDV calibration curve denoted by the solid line in Figure 10, Ref. *1*. The difference between the apparent HDV calibration curve and the true HDV calibration curve in that figure was believed to be on the order of experimental errors, and the correction shown by Equations 20 and 21 was not made. Reevaluation of the experimental data indicates that the values of the HDV obtained from route 1, Z_{TFE} (true), are in very good agreement with the corresponding values of the HDV obtained from route 2, Z_{TFE} (corr.), by using Equations 20 and 21 on the apparent HDV, Z_{TFE} (app). This is shown in Table I. If the correction denoted by Equations 20 and 21 is not made, the apparent HDV calibration curve can still be used to generate secondary molecular weight calibration curves by the previously discussed method of Provder and co-workers (*1, 2*) if the difference between the two HDV curves is small as is shown in Figure 10, Ref. *1*, and if the two HDV curves are reasonably parallel to each other over the retention volume range of interest. These small differences will be reflected along with instrument spreading effects and experimental errors in the effective values of ϵ and K obtained by the mathematical fitting procedures.

Materials and Methods

Samples. Fourteen PMMA whole polymers were prepared by routine free-radical bulk and solution polymerization methods to cover a wide molecular weight range. Reagent grade methyl methacrylate was polymerized without removing inhibitor, according to the specifications described in Table II. The 14 polymer samples were recovered from the reaction mixture by standard techniques (*10*). The recovered samples were combined in amounts specified under "blend" in Table II and were ground to form a physically homogeneous polymer blend for preparative GPC fractionation. The molecular weight characterization data of the whole polymers and blend are shown in Table III.

Solvents. Reagent grade THF ($n_D^{25} = 0.888$, bp = 64–66°C) containing 0.025 wt-vol % di-*tert*-butyl-*p*-cresol which served as an antioxidant was used for the preparative GPC fractionation. The solvent TFE ($n_D^{20} = 1.2907$, $d^{25} = 1.3823$, bp = 76°C, ionization constant $K_a = 4.3 \times 10^{-13}$) was obtained from Halocarbon Products Corp., Hackensack, N. J., and was used for both analytical GPC and viscometry. The recovery and

Table II. Synthesis Conditions for Experimental PMMA Samples

Sample	*BPO*[a], *mole/liter* × *10*^{3}	*OM*[b], *mole/liter* × *10*^{2}	*Reaction Time at 75°C, hours*	*Conv. %*	*Blend*[c], *grams*
122-1	1.8	—	70.7	89	8.02
122-2	1.8	0.13	14.7	86	8.00
122-3	1.8	0.33	14.7	80	8.01
122-4	1.8	0.66	14.7	69	5.02
122-5	1.8	3.3	70.7	89	8.01
122-6	1.8	6.6	70.7	53	5.00
122-7	0.72	—	70.7	91	5.06
122-8	0.72	0.13	14.7	95	5.02
123-9	0.72	0.33	14.7	99	5.02
123-10	0.72	0.66	14.7	87	8.00
123-12	0.72	6.6	14.7	44	3.00

Sample	*AIBN*[d] *mole/liter* × *10*^{2}	*Solvent*[e] *Concn.*	*Reaction Time at 60°C, hours*	*Conv. %*	*Blend*[c], *grams*
140-1	4.6	19.4	4.	45	7.68
140-2	3.0	9.7	4.	53	8.00
140-3	3.0	4.8	4.	40	5.12

[a] Benzoyl peroxide.
[b] Octyl mercaptan.
[c] Components of blend for preparative GPC fractionation.
[d] Azobisisobutyronitrile.
[e] Molar ratio, benzene: methyl methacrylate.

purification of the GPC-eluted, polymer-contaminated TFE has been described previously (*1*). Reagent grade benzene (n_D^{25} = 1.4979, bp = 80.9°C) also was used for viscometry.

Gel Permeation Chromatography. A Water Associates model 200 gel permeation chromatograph fitted with five Styragel columns having nominal porosity designations 10^7, 10^7, 10^6, 1.5×10^5, and 1.5×10^4 A was used for the analysis of molecular weight distribution in TFE at a temperature of 50.0 ± 0.5°C and a flow rate of 1.00 ± 0.05 ml/min. Further details concerning instrumental and operational parameters, sample preparation and injection, and data acquisition and reduction have been reported elsewhere (*1*).

A Waters Associates Anaprep GPC fitted with one 4 ft × 2.4 inches od Styragel column having a nominal porosity of 10^4 A was used for the preparative fractionation of the PMMA blend in tetrahydrofuran at a temperature of 25°C and at a flow rate of 30 ml/min. The degasser and differential refractometer were operated at 35° and 25°C, respectively. Samples having concentrations of 0.25 wt-vol % were respectively, automatically injected from a 100 ml loop over a 5-minute period. Ten 125 ml fractions were automatically collected for each sample injection. Upon evaporation of the tetrahydrofuran, eight fractions containing significant amounts of polymers were obtained and denoted as C, D, E, F, G, H, I, and J. Fractions E through J were purified to remove the antioxidant and peroxides of tetrahydrofuran by twice dissolving the fractions in acetone, reprecipitating with methanol, and then drying under vacuum. Fractions C and D first were extracted with methanol, and then the swollen polymer was extracted with cyclohexane and dried under vacuum. The baseline-adjusted retention volume curves of these fractions in TFE are shown in Figure 3. All fractions are reasonably bell-shaped except E. As a result of some inadvertent mixing, fraction E had a high molecular weight tail. This fraction was not used in the construction of the primary molecular weight calibration curve. However, it was used in establishing intrinsic viscosity relationships.

Membrane Osmometry and Viscometry. Number-average molecular weights of PMMA were determined with a Mechrolab model 501 high-speed membrane osmometer in toluene at 60°C except for samples 122-4 and 122-6 which were determined at 40°C.

Viscometry measurements were made in benzene at 30°C and in TFE at 50°C with uncalibrated Cannon-Ubbelohde dilution viscometers which gave solvent times greater than 100 seconds. The viscometers used had centistoke ranges denoted by viscometer sizes of 50 and 75 for benzene and TFE, respectively. Stock solutions were made up on gram solute/100 gram solution basis and converted to gram/deciliter *via* the solvent density at the temperature of measurement. The solvent densities used were $d^{30°C}_{\text{benzene}} = 0.8686$ (*11a*) and $d^{50°C}_{\text{TFE}} = 1.3429$ obtained from pycnometric measurements (*12*). The density–temperature relationship for TFE obtained from regression analysis of the experimental pycnometric data is

$$d_4^{\text{TFE}} = 1.4229 - 0.0016t, \quad 20°\text{C} \leq t \leq 55°\text{C} \tag{22}$$

The solvent and solution efflux times were determined by means of the Hewlett-Packard Autoviscometer system. The intrinsic viscosity $[\eta]$ was

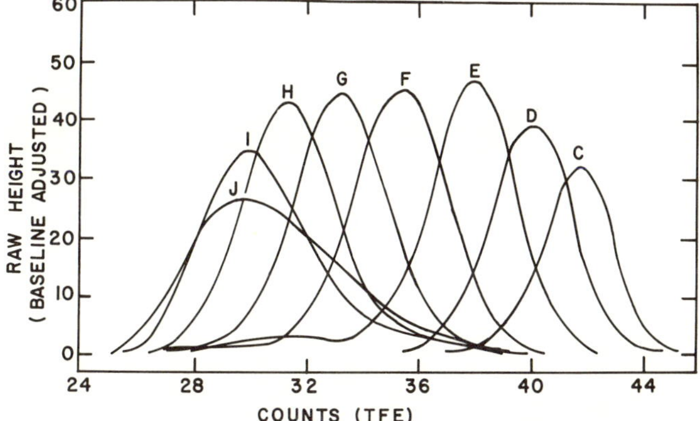

Figure 3. Baseline-adjusted chromatograms of PMMA fractions C, D, E, F, G, H, I, and J

determined from an equivalent form of the Schulz-Blaschke equation (*13*) derived by Heller (*14*) and Ibrahim (*15*). The intrinsic viscosity was the reciprocal of the intercept obtained from a linear least-squares fit of ($C, C/\eta_{sp}$) where η_{sp} is the specific viscosity and C (gram/deciliter) is the concentration and the abscissa of the parameter set. In most cases the experimental error in $[\eta]$ was less than 0.5%. Further details concerning instrumental and operational parameters of the Autoviscometer system and membrane osmometer as well as sample preparation techniques and data reduction have been reported elsewhere (*1*).

Results and Discussion

Evaluation and Preparative GPC Fractionation of PMMA. The baseline-adjusted retention volume curves of the preparative GPC fractions are shown in Figure 3. The characterization data are shown in Table III.

Table III. Molecular Weight Characterization Data for PMMA Whole Polymers and Fractions

Sample	*GPC Peak, counts*	$\overline{M}_n \times 10^{-3}$ [a]	$\overline{M}_v \times 10^{-3}$ [b]	$\overline{M}_v/\overline{M}_n$	$[n]^{30°C}_{\Phi H}$	$[n]^{50°C}_{TFE}$	*Specific Area*
122–1	29.25	547	1680	3.08	2.798	4.829	34.7
122–2	30.13	479	1250	2.61	2.213	4.262	33.6
122–3	29.75	415	1460	3.52	2.464	4.376	—
122–4	32.25	234	585	2.50	1.255	2.522	34.4
122–5	38.00	50	106	2.12	0.355	0.567	32.7
122–6	40.08	33.4	45.3	1.36	0.189	0.358	32.3
122–7	33.13	140	565	4.04	1.224	2.193	32.7
122–8	34.00	194	448	2.31	0.981	1.717	—
123–9	33.38	234	454	2.37	1.041	1.745	31.5
123–10	35.13	121	270	2.14	0.707	1.196	33.4
123–12	40.00	48.5	55.6	1.15	0.217	0.336	34.5
140–2	40.60	17.5	32.7	1.86	0.147	0.217	27.2
140–3	39.00	32.3	74.6	2.32	0.273	0.440	—
Blend	—	19.5	358	18.3	0.871	1.673	31.6
C	41.88	15.3[c]	16	1.05	0.104	0.152	18.8
D	40.15	30.8	36.4	1.18	0.160	0.225	27.9
E	38.13	68.9	196	2.85	0.559	0.908	31.4
F	35.30	130	281	2.15	0.729	1.179	33.7
G	33.30	227	513	2.26	1.139	1.985	34.2
H	31.40	513	918	1.78	1.757	3.032	31.6
I	30.00	671	1490	2.24	2.510	4.328	30.7
J	29.30	—	1370	—	2.349	4.187	33.9
Methyl methacrylate	46.60	0.1001	—	—	—	—	—

[a] Measurements made in toluene at 60°C except for samples 122-4 and 122-6 which were made at 40°C in toluene.

[b] $\overline{M}_v$ values obtained from $[n]^{30°C}_{\Phi H}$ data.

[c] $\overline{M}_n$ value has been diffusion corrected.

The $\overline{M}_v/\overline{M}_n$ values of Table III and the GPC traces of Figure 3 indicate that a reasonable fractionation was achieved with a single 10^4 A porosity preparative GPC column. Narrower molecular weight distribution fractions could have been achieved if another preparative column was used in series (*e.g.*, 10^6 A porosity column) with the 10^4 A column. Because of the high molecular weight and high viscosity of the PMMA polymers, we found that 0.25 wt-vol % was an upper limit to the sample concentration that could be injected in the preparative GPC without producing severe column overloading. Again, narrower molecular weight distribution fractions with less tailing could have been achieved by lowering the injected sample concentration with a concomitant reduction in sample through-put. It is recommended that tetrahydrofuran not be used as a preparative GPC solvent because the annoying presence of the antioxidant and peroxides in tetrahydrofuran makes recovery of the polymer fractions unnecessarily difficult. A good room temperature preparative GPC solvent substitute for tetrahydrofuran is methylene chloride which does not absorb water as does tetrahydrofuran.

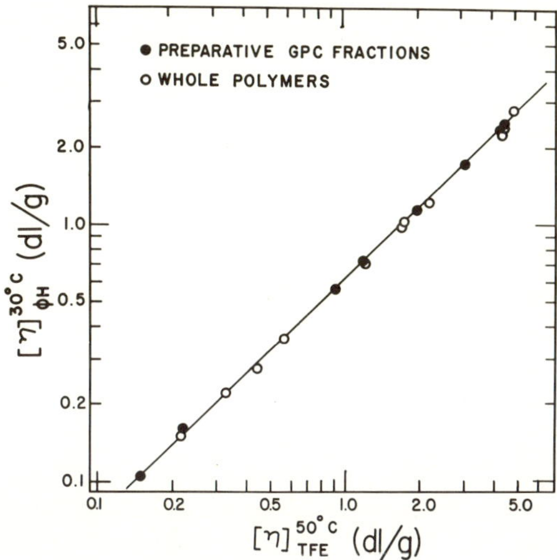

Figure 4. Relationship between $[\eta]_{\Phi H}^{30°C}$ *and* $[\eta]_{TFE}^{50°C}$ *for PMMA whole polymers and fractions*

Intrinsic Viscosity–Molecular Weight Relationship for PMMA in TFE. The intrinsic viscosities of the PMMA preparative GPC fractions and whole polymers in TFE at 50°C and in benzene at 30°C are shown in Table III and plotted in Figure 4. A least-squares analysis of the data plotted in Figure 4 yields the relation

$$\log_{10}[\eta]^{50°C}_{TFE} = 0.2321 + 1.070 \log_{10}[\eta]^{30°C}_{\Phi H},\ 0.152 \leq [\eta]^{50°C}_{TFE} \leq 4.765 \quad (23)$$

where the standard errors in the slope and intercept are ± 0.010 and ± 0.0013, respectively. The Mark-Houwink intrinsic viscosity–molecular weight relationship for PMMA in benzene at 30°C can be obtained from the data of Cohn-Ginsberg, Fox, and Mason (*16*). Their data consist of 14 well characterized narrow molecular weight distribution polymer fractions. A least-squares analysis of their data yields the relations

$$[\eta]^{30°C}_{\Phi H} = 6.83 \times 10^{-5}\overline{M}_w{}^{0.739},\ \overline{M}_w \geq 31{,}000 \quad (24)$$

$$[\eta]^{30°C}_{\Phi H} = 166 \times 10^{-5}\overline{M}_w{}^{0.431},\ \overline{M}_w < 31{,}000 \quad (25)$$

where the errors in the $\log_{10}$ of the pre-exponential factors and in the exponents of Equations 24 and 25 are ± 0.04 and ± 0.009, respectively. [The least-squares relationships represented by Equations 24 and 25 differ somewhat from the relationships originally reported by Cohn-Ginsberg, Fox, and Mason (*16*).] Use of Equations 24 and 25 with Equation 23 yields the intrinsic viscosity–molecular weight relationship for PMMA in TFE at 50°C:

$$[\eta]^{50°C}_{TFE} = 5.95 \times 10^{-5}\overline{M}_v{}^{0.791},\ \overline{M}_v \geq 31{,}000 \quad (26)$$

$$[\eta]^{50°C}_{TFE} = 181 \times 10^{-5}\overline{M}_v{}^{0.461},\ \overline{M}_v < 31{,}000 \quad (27)$$

Calibration Curves. In Figure 5 is shown a direct PMMA molecular weight calibration curve constructed from a plot of $\overline{M}_v$ *vs.* PRV. Also shown in Figure 5 is an indirect molecular weight calibration curve constructed according to the indirect method (*1*) as summarized in the theory section in steps A through D, using route 1. Within the experimental errors associated with the construction of the two molecular weight curves, the curves agree reasonably well over the retention volume ranges of the PMMA whole polymers and fractions. Some of the possible sources of errors involved with the construction of the direct molecular weight calibration curve are 1) experimental errors in the evaluation of $\overline{M}_v$ and 2) errors associated with relating $\overline{M}_v$ values with PRV. Possible sources of errors associated with the construction of the indirect molecular weight calibration curve involve experimental errors in the construction of the molecular weight calibration curve as outlined in steps A through D, using route 1 in the theory section. The largest source of error is involved with step C in the construction of the v_{THF} *vs.* v_{TFE} retention volume curve. As indicated previously (*1*), two different column sets were used to construct the v_{THF} *vs.* v_{TFE} retention volume curve. The column set used with tetrahydrofuran, which had nominal porosity designations 10^6, 10^5, 10^4, 10^3, 250 A, had less resolving power at high molecular weights and more resolving power at low molecular weights than the column set used with TFE. In addition to this difference in resolving power between both column sets, there would be differences in instrumental spreading effects (*2*) such as axial dispersion, skewing, and flattening of the observed chromatograms of the test polymers used with both column sets. The differences in resolving power and instrumental spreading effects primarily account for the slight differences between the two molecular weight calibration curves shown in Figure 5 over the retention volume ranges of the PMMA samples and the larger differences at very high molecular weights.

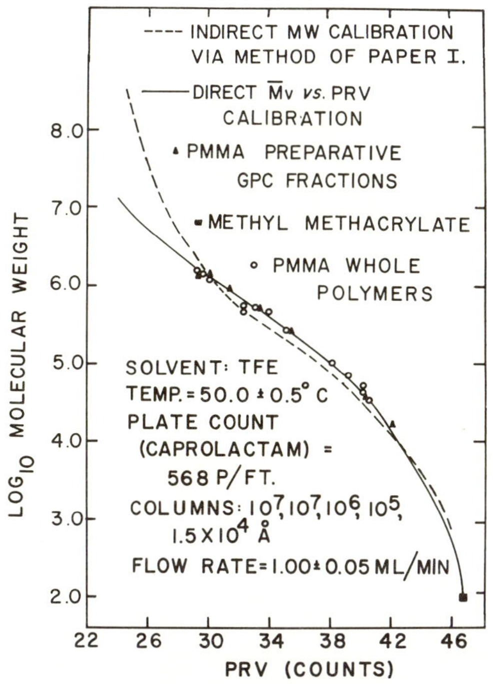

Figure 5. Molecular weight calibration curves for PMMA in TFE at 50°C

In Table IV, the values of $\overline{M}_n$, $\overline{M}_w$, $[\eta]$, and $\overline{M}_w/\overline{M}_n$ for the PMMA fractions D and H and whole polymers 122-7 and 122-9 calculated from the direct $\overline{M}_v$ *vs.* PRV calibration curve are compared with the available corresponding true values. In general, the calculated values of $[\eta]$ from the direct calibration curve are larger than the true values at low molecular weights as indicated in the values for fraction D and tend to converge to the true values as the molecular weight increases as indicated

Table IV. Comparison of $\overline{M}_n$, $\overline{M}_w$, $[\eta]$, *and* $\overline{M}_w/\overline{M}_n$ Values for PMMA Samples

Method	$\overline{M}_n \times 10^{-3}$	$\overline{M}_w \times 10^{-3\,a}$	$[n]$	$\overline{M}_w/\overline{M}_n$ [b]
	Fraction D			
True value	30.8	(36.4)	0.23	(1.18)
Direct $\overline{M}_v$ *vs.* PRV curve	27.2	47.5	0.29	1.75
	122–7			
True value	140	(565)	2.19	(4.04)
Direct $\overline{M}_v$ *vs.* PRV curve	183	650	2.18	3.56
	122–9			
True value	234	(454)	1.75	(2.37)
Direct $\overline{M}_v$ *vs.* PRV curve	251	511	1.85	2.03
	Fraction H			
True value	513	(918)	3.03	(1.78)
Direct $\overline{M}_v$ *vs.* PRV curve	582	930	3.06	1.60

[a] Values in parenthesis are $\overline{M}_v$ values, $\overline{M}_v \leq \overline{M}_w$.
[b] Values in parenthesis are $\overline{M}_v/\overline{M}_n$ values.

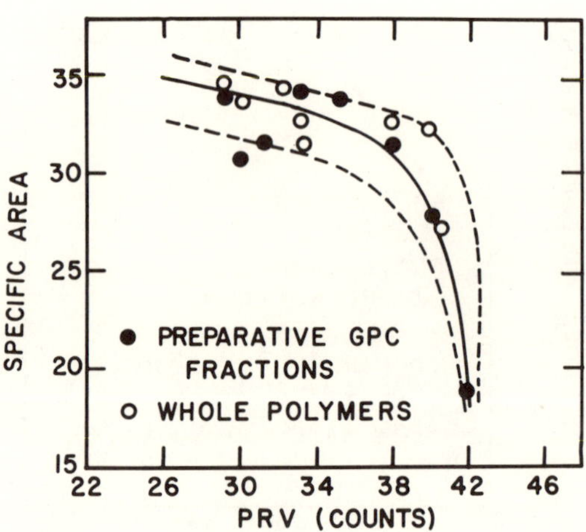

Figure 6. Specific area vs. *PRV in TFE for PMMA whole polymers and fractions*

by the values for fraction H. On the other hand, calculated $\overline{M}_n$ values from the direct calibration curve are in reasonable agreement with the true values at low molecular weights as indicated by the values for fraction D and tend to become larger than the true values at high molecular weights as indicated by the values for fraction H.

In part, this shift in $[\eta]$ and $\overline{M}_n$ can be attributed to the errors involved in the correlation of $\overline{M}_v$ with PRV. Also it can be attributed, in part, to a dependence of detector response upon the retention volume of the eluting species. Since the detector is a differential refractometer, this would require a dependence of the specific refractive index increment for PMMA in TFE at 50°C upon molecular weight. The total sample weight injected was in the 1 to 2 mg range. Thus, the concentration was low enough to rule out non-linear detector response over the retention volume range of interest in TFE. The specific area of the chromatogram is defined to be the area of the chromatogram divided by the product of the GPC sensitivity setting and the milligrams of sample injected on the column set. A plot of the specific area of the chromatogram *vs.* PRV for the PMMA samples in shown in Figure 6. The dotted lines indicate a scatter envelope corresponding to reasonable experimental errors attributable to uncertainties in chromatogram baseline and concentration errors. Since the chromatogram heights H are proportional to the true concentration C of polymer at a given retention volume,

$$H = K_1 C \tag{28}$$

it follows that the true polymer concentration is given by

$$C = H/K_1 = H \times R \tag{29}$$

where R is the response factor in units of mg/area required to correct the heights of the chromatogram. The response factor curve, which is the reciprocal of the specific area curve, for PMMA in TFE at 50°C is shown in Figure 7. The refractive index difference, Δn, measured by the differential refractometer is directly proportional to the polymer concentration,

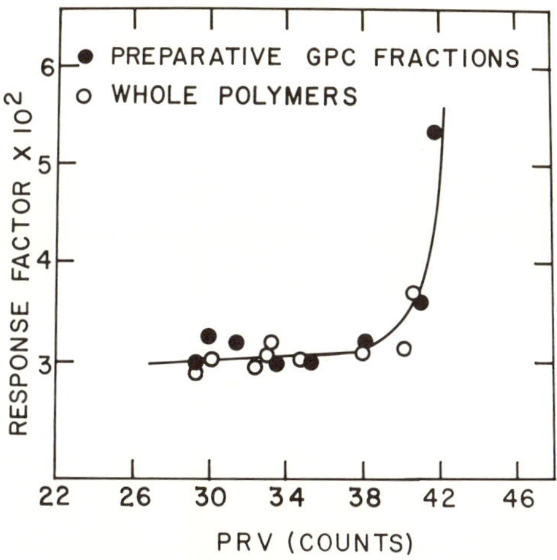

Figure 7. Response factor vs. *PRV in TFE for PMMA whole polymers and fractions*

$$\Delta n = n_D^{50} \text{ polymer} - n_D^{50} \text{ solvent} = K_2C \quad (30)$$

Also, the refractive index of the polymer is linear in the reciprocal of $\overline{M}_n$ (*17, 18*) as

$$n_D^{50} \text{ polymer} = a + b/\overline{M}_n \quad (31)$$

Equations 29, 30, and 31 can be combined to yield an equation similar to Equation 31 relating the response factor R to the reciprocal of $\overline{M}_n$

$$R = A + B/\overline{M}_n \quad (32)$$

Thus, a plot of R *vs.* $1/\overline{M}_n$ should be linear if, indeed, there is a molecular weight dependence of the specific refractive index increment. Figure 8 shows a plot of the response factor *vs.* the reciprocal of $\overline{M}_n$ for the fractions and whole polymers of PMMA used in Figure 6. The scatter about the least-squares line in Figure 8 can be attributed to uncertainties in chromatogram baseline, and concentration errors. Also, the sample polydispersity or sample retention volume range will effect the goodness of the correlation of R with $1/\overline{M}_n$. The seventeen data points are plotted about the least-squares line in Figure 8. The calculated correlation coefficient for the least-squares fit is 0.6413. If the correlation coefficient

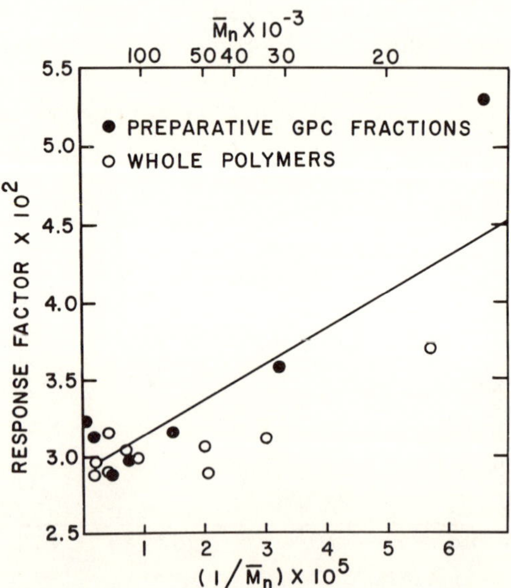

Figure 8. Response factor vs. $1/\overline{M}_n$ *in TFE for PMMA whole polymers and fractions*

were zero, then the probability level that a correlation coefficient as large as 0.641, for fifteen degrees of freedom, would be observed is 0.005 (*19, 20*). Therefore the linear correlation between the response factor and $1/\overline{M}_n$ is statistically significant.

Figure 8 shows that the response factor depends strongly upon the reciprocal of $\overline{M}_n$ for values of $\overline{M}_n$ less than 40,000. A similar trend of the dependence of the response factor upon the reciprocal of $1/\overline{M}_n$ was observed in tetrahydrofuran for these same PMMA samples. However, the magnitude of the trend over the molecular weight range was not as large in tetrahydrofuran as in TFE. This is expected since the specific refractive index increment value of 0.220 in TFE (*1*) is larger than the value of 0.0871 in tetrahydrofuran (*11b*) for PMMA. The strong specific refractive index increment dependence upon $\overline{M}_n$ at lower molecular weights to some extent can be associated with end-group functionality differing from the backbone functionality resulting from the presence of initiator or chain-transfer molecule fragments.

Figure 6 shows that the detector response, expressed as specific area, weights the higher molecular weight species below PRV of 38 considerably more than the molecular weight species above PRV of 38 (molecular weight $< 100{,}000$). Examination of Figure 3 shows that all the fractions have substantial low molecular weight tails above PRV of 38. The same is true of the whole polymers. Therefore it is not surprising that $\overline{M}_n$ values calculated from the direct molecular weight calibration curve tend to become larger than the true values at higher molecular weights since the chromatogram heights at low molecular weights are being weighted much less than the chromatogram heights at high molecular weights. To correct this situation, all chromatogram heights should be multiplied by the appropriate response factor, shown in Figure 7, prior to the calculation of the molecular weight averages. If the molecular weight dependence of the specific refractive index increment is not corrected for, anomalies such as $(\overline{M}_w/\overline{M}_n)$ GPC $<$ $(\overline{M}_w/\overline{M}_n)$ true can result. For example, for fraction H $(\overline{M}_w/\overline{M}_n)$ GPC $= 1.60$ while $(\overline{M}_v/\overline{M}_n)$ true $=$ 1.78. If instrument spreading corrections are attempted without prior corrections for the molecular weight dependence of the specific refractive index increment, improper values will result.

Using the PMMA direct molecular weight calibration curve shown in Figure 4 and the Mark-Houwink parameters for PMMA in TFE shown in Equations 26 and 27, an HDV calibration curve can be constructed as described in the theory section. Then secondary molecular weight curves can be constructed for other polymers of interest by the methods discussed in the theory section as was done in Ref. 1, using the indirect PMMA molecular weight calibration curve.

Conclusions

Characterized PMMA polymers can be used as primary standards for constructing a direct $\overline{M}_v$ *vs.* PRV molecular weight calibration curve in TFE at 50°C much as characterized polystyrene standards can be used in tetrahydrofuran for constructing primary molecular weight calibration curves. The direct PMMA molecular weight calibration curve compares quite well with the indirect molecular weight calibration curve constructed according to the method previously discussed (*1*). It also has been shown that the apparent anomalies $(\overline{M}_n)$ GPC $>$ $(\overline{M}_n)$ true and $(\overline{M}_w/\overline{M}_n)$ GPC $<$ $(\overline{M}_w/\overline{M}_n)$ true are directly attributable to a strong dependence of the specific refractive index increment upon the molecular weight at low molecular weights up to values of 100,000. Mark-Houwink parameters for PMMA have been established in TFE at 50°C. Thus, PMMA polymers can now be used as primary standards for constructing HDV curves in TFE from which secondary molecular weight calibration curves can be obtained for other polymers of interest.

Acknowledgments

The authors acknowledge the assistance of Masao Ohta in the polymerization of the PMMA whole polymers and in the recovery of the preparative GPC fractions and McClinton Rayford for some of the solution property characterization work. Also acknowledged is James E. Kurz for many helpful discussions.

Literature Cited

1. Provder, T., Woodbrey, J. C., Clark, J. H., *Separ. Sci.* (1971) **6**, 101.
2. Provder, T., Rosen, E. M., *Separ. Sci.* (1970) **5**, 437.
3. Coll, H., Prusinowski, L. R., *J. Polym. Sci., Part B* (1967) **5**, 1153.
4. Meyerhoff, G., *Separ. Sci.* (1971) **6**, 239.
5. Goedhart, D., Opschoor, A., *J. Polym. Sci., Part A-2* (1970) **8**, 1227.
6. Billmeyer, F. W., *J. Polym. Sci.* (1949) **4**, 83.
7. Huggins, M. L., *J. Amer. Chem. Soc.* (1942) **64**, 2716.
8. Marquardt, D. W., *J. Soc. Ind. Appl. Math.* (1963) **2**, 431.
9. Nelder, J. A., Mead, R., *Computer J.* (1965) **7**, 308.
10. Fox, T. G., Kinsinger, J. B., Mason, H. B., Schuele, E. M., *Polymer* (1962) **3**, 71.
11. Brandrup, J., Immergut, E. H., "Polymer Handbook," Interscience Publishers, New York, 1966, a) Vol. VIII p. 53, b) Vol. IV p. 288.
12. Daniels, F., Williams, J. W., Bender, P., Alberty, R. A., Cornwell, C. D., "Experimental Physical Chemistry," McGraw-Hill Book Co., Inc., New York, 1962, p. 452.
13. Schulz, G. V., Blaschke, E., *J. Prakt. Chem.* (1941) **158**, 130.
14. Heller, W., *J. Colloid Sci.* (1954) **9**, 547.
15. Ibrahim, F. W., *J. Polym. Sci., Part A* (1968) **3**, 469.
16. Cohn-Ginsberg, E., Fox, T. G., Mason, H. F., *Polymer* (1962) **3**, 97.

7. Geczy, I., *Vysokomol. soyed* (1965) **7**, 642.
8. Rhein, R. A., Lawson, D. D., *Chem. Tech.* (Feb. 1971) 122.
9. Dixon, W. J., Massey, F. J., "Introduction to Statistical Analysis," McGraw-Hill Book Co., Inc., New York, 1957, p. 468.
0. Volk, W., "Applied Statistics for Engineers," McGraw-Hill Book Co., Inc., New York, 1969, p. 266, ff.

ECEIVED February 14, 1972.

12

Gel Permeation Chromatography—Data Acquisition and Processing System Using a Minicomputer

A. E. HAMIELEC, G. WALTHER, and J. D. WRIGHT

Department of Chemical Engineering, McMaster University, Hamilton, Ontario, Canada

A low-cost minicomputer has been developed for data acquisition and reduction in gel permeation chromatography. The system includes a Nova 1200 CPU, Teletype, and special-purpose interface. With efficient assembly-language programming, a minimum configuration system containing 4k words of 16-bit memory was adequate for simultaneous data acquisition from four GPC instruments or for data acquisition and reduction from one. Monitored data include peak heights, elution volumes, and sample injections. Molecular weight averages and distributions are calculated and are available within a few minutes of the completion of a sample run. The user can monitor and direct the operation of the system using simple Teletype commands. The computer system is modular and is flexible enough to handle other laboratory equipment.

Data acquisition and reduction in gel permeation chromatography (GPC) can be most tedious and time-consuming work. The analysis of high polymers, unlike the analysis of oligomers and small molecules by liquid and gas chromatography, requires several integrations of the detector response to obtain higher moments and a molecular weight distribution (MWD). Often the moments and MWD must be corrected for axial dispersion. The latter calculations must be done with a digital computer. Manual acquisition of chromatogram heights is very costly in manpower and is usually not done on a routine basis. Some analytical laboratories have automated data acquisition with an analog-to-digital converter (*1*, *2*). The raw chromatogram heights are punched on paper tape and later processed on a digital computer. A few commercial laboratories have developed complete data acquisition and reduction systems for GPC using a dedicated computer (*3, 4*); however their systems have been rather costly.

At McMaster University we have four GPC instruments being used by the polymer research and waste water groups. It was worthwhile, therefore, to develop a low-cost, dedicated minicomputer system to acquire and process GPC data internally. In addition, the minicomputer system was to be used as part of the chemical engineering graduate program for general purpose, real time, data acquisition. The chemical engineering department, in addition to its GPC instruments, has in use many measuring devices commonly found in the chemical laboratory and pilot plant. Most of the signals are low level (in the 0–100 mv range) and require sampling rates not in excess of 10,000 samples/second (most may be sampled at relatively low rates). These latter specifications required a system that combined accuracy with flexibility and modularity.

Funds available for hardware for one year were about $11,000. This limited the size of our basic system and made the requirement for modularity especially important. We plan to expand our system to increase the number of detectors that may be sampled simultaneously and increase the complexity of data reduction calculations. The expansion of core size to permit the use of Fortran or Algol would also be desirable at some future date.

We will now discuss our experiences in interfacing and implementing a system, which we have called DARS (Data Acquisition and Reduction System), that fulfills all of the above requirements. In particular, we will examine the following questions:

What is a minicomputer system?
How is a minicomputer programmed?
What software has been developed?
What are the hardware costs?

What is a Minicomputer System?

A minicomputer is generally defined as a system comprising a central processor with a minimum configuration of 4k words of memory which can be purchased for less than $10,000. A basic minicomputer system, adequate for monitoring and doing all calculations encountered in GPC, consists of a central processing unit (CPU), a real-time clock, an interface, and a Teletype (Figure 1). Auxillary equipment might include a high-speed paper tape reader and punch, a disk, a cassette tape recorder, a line printer, etc. Detailed below are a few features of each component as it applies to the DARS.

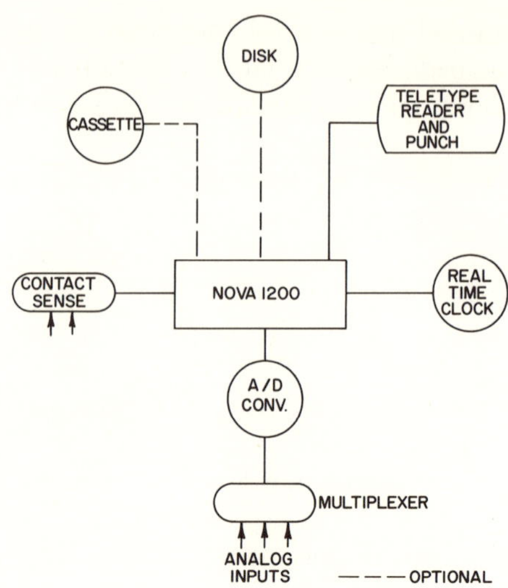

Figure 1. DARS system

CPU. The central processing unit for the DARS is a Nova Model 1200 by Data General in the United States and Datagen in Canada. The Nova 1200 has a 16-bit word length and a cycle time of 1.2 microseconds. As many as 62 separate devices may be interfaced to the CPU. The basic system contains 4k (4096) words of memory which is expandable up to a maximum of 32k words. Included in the in-out system are facilities for hardware interrupts or a real-time priority basis.

Real-Time Clock. The real-time clock produces a program interrupt at one of four frequencies, independent of processor timing. The real-time clock is used in GPC work to implement sampling at user specified intervals.

Teletype. The Teletype is an ASR-33, with a printing speed of 10 characters per second. All user communication, including output, in the basic system is implemented through the Teletype.

Interface. The interface includes a 10-bit analog-to-digital converter, a ten-channel multiplexer, and eight contact sense lines. The converter will operate at rates up to 20,000 conversions per second with a precision of 1/1000 for analog inputs with either 10 or 100 mv full scale values. The contact sense lines are used to communicate elution dumps and sample injections by signaling the computer on a circuit closure. The interface, with eight contact sense lines and four active A/D channels, is capable of handling data from four GPC instruments simultaneously.

Auxiliary Equipment. A high-speed reader, high-speed punch, disks, cassette recorder, or line printer provide greater input-output efficiency. While they are unnecessary with the basic system for day-to-day operation with analytical instruments in a dedicated mode, they are essential for general purpose computation.

Why a Minicomputer System?

A minicomputer system that collects and processes data offers many advantages to the GPC user. First, molecular weight averages are calculated and available within a minute of the completion of an analysis. Second, any errors that arise, either human or mechanical, in preparing data for processing are eliminated because the data are stored and processed internally. Third, there are no interference effects when two events occur simultaneously (*i.e.*, elution volume and height conversion) as has been reported by Brussau (*1*), because the two servicing systems are completely independent. Fourth, sample processing with a minicomputer is economically competitive with sample processing on larger computers. (Processing similar to what can be done on the minicomputer costs about $2.00 per sample on a CDC 6400.) The minicomputer requires no special system's support, and is able to operate in most laboratory environments without any protective enclosure, eliminating costly maintenance. Finally, the minisystem is very flexible in that it may be expanded to handle data from additional GPC instruments or programmed to operate with a variety of other laboratory instruments when not on-line with a GPC system.

How Is a Minicomputer Programmed?

The minicomputer can be programmed in several languages including Fortran, Algol, Basic, and Assembler. All GPC programming was done in Assembler language because it is the most efficient language when attempting to minimize the amount of core required. Compiler level languages such as Fortran generally involve large overheads in core space and hardware in which to operate and are uneconomical for a small dedicated real-time system.

The GPC programs are structured around a priority interrupt system which allows the program to ensure that events are acknowledged in their proper sequence. Each device is assigned a priority. Any device of higher priority may interrupt programs or devices being serviced at lower priorities. Once the higher priority device has been serviced, control is returned to the point at which the lower priority routine was interrupted. All interrupts for lower priority devices are deferred (masked out) until all higher priority devices have been serviced. For example, Figure 2 is the priority structure for the GPC system. The real-time clock may

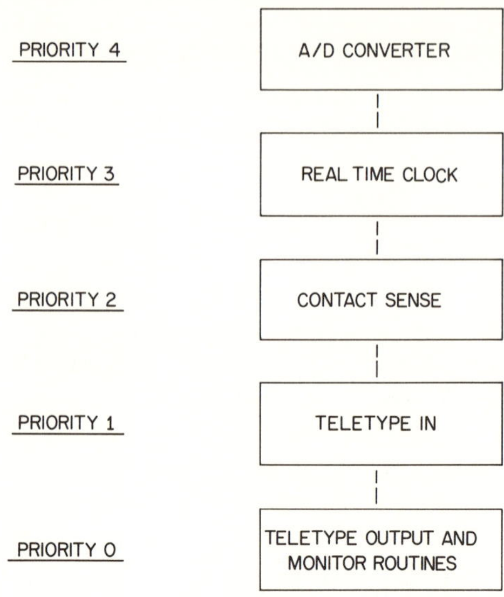

Figure 2. Priorities for GPC system

interrupt the Teletype but it itself would be interrupted by the A/D converter.

The basic procedure for programming a minicomputer in Assembler language is not unlike programming a large computer in Fortran. Once a flow chart is established, each subroutine is written, assembled, tested, and corrected before constructing the entire system. Most minicomputer manufacturers provide some system software which may include editors, assemblers, debuggers, and binary loaders with their machine. It is particularly useful to have system debugging programs which allow on-line, dynamic monitoring and altering of programs in the development stage.

What Programs Have Been Developed?

The programs developed for GPC include an operating system for control of data acquisition and a library of calculational routines for data reduction. Each program was developed with the single purpose of minimizing core requirements. This was necessary to ensure that a workable system could be constructed in the minimum configuration system of 4k. To accomplish this goal it was necessary to rewrite many of the standard routines, eliminating those features which are not required in GPC calculations.

The operating system is the main channel of communication between the user, computer, and GPC. Included in the operating system are routines to read and store a chromatogram height, acknowledge an elution dump, out-put data, etc. The user may access one of eight input-output control routines by entering a command at the Teletype. Figure 3 contains a summary of the basic commands and their use. to obtain the status of the system at any time, the user would type "LOOK (ESC)," to which the computer would respond by printing out current operating information.

```
LOOK   -   OUTPUT CURRENT SYSTEM INFORMATION.

DATA   -   ENTER AUTOMATIC DATA COLLECTION AND PROCESSING PARAMETERS.

BEGIN  -   ENTER DATA COLLECTION PARAMETERS FOR INDIVIDUAL SAMPLES.

STOP   -   MANUAL STOP

CALC   -   MANUAL INITIATED CALCULATION

ATD    -   ANALOG TO DIGITAL TEST ROUTINE

INJ    -   KEYBOARD ACKNOWLEDGED INJECTION

TYPE   -   MANUAL OUTPUT OR DATA

      (All commands are followed by a delimiting signal (ESC key)
```

Figure 3. Command summary

Shown in Figure 4 is a start-up for a system with one GPC instrument which has an automatic sample injection valve. The user has pressed the start switch which initiated the system and returned a restart message. He then entered the "data" command which requested sample control information. The system will now acknowledge continuous sample injections, beginning with sample number 1500. When the 45th elution dump following the injection is reached, all data will be typed out and a calculation will be executed. A record of all output will be punched on paper tape if the punch is turned on at the keyboard. User options include facilities to process samples either manually or automatically, overlap samples, update data collection parameters during a run, or reprocess an old data tape while simultaneously acquiring data for new samples.

The calculation library includes programs to process all data encountered in normal GPC operation for standards and unknowns. It includes routines to decode data, correct for baseline drift, interpolate for flow variation, calculate mean elution volume, calculate molecular weight averages, and calculate corrected molecular weight averages and distributions. Molecular weight averages are obtained from the moments of the chromatograph as calculated by Simpson's rule. Corrected molecular averages are found by adjusting the uncorrected averages according to the method of Balke and Hamielec (5). The corrected molecular weight distribution can be calculated by Method 2 of Ishige, Lee, and Hamielec (6). This method of determining the corrected molecular weight distribution was chosen for its accuracy and small intermediate storage requirements.

```
RESTART

*


DATA
TYPE Ø-OFF, 1-ON

AUTO INJ  1
AUTO TYPE 1
AUTO CALC 1


SAMPLE NO. 15ØØ
RATE CODE  1
ON AT EV  2Ø
OFF AT EV 45

BEGIN BASE 22
END BASE   42
BEGIN CALC 26
END CALC   38

D1 = 3.654E13
D2 = .63457


INJECTION NOTED FOR Ø15ØØ
```

Figure 4. Start up of GPC system

One of the major advantages of the minicomputer is the speed and precision of data analysis. Data reduction can begin immediately upon completion of a run. Uncorrected molecular weight averages are usually available within a minute. Corrected molecular averages and MWD values take longer, depending on the method employed. Normal calculation time using the above Method 2 was about 5 minutes. Molecular weight averages obtained from a minicomputer and a CDC 6400 were in almost exact agreement.

```
CALC
0-TAPE, 1-MEM  0
SSAMPLE NO. 1500
BEGIN BASE 22
END BASE    42
BEGIN CALC 24
END   CALC 30


D1 = 3.6543E13
D2 = 0.63457


MEAN ELUTION VOLUME = +.2645761E+02


MOLECULAR WEIGHT AVGS.

MN = +.1484059E+06
MW = +.1995748E+06
MZ = +.2517811E+06

  MOL  WT.          FRACTION

.1200000E+04      .7648611E-06
.1300000E+04      .1657890E-05
.1400000E+04      .2416780E-05
```

Figure 5. Processing a data tape

Figure 5 shows the communication necessary for manually initiating a calculation to reprocess an old data tape. This situation could arise if, for example, the operator decided to try new limits for the calculation. The user has entered the calculation routine by entering "CALL (ESC)" at the Teletype. The system responds with questions about the source of the data and calculation parameters to be used. The calculation will begin when the user places the data tape in the paper tape reader.

The number of programs which reside in core at any one time is dependent on the memory size of the system. Figure 6 is a core map for a 4k and an 8k system. As Figure 6 indicates, a 4k system is adequate to acquire data and calculate molecular weight averages and MWD values for one GPC system, corrected for axial dispersion. A 4k system includes a data storage capability of 400 fixed point numbers for raw data, and storage of 61 floating point numbers for adjusted data. The 4k system could also be used for data acquisition only, for four GPC

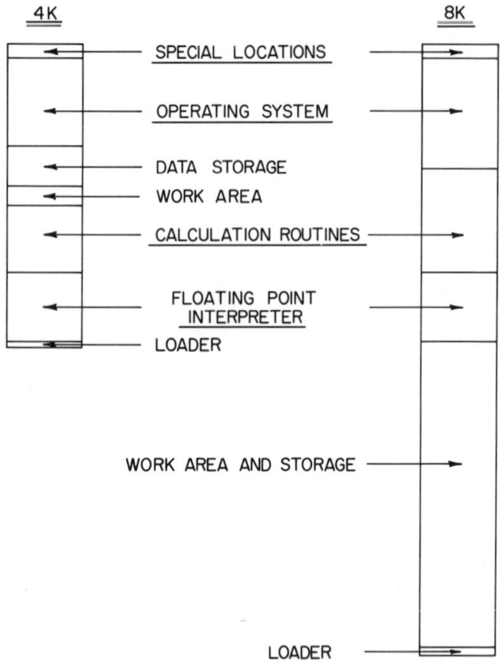

Figure 6. Core map

systems simultaneously. Output from this data-taking system could be processed later, or used as raw data for processing by a central computer or time sharing. An 8k system, on the other hand, can be used to acquire and process data from 4 GPC instruments simultaneously. Data storage buffers for fixed point numbers could be as large as 800 words, depending upon the complexity of additional calculational routines.

What Are the Hardware Costs?

The hardware costs for the DARS are summarized below (quoted without Federal or Provincial taxes):

Nova 1200 CPU	$ 2,665
4k words, 16-bit memory	3,000
Real-time clock	445
Power monitor and autorestart	445
Teletype interface	220
ASR 33 Teletype	1,300
Interface (4 channels low-level A/D—10 bits, 8 contact sense inputs)	2,500
	$10,575 (1972)

An 8k word system involves only the purchase of an additional memory board.

Conclusion

We have developed a minicomputer system and found it to be ideally suited to the problems associated with data acquisition and reduction in gel permeation chromatography. Such a system eliminates tedious work, while providing fast and accurate molecular weight averages and MWD values. With efficient programming, a 4k word system costing less than $11,000 can be used for simultaneous data acquisition from four GPC systems or for data acquisition and reduction from one. Data acquisition and reduction for four systems operating simultaneously is possible by expanding memory to 8k at an additional cost of $3,000.

Literature Cited

1. Brussau, R. J., *J. Chromatogr. Sci.* (1971) **9**, 1.
2. Moore, L. D., Overton, J. R., *9th Int. GPC Seminar Preprints* (1970) 399.
3. "Nuclear Data Publication," Nuclear Data Inc., Palatine, Ill.
4. "Problematic Preliminary," Problematics Inc., Waltham, Mass.
5. Balke, S. T., Hamielec, A. E., *J. Appl. Polym. Sci.* (1969) **13**, 1381.
6. Ishige, T., Lee, S. I., Hamielec, A. E., *J. Appl. Polym. Sci.* (1971) **15**, 1607.

RECEIVED January 17, 1972.

13

Molecular Weight Averages from Gel Permeation Chromatography Using the Universal Calibration Method

EDGAR NICHOLS[1]

Gulf Oil of Canada Ltd., Montreal, Quebec, Canada

In GPC, the product $[\eta]$M, *(or the hydrodynamic radius* R_e) *has been widely accepted as a universal calibration parameter. In the Ptitsyn-Eizner modification of the Flory-Fox equation the quantity* Φ, *which relates the dimensional parameters to the above product, is taken as a variable. The value of* Φ *depends upon molecular expansion in solution as represented by a function* f(e). *Because of this dependence polymeric species having the same* $[\eta]$M *value cannot have the same statistical dimensions (radius of gyration or end-to-end distance) unless they have the same* e *value. Thus, if* $[\eta]$M *is a universal calibration parameter, the statistical parameters cannot be used as such. A method is presented for obtaining the* M_w/M_n *ratio from GPC data even though universal calibration is used.*

In GPC the product $[\eta]M$ has been widely accepted as a universal calibration parameter, where $[\eta]$ is the intrinsic viscosity and M is the molecular weight. This product is defined by the Einstein-Simha viscosity expression (*1*) as

$$[\eta]M = \Phi_o R_e^3 \tag{1}$$

where R_e is the radius of a hydrodynamic sphere and Φ_o is a constant having the value 6.308×10^{24} (cgs units). Flexible polymers in dilute

[1] Present address: NL Industries, W. Caldwell, N. J. 07006

solution behave as spheres of radius $R_e = R\xi$, where R is the root-mean-square radius of gyration. The $[\eta]M$ product for such polymers, then, is given by

$$[\eta]M = \Phi_o \xi^3 R^3 \tag{2}$$

and for the end-to-end distance $\overline{h}$,

$$[\eta]M = \Phi_o \xi^3 \overline{h}^3/(6)^{3/2} \tag{3}$$

The value of ξ can be expressed by the Ptitsyn-Eizner relation (*2*) as

$$\xi = [\xi_\theta{}^3 f(e)]^{1/3} \tag{4}$$

where ξ_θ refers to an ideal or theta solvent and

$$f(e) = 1\text{-}2.63e + 2.86e^2 \tag{5}$$

The parameter (e) is the exponential dependence of α^2 on M where α is the molecular-expansion coefficient. From the value of 2.86×10^{23} found by Ptitsyn and Eizner for Φ_θ, it follows that $\xi_\theta = 0.875$.

The ratio $(R_e/R)^3 = (\xi)^3$ is thus implicit in the value of Φ in the Flory-Fox equation and has a value of 0.49, corresponding to the Flory-Fox Φ value of 2.1×10^{23}. It is clear from Equations 1, 2, and 3 that $[\eta]M$ cannot be related to the statistical polymer dimensions h and R without a knowledge of ξ, *i.e.*, Φ, which varies with solvent for a given polymer. It follows, that if all species having the same $[\eta]M$ elute together from the GPC columns, then only R_e can be the universal parameter, since ξ will not be the same for all solute-solvent pairs and $\overline{h}$ and R will not be equally correct for universal calibration.

Polydispersity

It can be shown following the arguments of Newman *et al.* (*3*), that only the number-average molecular weight and hydrodynamic radius are valid when the Einstein-Simha viscosity expression is applied to whole polymers. Higher moments require a polydispersity factor. Thus,

$$[\eta] = \Phi_o \frac{\langle R_e \rangle^3{}_n}{\langle M \rangle_n}, \text{ but} \tag{6}$$

$$[\eta] = q\Phi_o \frac{\langle R_e \rangle^3{}_w}{\langle M \rangle_w} \tag{7}$$

where q is a heterogeneity factor having the definition in this case of

$$q = \left[\frac{\langle R_e\rangle_n}{\langle R_e\rangle_w}\right]^3 \frac{\langle M\rangle_w}{\langle M\rangle_n} \tag{8}$$

Values of R_e can be calculated from Equation 6 for narrow fractions of known molecular weights and plotted as a function of V_r, the retention volume. $<R_e>_n$ and $<R_e>_w$ can then be calculated from the GPC molecular-size distribution curve for unknown whole polymers. To obtain $<M>_w/<M>_n$ from Equation 8 requires a value for q.

By use of both the appropriate value for ξ in Equation 2 and the Mark-Houwink viscosity expression, one may write

$$\Phi \langle R\rangle^3{}_n = Q_n K_v \langle M\rangle_n{}^{a+1} \tag{9}$$

where Φ is equivalent to $\Phi_0\xi^3$, Q is a heterogeneity factor, and K_v is a constant for a given polymer-solvent system and $a = (3e + 1)/2$. Thus one may write

$$\langle R\rangle^3{}_n = \frac{Q_n K_v \langle M\rangle_n{}^{1+a}}{(\xi)^3 6.308 \times 10^{24}} \tag{10}$$

From solution theory it can be shown that K_v can be written as

$$K_v = \frac{4.291 \times 10^{23}\xi^3\alpha^3(B/M_o{}^{1/2})^3}{M^{3e/2}} \tag{11}$$

where $(B/M_o{}^{1/2})$ is an intrinsic constant which will be represented by A. Since $\alpha \sim M^{e/2}$, equations 10 and 11 can be combined as

$$\langle R\rangle^3{}_n = Q_n \left[\frac{C_T A^3}{6^{3/2}}\right] \langle M\rangle_n{}^{a+1} \tag{12}$$

where C_T is a temperature-dependent constant. It follows then that

$$q\left[\frac{\langle R\rangle_w}{\langle R\rangle_n}\right]^3 = \left[\frac{Q_w}{Q_n}\right]\left[\frac{\langle M\rangle_w}{\langle M\rangle_n}\right]^{3/2}\left[\frac{\langle M\rangle_w}{\langle M_n\rangle}\right]^{3e/2} \tag{13}$$

Equation 13 is valid only when the molecular-weight dependence of α is considered. When this is treated as insignificant, Equation 12 becomes

$$\langle R\rangle^3{}_n = Q_n \left[\frac{\alpha\langle M\rangle_n{}^{3/2} A^3}{6^{3/2}}\right] \tag{14}$$

so that if $<M>_w/<M>_n = D$, then

$$(\langle R\rangle_w/\langle R\rangle_n)^3/D^{3/2} = \frac{Q_w}{Q_n} \tag{15}$$

and therefore, from Equation 13,

$$q = D^{3e/2} \tag{16}$$

This means that the only effect of the q factor on the $[\eta]$–M relationship is that of the excluded volume due to coil expansion in a good solvent. It follows from Equations 8 and 16 that

$$\left[\frac{\langle R\rangle_w}{\langle R\rangle_n}\right]^{6/(2-3e)} = D \tag{17}$$

It can be seen then that one need not use Equation 7 at all, since $<M>_n$ can be obtained from Equation 6 and $<M>_w/<M>_n$ from Equation 17, thus permitting $<M>_w$ to be calculated.. If one is dealing with an unknown whole polymer, Equation 6 permits the determination of $<M>_n$ with no knowledge of solute-solvent interaction. To find $<M>_w/<M>_n$ and hence $<M>_w$ requires a knowledge of (e) for the application of Equation 17. This is easily obtained from GPC provided two or more samples of a different molecular weight can be found. Equation 6 permits the determination of $<M>_n$ for such samples from which the Mark-Houwink exponent (a) can be determined from the relationship in Equation 9, *i.e.*,

$$\log \langle R\rangle_n = \frac{(1 + a)}{(\;\;3\;\;)} \log \langle M\rangle_n + \text{constant} \tag{18}$$

Experimental

The calibration was established with polystyrene standards supplied by Waters Associates. The broad-distribution poly (vinyl chloride) standards were obtained from Arro Laboratories, Joliet, Ill. Other PVC samples studied were obtained from Shawinigan Chemicals Division, Gulf Oil of Canada Ltd. For all GPC analyses, small sample loads (*ca* 4 mg) and slow rates *ca* 0.76 ml/min) were used to maximize resolution.

Results and Discussion

Using the method outlined above, values of (a), (e), and $[\xi_\theta^3 f(e)]^{1/3} = \xi$ have been calculated for poly (vinyl chloride), poly (vinyl acetate), and polystyrene in tetrahydrofuran (THF) at 25°C, and are shown in Table I.

Table I. Viscosity Parameters—THF, 25°C

Polymer	*a*	*e*	ξ
PVC	0.76	0.179	0.748
PS	0.70	0.134	0.778
PVAc	0.64	0.090	0.805

Table II. A Comparison of Data Obtained for

PVC Resin Code No.	$\langle R_e\rangle_n$, *Ang.*	$\langle R_e\rangle_w$, *Ang.*	$\langle R_e\rangle_w / \langle R_e\rangle_n$	*From GPC and Equations 6 and 17* $\langle M\rangle_n$	$\langle M\rangle_w$	$\langle M\rangle_w / \langle M\rangle_n$
400–2	65.4	83.2	1.27	25,100	65,100	2.6
400–3	88.8	113.0	1.27	41,000	106,500	2.6
400–4	100.5	126.5	1.26	53,600	131,500	2.45

Table III. Comparison of Data Obtained for Some from a Mark-Houwink

PVC Resin	$\langle R_e\rangle_n$, *Ang.*	$\langle R_e\rangle_w$, *Ang.*	$\langle R_e\rangle_w / \langle R_e\rangle_n$
A	67.5	83.0	1.23
B	73.60	92.0	1.25
C	86.0	105.0	1.22
D	90.0	111.5	1.24

Substitution of the appropriate values of (e) into Equation 17 gives the following relationships for PVC, PS, and PVAc in THF:

$$\text{(PVC)}, \ [\langle R\rangle_w/\langle R\rangle_n]^{4.0} = D$$
$$\text{(PS)}, \ [\langle R\rangle_w/\langle R\rangle_n]^{3.75} = D$$
$$\text{(PVAc)}, \ [\langle R\rangle_w/\langle R\rangle_n]^{3.47} = D$$

and since $(e) = 0$ in a θ solvent,

$$(\phi), \ [\langle R\rangle_w/\langle R\rangle_n]^{3.0} = D$$

Thus, from a log-log plot of any of the above equations, and $<R>_w$ and $<R>_n$ obtained from the universal calibration curve and the GPC trace, the dispersity index D can be easily determined.

Table I compares results obtained in this manner with those reported by Arro Laboratories for 3 broad-distribution PVC standards. Table II does the same for several PVC commercial suspension resins from Shawinigan, using a Mark-Houwink expression as a basis for comparison of $<M>_w$ values.

PVC Standards by GPC and by Absolute Methods

By Light Scattering (L) or Osmometry (O)		
$\langle M \rangle_n$(O)	$\langle M \rangle_w$(L)	$[\eta]$ *25°C–THF, ml/gram*
25,500	68,600	70
41,000	118,000	113
54,000	132,000	125

PVC Suspension Resins by GPC and by Calculation Viscosity Expression

From GPC and Equations 6 and 17				
$\langle M \rangle_n$	$\langle M \rangle_w$	$\frac{\langle M \rangle_w}{\langle M \rangle_n}$	$\langle M \rangle_w$ *(calc)*[a]	$[\eta]$, *ml/gram, 25°C*
29,100	64,000	2.20	60,000	66.7
30,700	75,400	2.45	78,000	81.4
41,000	88,200	2.15	96,000	95.5
45,000	107,000	2,37	108,000	103.0

[a] $[\eta] = 1.55 \times 10^{-4} \langle M \rangle_w^{0.76}$, THF, 25°C.

Literature Cited

1. Einstein, A., "Investigations on the Theory of Brownian Motion," Dover Publications, New York, 1956.
2. Ptitsyn, O. B., Eizner, T. F., *Transl. Fr. Zhurnal Techniches Koi Fiziki* (1959) **24,** 1117.
3. Newman, S. *et al., J. Polym. Sci.* (1954) **14,** 451.
4. Baijal, M. D., *J. Macromol. Sci. Chem.* (1963) **A-2,** 1055.

RECEIVED November 29, 1972.

14

GPC Analysis of Block Copolymers

FRANKLIN S. C. CHANG

Borg-Warner Corp., Roy C. Ingersoll Research Center, Des Plaines, Ill. 60018

The GPC analysis of block copolymers is handicapped by the difficulty in obtaining a calibration curve. A method has recently been suggested to circumvent this difficulty by using the calibration curves of homopolymers. This method has been extended so that the calibration curves of block copolymers of various compositions can be constructed from the calibration curve of one-component homopolymers and Mark-Houwink parameters. The intrinsic viscosity data on styrene–butadiene and styrene–methyl methacrylate block polymers were used for verification. The average molecular weight determined by this method is in excellent agreement with osmometry data while the molecular weight distribution is considerably narrower than what is implied by the polydispersity index calculated from the GPC curve in the customary manner.

The GPC analysis of block copolymers is handicapped by lack of a calibration curve and the complication of a possible composition distribution. Calibration of elution volume needs monodispersed copolymers which are difficult to prepare. Without knowing the composition distribution the chromatogram height cannot be converted to weight distribution. Runyon and co-workers (*1*) have demonstrated that the use of an ultraviolet detector is very helpful in determining the composition distribution of styrene (S)–butadiene (B) block copolymer. In a previous paper (*2*) we have suggested that the molecular weight of a block copolymer can be calculated from its composition and the molecular weight of the constituent homopolymer at the same elution volume. Here we show that Mark-Houwink parameters of the homopolymers can be used to calculate the molecular weight of a block copolymer. The adequacy of this method is shown by its application to the intrinsic viscosity data of various block copolymers in a number of solvents. This method of analysis has been applied to laboratory prepared block copolymer samples. The molecular weight obtained is in excellent agreement with osmometry result. In addition the results show that the molecular weight distribution (MWD) of anionic block copolymer can be represented by a Poisson distribution, considerably narrower than the apparent polydispersity index derived from the GPC curve without considering the instrument spreading.

The Equivalence Ratio

The linear segment of the universal calibration curve is customarily expressed as

$$\log [\eta] M = b_1 V + b_0 \tag{1}$$

where $[\eta]$ and M are the intrinsic viscosity and molecular weight (MW) of a polymer eluted at solvent volume V; b_0 and b_1 are constants. The intrinsic viscosity $[\eta]$ is related to M by the well known Mark-Houwink equation: $[\eta] = KM^a$. Substituting this relation into Equation 1 and transposing

$$\log M = (b_1 V + b_0 - \log K)/(1 + a) \tag{2}$$

The ratio of the molecular weights of two homopolymers eluted at the same solvent volume is defined as the hydrodynamic volume equivalence ratio (equivalence ratio for short).

$$r = M_A/M_B = \text{antilog}\left[(b_1 V + b_0)\left(\frac{1}{1 + a_B} - \frac{1}{1 + a_A}\right) + \frac{\log K_B}{1 + a_B} - \frac{\log K_A}{1 + a_A}\right] \tag{3}$$

where the subscripts designate the individual polymers. When $a_A = a_B = a$, the relation is simplified to

$$r = \left(\frac{K_B}{K_A}\right)^{1/(1+a)} \tag{4}$$

This is the case of parallel calibration curves discussed in the previous paper. Equation 4 shows that when the calibration curves are parallel, the equivalence ratio, r, is constant to the elution volume V. It varies with the latter when $a_A \neq a_B$—*i.e.*, the calibration curves are not parallel. For that case Equation 3 would have to be used. Equation 4 also shows that the equivalence ratio can be calculated from the Mark-Houwink parameters K and a. It offers a way to determine r in addition to obtaining it from the GPC calibration curves of homopolymers.

Assuming that this equivalence ratio applies to the individual blocks in a block copolymer, the molecular weight M_c of the block copolymer of the AB and ABA type is (2):

$$M_c = M'_A/[1 + (r - 1)W_B] \tag{5}$$

where M_c and M_A' are the MW's of the copolymer and the component homopolymer A at the same elution volume, and W_B is the weight fraction of component B. Thus the calibration curves for block copolymers are readily established from those for the homopolymers by using Equation 5.

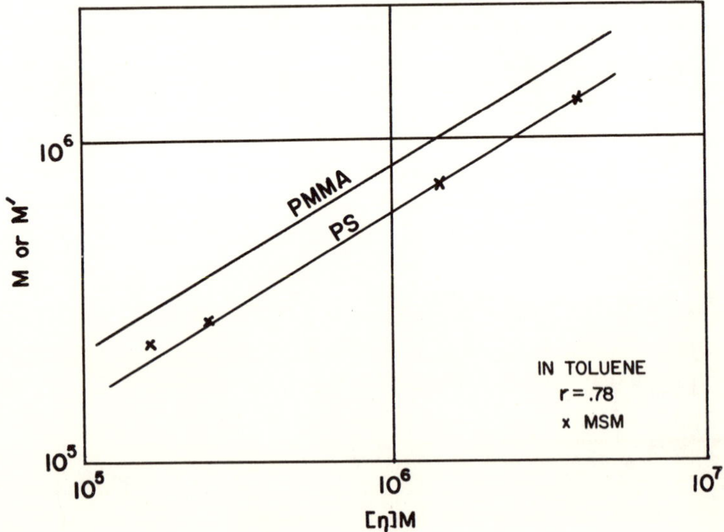

Figure 1. M' vs. [η] *M curve for MSM in toluene*

To test the adequacy of Equation 5 the $[\eta]$–$M_W^{1/2}$ curve published by Kotaka and co-workers (3) on styrene (S)–methyl methacrylate (M) block copolymers in toluene is replotted as M'-$[\eta]M$ curve in Figure 1. When a value of $r = 0.78$ is used, the data points for MSM copolymers (shown as crosses) coincide with the curve for polystyrene. According to Equation 4, r may be calculated from the coefficient K for homopolymers. When applied to the data on polystyrene and poly(methyl methacrylate) (3), a value of 0.72 is obtained, in good agreement with the above value of 0.78. The slight difference is probably the result of the polydispersity in the copolymers since the Mark-Houwink equation requires monodispersed polymers.

The intrinsic viscosity data on styrene (S)–butadiene (B) block copolymer measured by Utracki and co-workers (4) are replotted in

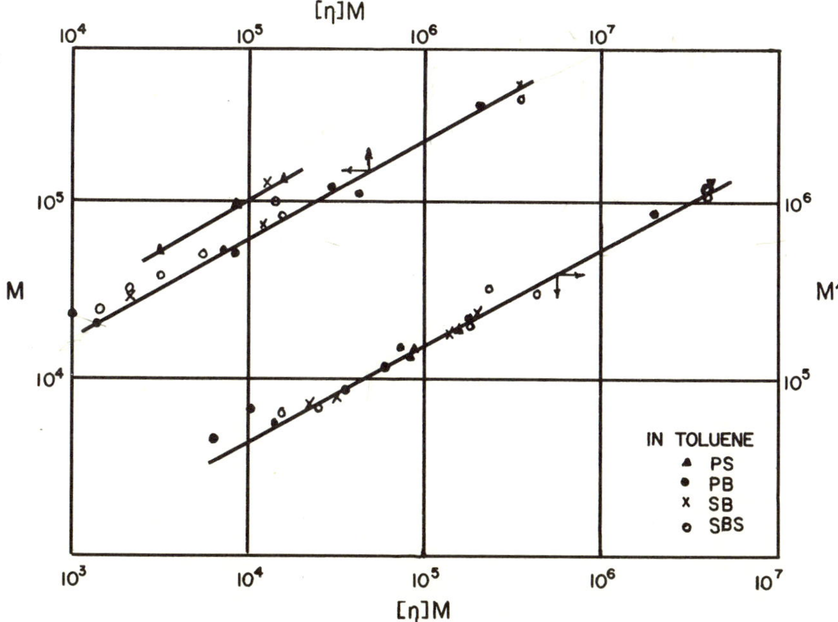

Figure 2. M′ vs. $[\eta]M$ *curve for SB and SBS in toluene*

Figures 2 and 3 as additional proof to Equation 5. The two solid lines in the upper left corner of Figure 2 are the $M'-[\eta]M$ curves for styrene and butadiene homopolymers. The data points for block copolymers shown as cross and open circle fall in between these two curves. When plotted as $M'-[\eta]M$ curve, all points fall on or near the curve for polystyrene shown as a solid line in the lower right part of Figure 2. The behavior of these copolymers in toluene and dioxane is shown in Figure 3. Since these block copolymers cover a wide range of composition (%S = 3.6–45.9) as well as molecular weight (M = 34,000–620,000), these results prove unequivocally the adequacy of Equation 5. This equation will make it possible to interpretate the chromatogram of block copolymer without preparing monodispersed copolymers which is something difficult, if not impossible.

Analysis of Block Copolymer

The mole fraction or probability P of anionic polymer of degree of polymerization is described by the Poisson distribution

$$P = e^{-\nu}\nu^{n}/n!$$

where ν is the average degree of polymerization. For i blocks built in sequence the probability is

$$P = \prod_i P_i$$

For instance, the mole fraction N of a triblock polymer with n_1, n_2, and n_3 monomer units in the respective blocks is

$$N = P_1P_2P_3 = \exp[-(\nu_1 + \nu_2 + \nu_3)]\nu_1^{n_1}\nu_2^{n_2}\nu_3^{n_3}/n_1!n_2!n_3!$$

and the *MWD* is

$$W_M = M \sum_{n_1n_2}\sum N / \sum_M M \sum_{n_1n_2}\sum N \qquad (6)$$

with the restriction that

$$n_3 = (M - n_1m_1 - n_2m_2)/m_3$$

where m_i designate the *MW* of the monomers. In GPC the separation is performed according to the hydrodynamic volume of the molecules which may be specified by the values of M'. Thus the expression *MWD* takes the following form:

$$W_{M'} = \sum_{n_1n_2}\sum NM / \sum_{M'n_1n_2}\sum\sum N\,M \qquad (7)$$

under the restriction

$$n_3 = (M' - n_1m_1 - r_2n_2m_2)/r_3m_3$$

where the equivalence ratio $r_i = (M_1/M_i)_v$. This situation of different *MWD* from mass and volume separations does not arise with homopolymers; it is unique to block copolymers.

Calculating programs have been written to carry out the summations according to Equations 6 and 7 and to take care of the instrument spreading by the equation after Tung (5). Calculations have been done on a Wang calculator model 700 on various diblock and triblock copolymer structures. A typical set of results obtained from S(19,250)B(68,000)S(19,250) is shown in Tables I and II, the headings are self-explanatory, except that each point of data represents a mixture of structures of the same value of M (in Table I) or M' (in Table II). Therefore, all the tabulated figures except weight % are the average values for each fraction.

When these two tables are compared, there is indeed a composition distribution superimposed on the *MWD* in the case of Poisson block copolymer, but the range is not very wide (Table I). In volume separation, however, the range is drastically reduced. For all practical purposes,

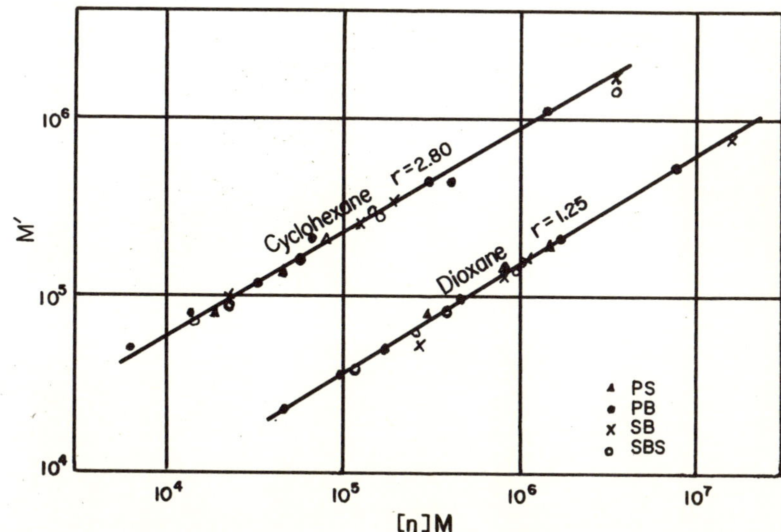

Figure 3. M′ vs. [η] M *curves for SB and SBS in benzene and in dioxane*

the composition of volume fractions can be considered constant. The weight distributions in these two modes of fractionation are different although not by much. The values of $\overline{M}_n$ and $\overline{M}_w$ are the same, as they should be. The *MWD* is very narrow in conventional sense.

The peak point (marked by P in the first columns of Tables I and II) in both modes of fractionation is the same and coincides with the nominal or average structure of the polymer sample. These observations hold true for all the structures studied. They also simplify the work needed to determine the structure. All that is necessary is to analyze the peak point in the chromatogram. When this is done, the structure is known, and the *MWD* can be calculated by use of the Poisson model.

Table I. *MWD* of S(19,250)B(68,000)S(19,250)

$M \times 10^{-5}$	*% S*	*% wt*	$M \times 10^{-5}$	*% S*	*% wt*
.983	34.84	.16	1.076	36.28	14.37
.993	35.00	.46	1.087	36.44	11.55
1.003	35.13	1.13	1.098	36.60	7.98
1.013	35.32	2.47	1.109	36.76	4.72
1.023	35.48	4.69	1.120	36.92	2.37
1.033	35.63	7.79	1.131	37.08	1.02
1.044	35.79	11.27	1.142	37.27	.36
1.054	35.95	14.15	1.154	37.41	.11
1.065P	36.12	15.41			

$\overline{M}_n = 1.064 \times 10^5$ $\overline{M}_w = 1.065 \times 10^5$

Table II. Volume Fraction of S(19,250)B(68,000)S(19,250)

$M' \times 10^{-5}$	$\overline{M}_n \times 10^{-5}$	$\overline{M}_w \times 10^{-5}$	% *S*	% *wt*
1.612	.984	.984	36.17	.07
1.628	.994	.994	36.20	.23
1.645	1.004	1.004	36.17	.63
1.661	1.014	1.014	36.17	1.47
1.677	1.024	1.024	36.16	2.96
1.694	1.034	1.034	36.15	5.13
1.711	1.044	1.044	35.14	7.64
1.728	1.055	1.055	36.13	9.75
1.746P	1.065	1.065	36.12	10.61
1.763	1.076	1.076	36.12	9.83
1.781	1.089	1.087	36.12	7.73
1.799	1.097	1.097	36.10	5.13
1.816	1.108	1.108	36.09	2.87
1.835	1.119	1.119	36.08	1.35
1.853	1.130	1.130	36.07	.53
1.872	1.142	1.142	36.07	.17
1.890	1.153	1.153	36.05	.04
Average	1.064	1.065	36.12	

To analyze a diblock copolymer being characterized completely by two parameters—*e.g.*, the molecular weights of both blocks, or the molecular weight of the copolymer and its composition—two pieces of data must be measured. The value of M' and the composition determined by ultraviolet or infrared spectroscopy at the peak point are sufficient. For the complete analysis of triblock copolymers, one additional piece of data is necessary. Theoretically the intensity or the height of the chromatogram at the peak point can be used for this purpose, but the process is involved, and after correcting for the refractive index the weight percentage is not very sensitive to structure. It is more convenient to use the homopolymer peak in the chromatogram, which is usually present. It has been found in the analysis of SB type copolymer, the secondary peak at large elution volume is the same as the first block. In case this peak is absent or too weak, the analysis of precursor is called for. If precursors are not available, the number average molecular weight of the whole copolymer can be determined by osmometry, and a trial-and-error process must be followed. This method of peak analysis has been successfully applied to a host of styrene–butadiene block copolymers. One of them is discussed here as an example.

Experimental

The anionic catalyzed SBS sample was prepared in our laboratory. A column combination of 10^5, 10^4, 3×10^3, 10^3, and 10^2A was used in the GPC with THF flow rate at 1.0 ml/min at room temperature. The

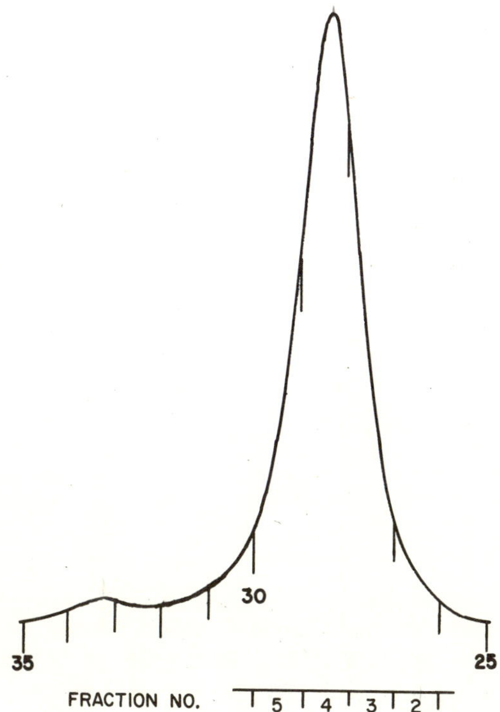

Figure 4. Chromatogram of an SBS block copolymer

resulting chromatogram is shown in Figure 4. Fractions of 5 ml each were collected for ultraviolet determination of styrene content at 260 mμ on a Beckman DU spectrometer. A correction was applied for the volume (*ca.* 1 ml) between the differential refractometer and the siphon.

Results and Discussion

The results of the analysis of one SBS sample are summarized in Table III, in which the ultraviolet optical density and the average height in GPC obtained from the area under curve are listed for each fraction in columns 2 and 3, respectively. A conversion factor of 7.2 obtained from polystyrene (PS) calibration was used to convert the optical density to the styrene contribution to the average chromatogram height which is tabulated in column 4. The difference between columns 3 and 4 was taken as the relative weight of butadiene after being multiplied by 1.37, a correction for the difference in refractive indices of styrene and butadiene (column 5).

Although a composition distribution is indicated by the values of percentage of styrene for the fractions shown in the last column in Table

Table III. Analysis of SBS

Fraction	OD	Average H	Relative Weight S	Relative Weight B	% S
2	.04	0.78	.29	.67	30.0
3	.15	2.28	1.07	1.66	39.2
4	.39	7.28	2.81	6.13	31.4
5	.20	3.94	1.45	3.41	29.6
				Average	32.6

	Elution Volume	M′	M
PS (= first block)	33.3		15,000
SBS (32.6% S)	28.4	150,000	90,000
Whole polymer, $\overline{M}_n$ calculated			80,000
Whole polymer, $\overline{M}_n$ osmometry			90,000
SB calculated		137,000	
SB GPC	28.5	140,000	

I, the extremely low concentration of polymer in the fractions collected reduced the precision in the ultraviolet optical density values measured, the influence of which is amplified in the figures in the last column. Therefore, the average styrene content was used as the composition of the peak point to calculate the *MW* of the polymer by Equation 5. The *MW* of the styrene homopolymer which is present in small amounts was taken as the *MW* of the first block, as explained above. In this manner the molecular structure of the polymer was found to be S(15,000) B(61,000)S(14,000) the figures given in parentheses being the *MW*'s of the successive blocks.

For verification the amount of styrene homopolymer was estimated from the area under the secondary peak in the chromatogram as 2.4 wt % in the whole polymer. From this a value of 80,000 was obtained for M_n of the whole polymer, in good agreement with the osmometry value of 90,000. The calculated styrene content of the whole polymer is 34.2%, comparing very well with the value determined by infrared spectroscopy (36.0%). As further verification the value of *M′*, the *MW* of polystyrene that would have been eluted at the same solvent volume as the diblock copolymer consisting of the first two blocks, was calculated as 137,000 in excellent agreement with the value (140,000) read from the calibration curve against the experimentally determined elution volume. The polydispersity index calculated for a Poisson polymer of the structure of S(15,000) B(61,000) S(14,000) is 1.002—much smaller than the apparent value of 1.64 derived directly from the chromatogram without any corrections for composition and instrument spreading.

The polydispersity index of the preparation is probably between these two values. The characterization of a block copolymer by *MWD* deserves careful consideration which is beyond the scope of this paper.

Conclusions and Summary

The equivalence ratio can be calculated from the Mark-Houwink coefficient, *K*, of component homopolymers. The composition distribution in the chromatogram of a block copolymer is negligible. The peak point of a block copolymer chromatogram corresponds to the average structure of the polymer. Thus, analysis of block copolymers is reduced to analysis of the peak point. Analyses of anionic block copolymer structures have been successfully accomplished by this peak analysis technique with the aid of equivalence ratio.

Acknowledgment

The author is indebted to Borg-Warner Corp. for permission to publish this paper and to J. F. Pendleton and R. L. Gay for supplying the block copolymer samples.

Literature Cited

1. Runyon, P., *et al.*, *J. Appl. Poly. Sci.* (1969) **13**, 2359.
2. Chang, F. S. C., *J. Chromatog.* (1971) **55**, 67.
3. Kotaka, T., *et al.*, *Polymer J.* (1970) **1**, 265.
4. Utracki, L. A., *et al.*, *J. Poly. Sci., Pt. A-2* (1968) **6**, 2051.
5. Tung, L. H., *J. Appl. Poly. Sci.* (1966) **10**, 375.

RECEIVED January 17, 1972.

15

Treating a Gel Permeation Chromatogram as a Summation of Narrow–Fraction Chromatograms

B. S. EHRLICH[a] and W. V. SMITH

Uniroyal, Inc., Oxford Management & Research Center,
Middlebury, Conn. 06749

A slightly modified version of the chromatogram analysis procedure of Pickett, Cantow, and Johnson (1) *is used to obtain molecular weight distributions from polystyrene chromatograms. Results for combined-fraction chromatograms of known distributions indicated that this method greatly increased chromatogram resolution. For NBS Standards 705 and 706, results showed that good values of* $\overline{M}_w$, $\overline{M}_n$, *and* $[\eta]$ *could be calculated from the corrected chromatogram. We also show that for a series of different polymer fractions reverse-flow and normal narrow-fraction chromatogram widths are independent of polymer type and depend only on peak elution volume. Chromatogram widths of fractions of narrow-distribution polymers obtained by elution chromatography approach reverse-flow chromatogram widths confirming that reverse flow widths are representative of monodisperse polymers.*

Several methods of obtaining the molecular weight distribution from a gel permeation chromatogram have been presented in the literature. Most of these methods (*2–4*) try to correct for the finite resolution of gel permeation chromatography (GPC) columns by treating a chromatogram as a summation of single-species chromatograms which are usually approximated by a suitable function such as a Gaussian. One method that eliminates this dependence on an assumed single species chromatogram has been proposed by Pickett, Cantow, and Johnson (*1*). They consider an experimental chromatogram as a summation of narrow-fraction chromatograms of presumably known distribution. A closely

[a] Present address: Uniroyal, Inc., Chemical Division, Naugatuck, Conn. 06770.

packed set of narrow-fraction chromatograms (basis) is obtained by linear interpolation from a series of experimentally obtained Pressure Chemical polystyrene standard chromatograms which span the entire GPC range of interest. A modified least-squares computational procedure is used to generate the group of basis chromatograms and their respective amounts (limited to positive contribution) which give the best fit to the experimental chromatogram. From the calculated coefficients of the basis chromatograms, a reshaped experimental chromatogram (corrected for column spread) and a molecular weight distribution curve for the experimental chromatogram are generated. This procedure requires a knowledge of the molecular weight distribution of the basis chromatograms. These are approximated by assuming that the polystyrene standard chromatograms would be Gaussian if the column had infinite resolution, and using the values of $\overline{M}_w$ and $\overline{M}_n$ supplied with the standards along with the GPC column calibration curve to obtain the standard deviations of these Gaussians.

In this paper we have used a slightly modified version of this method. The modification consists of replacing the Pressure Chemical standards by sharp fractions of the same, obtained by elution chromatography. Making the approximation that these sharp fractions are monodisperse, we are able to generate a basis set of monodisperse chromatograms. This eliminates the need for correcting for the distribution of the PS standards and thereby the necessity for assuming a function for their MWD's. In addition, depending on experimental values of $\overline{M}_w$ and $\overline{M}_n$ of the standards to calculate their distributions is not necessary.

This procedure has been tested by applying it to known distribution chromatograms formed by blending of fractions and to NBS Standards 705 and 706 for which experimentally measured values of $\overline{M}_w$, $\overline{M}_n$, and $[\eta]$ are available.

Results and Discussion

Fractionation of the Pressure Chemical Polystyrene Standards. Five polystyrene standards (molecular weights 5.1, 16, 41, 86, and 180 $\times$ 10^4) were fractionated by elution chromatography. Each sample (1 g) was fractionated by eluting from a column of polymer-coated glass beads using a THF–acetone solvent mixture (continually increasing the THF concentration) with no temperature gradient. Fractions were collected, and GPC chromatograms were obtained for selected ones. A typical set of fractionation data (PS 41 $\times$ 10^4) is given in Table I. Chromatograms were obtained using a 3.2 ml/min flow rate and injecting 1 mg of polymer (0.05% solution, 2-ml loop).

The data in Table I show that fraction chromatogram peak positions and widths appear to approach limiting values (7.64–7.68 and 0.92

Table I. Fractions of Pressure Chemical Polystyrene Standard (41×10^4)

Fraction No.	*GPC Peak,[a] counts*	W,[b] *counts*	F_i[c]	$M \times 10^{-4}$
6	9.33	0.9	.001	3.0
8	8.98	0.81	.003	6.2
10	8.52	0.9	.007	13.0
12	7.75	1.15	.035	36.5
14	7.72	0.94	.170	38.0
16	7.68	0.92	.452	39.5
18	7.64	0.92	.329	41.0
20	7.64	0.9	.003	41.0
Overall Std	7.64	0.97		

From fractionation results $\overline{M}_w = 39.3 \times 10^4$

$\overline{M}_n = 38.3 \times 10^4$ $\qquad \overline{M}_n/\overline{M}_w = 1.025$

[a] Column system 10^6, 4×10^5. One count = 10 ml.
[b] W = width of chromatogram at 0.5 height.
[c] F_i = fraction of polymer collected present in each sample.

counts, respectively) at which point most of the polymer is eluted. This is expected for a polymer which has a distribution which is very sharply defined (nearly monodisperse) at the high molecular weight end. Fractionation results obtained for PS 8.6×10^5 and 1.6×10^5 indicate distributions which also are nearly monodisperse at the high molecular weight ends with slight tails extending to lower molecular weights. For the 1.8×10^6 PS standard, there is a small spread in the distribution at the high molecular weight end along with a significant low molecular weight tail. For this standard, fractionation results indicate $\overline{M}_w/\overline{M}_n \approx 1.49$. Even for this polymer, the high molecular weight end is nearly monodisperse. Therefore, the approach of Tung and Runyon (*5*) who suggest using the leading halves of high molecular weight PS standard chromatograms to obtain a calibration for column resolution (spreading) appears reasonable.

Comparison of Chromatogram Widths (Reverse Flow *vs.* Narrow Fraction). To obtain a column resolution (spreading) calibration for our two-column GPC chromatograph (10^6, 4×10^5) we have used the reverse-flow procedure (*6*). Figure 1 presents chromatogram half-width data obtained for several different polymer fractions by reverse flow. These widths are characteristic of monodisperse polymers. Data which have been obtained for polystyrenes, EPDM's, and branched and linear polybutadienes all fall on one smooth curve demonstrating the independence of reverse-flow chromatogram widths on polymer type, and their dependence on peak elution volume, and confirming Tung and Runyon's conclusions (*5*).

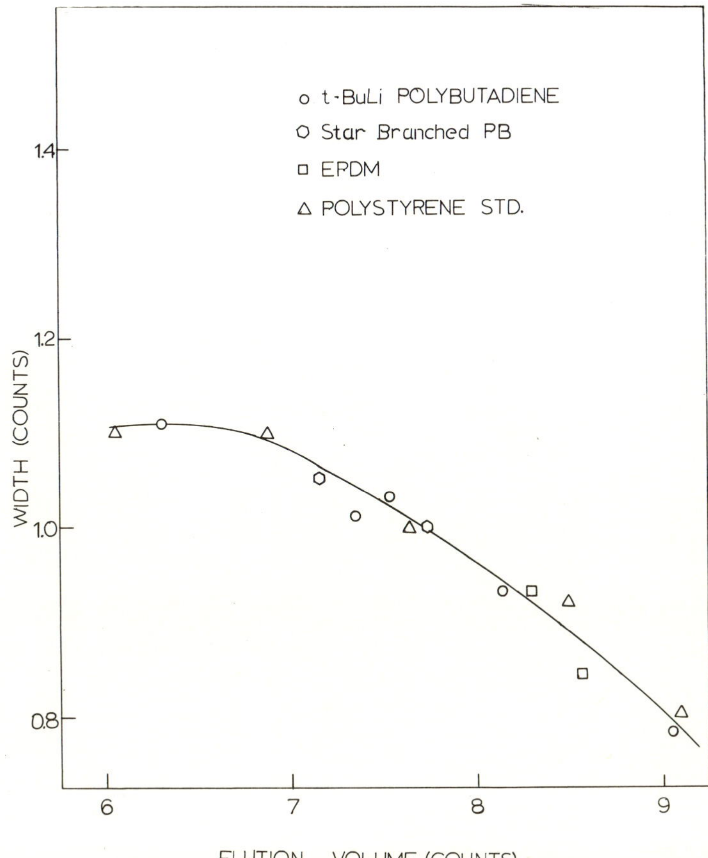

Figure 1. Reverse-flow chromatogram widths

In Figure 2 we compare some of our fractionated polymer chromatogram widths with the reverse-flow curve. For both *tert*-BuLi PB (circles) and the Pressure Chemical PS standards (triangles) we can see that the elution chromatographic fractionation of these polymers gives fractions having chromatogram widths approaching those obtained by reverse flow. This is expected if reverse-flow chromatogram widths can be assigned to monodisperse polymers. We can also see that chromatogram widths obtained for narrow polystyrene and polybutadiene fractions support the conclusion reached from reverse-flow data that chromatogram widths are independent of polymer type. If we assume that reverse-flow and monodisperse chromatogram widths are equal, and make the approximation that chromatograms of the reverse flow and the sharply fractionated PS standards can be approximated as Gaussian, then we can simply calculate the polydispersity indices for the narrow

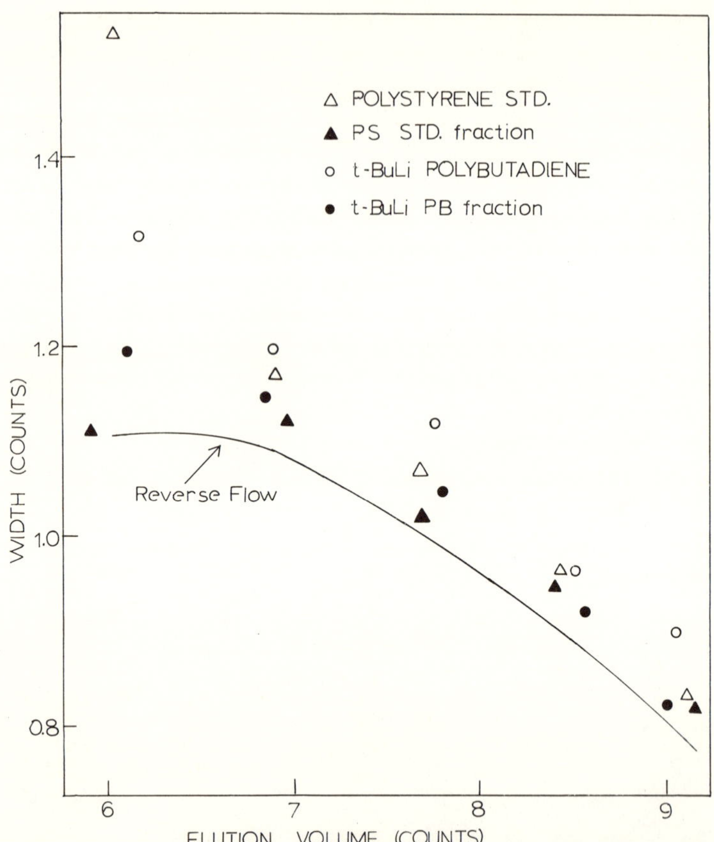

Figure 2. Comparison of narrow-fraction and reverse-flow chromatogram widths

fractionated PS standards. Such an approach gives values of $\overline{M}_w/\overline{M}_n \approx 1.01$ for fractions of PS standards (1.8×10^6, 8.6×10^5, 4.1×10^5, 1.6×10^5). For the 5.1×10^4 fractionated standard $M_w/M_n \approx 1.03$.

Generating a Basis Set of Monodisperse Chromatograms. We have used the five fractionated PS standards along with three unfractionated standards to generate an interpolated (*1*) basis set of monodisperse chromatograms. The basis-generating chromatograms are specified in Table II. A 10^6, 10^5, 10^4, 10^3 column system was used with a flow rate of 1 ml/min and sample injection concentrations of 0.05% (1 mg). The chromatograph is equipped with an on-line digitizer which gives the chromatogram height every minute. Samples 1–5 (Table II) are the fractionated standards. Samples 6–8 are unfractionated. Since samples 1–4 have polydispersity indices ≈1.01, it is quite reasonable to approximate them as monodisperse. Samples 5–8 have higher polydispersities, but these have also been assumed monodisperse to simplify calculations.

All chromatograms were normalized to the same area, and linear interpolation (*1*) was used to generate a closely packed basis set of approximately monodisperse chromatograms separated by either 0.2 or 0.4 count (1 count = 5 ml, 0.2 count corresponds to a 14% change in molecular weight for the column system used). This interpolated set of chromatograms (basis) was then used to resolve experimental chromatograms following the mathematical procedure of Pickett, Cantow, and Johnson (*1*). The chromatogram was treated as a summation of the basis chromatograms, and a modified least-squares fitting procedure was used to find the best combination of basis chromatograms limited to positive contributions required to generate the experimental chromatogram. Digitized data were used throughout with 0.4- or 0.2-count separation between data points.

Table II. Basis-Generating Chromatograms of Narrow Polystyrene Fractions

Fraction No.	$M \times 10^{-3}$ [a]	*Peak Position,* [b] *counts*
1	1800	24.00
2	860	25.35
3	411	26.58
4	160	28.07
5	51	29.85
6	20	31.22
7	10	32.28
8	5	33.63

[a] Fractions 1–5 were fractionated by elution chromatography. All molecular weights given are for original unfractionated Pressure Chemical standards.
[b] One count = 5 ml; column system 10^6, 10^5, 10^4, 10^3 (THF at room temperature). Sample injection: 0.05% solution (1 mg).

Testing the Chromatogram Resolution Procedure. KNOWN DISTRIBUTION CHROMATOGRAMS. Several chromatograms of known distribution were prepared by selectively blending some of the PS fractions (Table II). These chromatograms were used to test the resolution procedure.

In Table III are some typical results obtained from the combined fraction chromatograms. Results presented were obtained using a basis chromatogram set with peak positions separated by 0.4 count along with digitized data spaced at 0.4 count. To allow easy comparison of calculated and actual distributions, the weighted average calculated results are given in Table III. To obtain the weighted average results, intensities calculated for adjacent basis chromatograms, (0.40-count separation) were combined, and the total intensity assigned to an intermediate peak position was found by taking a weighted average of the contributions at the two adjacent peak positions. For example, if the calculated results indi-

Table III. Combined Fraction Chromatograms

	Known Distribution		*Calcd Distribution*		*Wt Av Calculation*	
Frac-tions	*Amount, %*	*Peak Position, counts*	*Amount, %*	*Peak Position, counts*	*Amount, %*	*Peak Position, counts*
			Sample 1			
			0.2	29.20	0.3	29.33
			0.1	29.60		
6	46 ± 2	31.22	3.4	30.80	39.8	31.17
			36.4	31.20		
7	54 ± 2	32.28	37.1	32.00	59.9	32.15
			22.8	32.40		
			Sample 2			
2	35 ± 2	25.35	27.4	25.20	34.7	25.28
			7.3	25.60		
4	33 ± 2	28.07	27.4	28.00	33.1	28.07
			5.7	28.40		
6	33 ± 2	31.22	30.2	31.20	32.2	31.22
			2.0	31.60		
			Sample 3			
4	23 ± 1	28.07	19.2	28.00	22.6	28.06
			3.4	28.40		
5	26 ± 1	29.85	5.8	29.60	27.4	29.92
			21.6	30.00		
6	23 ± 1	31.22	13.8	31.20	23.3	31.36
			9.5	31.60		
7	28 ± 1	32.28	7.9	32.00	26.7	32.28
			18.8	32.40		
			Sample 3 Values Calculated with Extended Basis Set			
			6.1	28.00		
4	23 ± 1	28.07	15.4	28.07	22.9	28.09
			0.7	28.40		
			0.8	28.80		
			20.1	29.85		
5	26 ± 1	29.85	5.1	30.00	25.6	29.89
			0.4	30.40		
6	23 ± 1	31.22	20.3	31.22	25.0	31.29
			4.7	31.60		
7	28 ± 1	32.28	20.3	32.28	26.5	32.31
			6.3	32.40		

cated equal intensities at 30.00 and 30.40 counts, the weighted average result would place the total intensity at 30.20 counts. Figure 3 shows the fits obtained between calculated and experimental chromatograms for samples 1, 2, and 3 (Table III). If chromatograms 1–8 (Table II) are added to the basis set and the calculations redone, the agreement between calculated and experimental chromatograms is improved. However, the calculated results still do not exactly replicate the known distribution as would be expected if the GPC chromatogram were a true sum of its components and GPC were perfectly reproducible. The calculated data obtained for sample 3 when the basis set is extended by adding chromatograms 1–8 are shown at the bottom of Table III. Calculated weighted average results obtained with the extended basis set and the normal basis set for sample 3 are nearly identical. The deviation of the calculated distribution (not weighted average) from the actual distribution (in the extended basis set calculation) is a measure of the

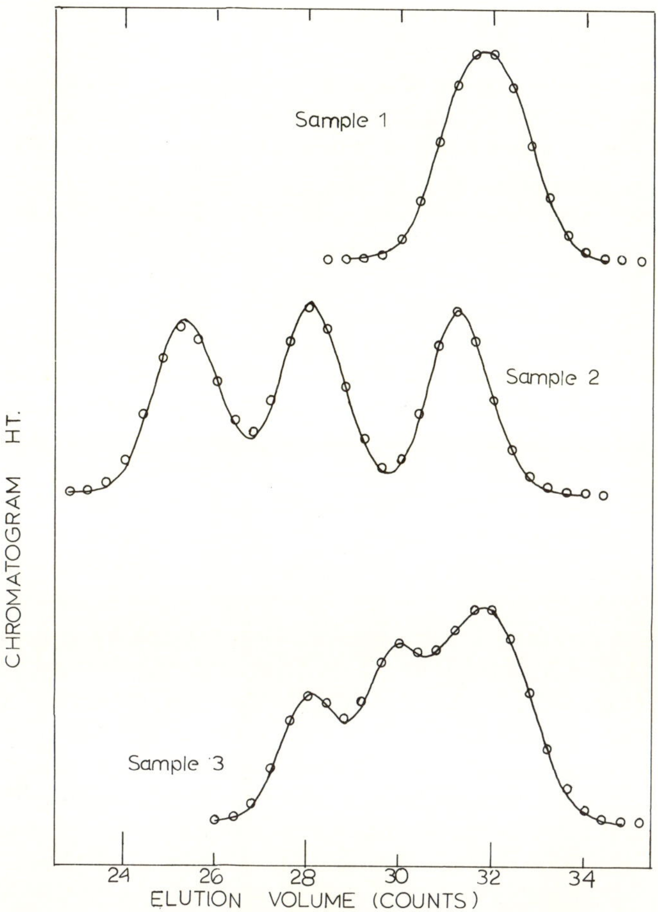

Figure 3. Fraction blends: experimental vs. *calculated chromatograms*

uncertainty of this method in generating a distribution. This uncertainty should have a much smaller effect on calculated average molecular weights than it does on the calculated distribution.

The results obtained for sample 1 (Table II) show some of the errors that can occur for this procedure. A small amount of intensity has been assigned to 29.33 counts (0.3%, weighted average calculation). This is probably from a slight fluctuation in the experimental GPC near the base line which would be treated as real, and is an indication of the uncertainty in GPC data. Also, the calculated weighted average results indicate the presence of two species with a ratio of 6/4 favoring the low molecular weight species. This ratio is higher than that of the known distribution (54/46), and both weighted average calculated peaks are shifted to lower than their known positions (31.17 eV *vs.* 31.22 eV) and (32.15 eV *vs.* 32.28 eV). These two errors tend to cancel when average molecular weights are calculated. Therefore, we can again see that uncertainties in molecular weight averages calculated by this procedure will be less than uncertainties in calculated MWD's. Results obtained using 0.2-count separations between digitized data points and interpolated basis chromatogram peaks are considerably closer to the known distribution. The calculated ratio is 55/45 *vs.* 54/46 for the known distribution, and peaks occur at 31.20 *vs.* 31.22 and 32.21 *vs.* 32.28. Results obtained using an extended basis set which contains chromatograms 1–8 (Table II) agree closely with the results found using the normal basis set with 0.2-count separations. Going to narrower separations between basis chromatogram peaks and digitized data points (0.2 count) improves the quality of the MWD results somewhat along with the fit to the experimental chromatogram; however, the computer time required to obtain a fit increases significantly. We have generally chosen to use the greater separations (0.4 count) to reduce the computer time requirement. The fits to experimental chromatograms for samples 1–3 presented in Figure 2 are adequate. Better fits are typically obtained for broad unimodal chromatograms.

Any uncertainties in GPC peak positions (probably of the order of ±0.1 count) might account for some of the difference between calculated and experimental peak positions. We also have the problem of spillover. For example, in sample 1 (Table II) if a small part of the intensity assigned to position 32.00 is split evenly between 31.60 and 32.40, the change in the fit to the experimental chromatogram will be slight (within experimental error of the chromatogram); however, the agreement between weighted average calculated results and known distribution will be improved. This could be part of the cause of the error in the calculated results for sample 1 and could also contribute to oscillations in the calculated MWD's which have been noted (7) for all methods examined where an attempt is made to obtain MWD by decomposing chromatograms into component elements.

The calculated distribution obtained for samples 2 and 3 appears to be quite reasonable, and our general conclusion for these combined-fraction chromatograms is that this procedure gives excellent calculated distributions.

NBS Standards. Obtaining $\overline{M}_n$, $\overline{M}_w$, $[\eta]$, and MWD. Chromatograms of two NBS standard polystyrenes, NBS 706 (broad distribution) and NBS 705 (narrow distribution), have been analyzed. Calculated results have been obtained using the two previously mentioned fitting procedures: (1) 0.4-count separations between basis chromatograms and digitized data points and (2) 0.2-count separations.

Table IV. Calculated Values of Average Molecular Weights and Viscosities for NBS Standards

		$\overline{M}_w$	$\overline{M}_n$	$[\eta]_{THF, 30°}$
NBS 705				
Calculated values	unresolved GPC	180,000	139,000	0.71
	resolved GPC	173,000	162,000	0.70
Experimentally determined values		173,000 [a]	164,000 [b]	0.73 [c]
NBS 706				
Calculated values	unresolved GPC	289,000	122,000	0.94
	resolved GPC	272,000	139,000	0.92
Experimentally determined values		257,800 [a]	136,500 [b]	0.92 [c]
		288,100 [d]		

[a] NBS value from light scattering.
[b] NBS value from osmometry.
[c] $[\eta]$ values measured in this laboratory.
[d] NBS value from sedimentation.

Results obtained for average molecular weights and $[\eta]$'s by the two fitting procedures are nearly identical. Data presented in Table IV are for procedure 1 (0.4 count separations) which uses one-fourth the computer time required by procedure 2. Viscosities have been found using our independently obtained realtionship between $[\eta]_{THF}$ 30°C and M for polystyrene along with the resolved chromatogram data. In Table IV the results calculated from the resolved chromatogram are compared with values calculated directly from the experimental chromatogram and experimentally determined values(NBS $\overline{M}_w$ and $\overline{M}_n$, our $[\eta]$) for the standards. The agreement between the calculated and measured molecular weight averages and viscosities for the NBS standards is excellent. The closeness of the values obtained for NBS 705 is somewhat fortuitous as approximations used in our calculations (such as the assumption that basis-generating chromatograms were monodisperse) should be more

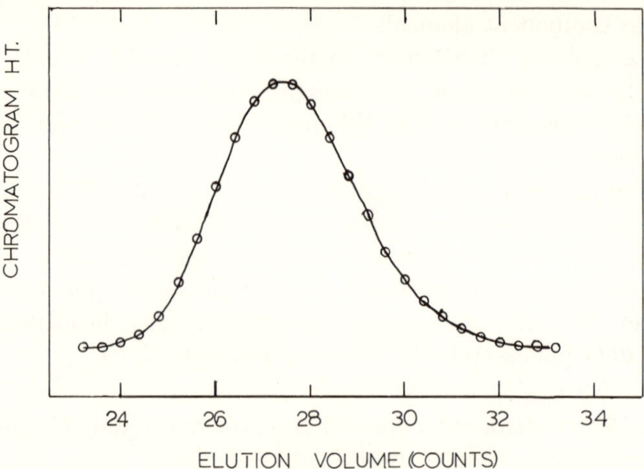

Figure 4. NBS 706: experimental vs. *calculated chromatograms*

significant in treating a narrow-distribution polymer. Figure 4 shows the experimental and calculated chromatogram for NBS 706. The quality of the fit is clearly well beyond the reproducibility of the GPC data. In Table V the resolved GPC data obtained for NBS 706 using the 0.4-count separation procedure are presented.

Figures 5 and 6 present integral MWD's obtained for NBS 706 and 705 by the two fitting procedures. We have used $\Sigma F_i - \frac{1}{2}F_i$ (Table V)

Table V. Calculated MWD of NBS 706

GPC Counts	$M \times 10^{-5}$	F_i[a]	ΣF_i	$\Sigma F_i - {}^1/_2 F_i$
31.60	0.157	0.0074	0.0074	0.0037
31.20	0.201	0.0031	0.0105	0.0090
30.80	0.260	0.0071	0.0176	0.0141
30.40	0.335	0.0303	0.0479	0.0328
29.60	0.56	0.0555	0.1034	0.0757
29.20	0.72	0.0257	0.1291	0.1163
28.80	0.93	0.0630	0.1921	0.1606
28.40	1.20	0.1077	0.2998	0.2460
28.00	1.56	0.0593	0.3591	0.3295
27.60	2.01	0.1590	0.5181	0.4386
27.20	2.60	0.1133	0.6314	0.5748
26.80	3.35	0.0867	0.7181	0.6748
26.40	4.3	0.1973	0.9154	0.8168
25.60	7.2	0.0767	0.9921	0.9538
25.20	9.3	0.0079	1.0000	0.9961

[a] F_i = fraction of total polymer.

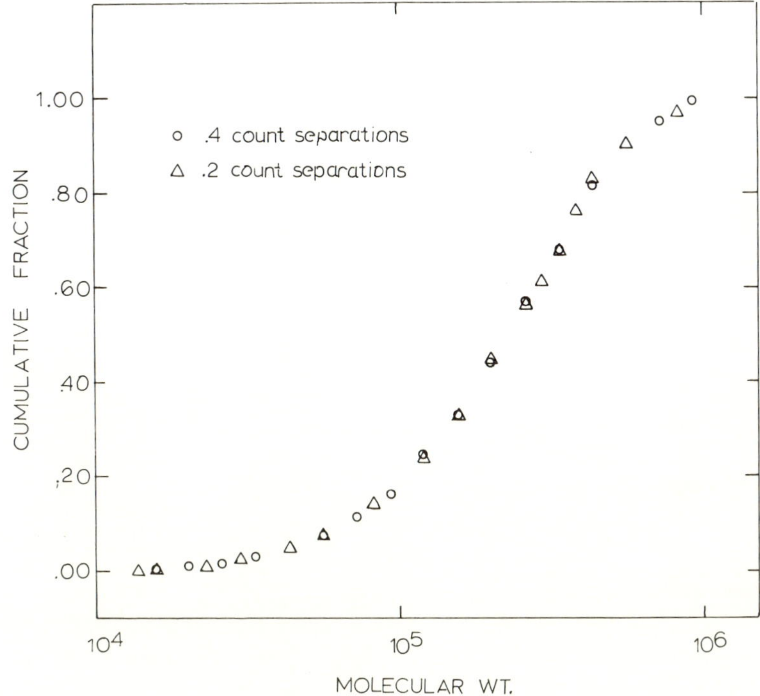

Figure 5. NBS 706: molecular weight distribution

as the cumulative fraction axis. This procedure assigns one-half the intensity calculated at any molecular weight to lower molecular weights and one-half to higher molecular weights. It considerably reduces data oscillations that occur in this chromatogram analysis treatment. If a differential distribution is desired, it would be best obtained from the best curve that can be drawn through the integral distribution data points. However, there is still a problem in data point oscillation which may interfere with obtaining accurate differential MWD's, particularly if fine details of the distribution are of concern.

We have also obtained some results using unfractionated Pressure Chemical standards to generate a basis set and not correcting for their distribution. For most cases this appears to be a reasonable approximation. Molecular weight averages obtained for NBS 706 using this approach are nearly indistinguishable from those given in Table IV. Some error might be introduced when examining higher molecular weight samples since the highest molecular weight Pressure Chemical standard (molecular weight 1.8×10^6) is of poor quality. Errors would also be expected in examining very narrow-distribution polymers ($\overline{M}_w/\overline{M}_n < 1.1$).

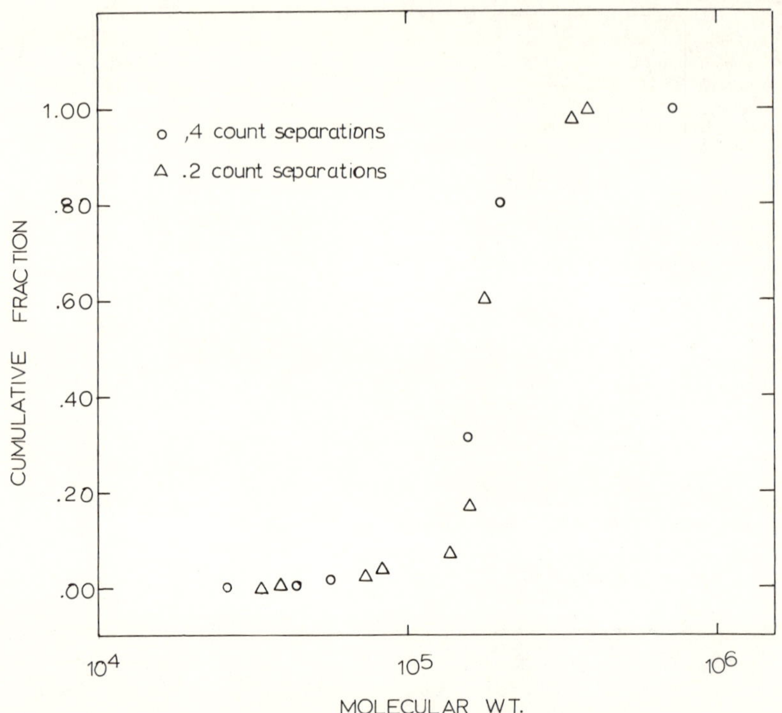

Figure 6. NBS 705: molecular weight distribution

Conclusion

The Pickett, Cantow, Johnson (*1*) method appears to be an excellent method for obtaining molecular weight averages and viscosities from GPC chromatograms when skewing is not significant (low concentrations). Within the limit of uncertainty generated by the oscillation of data points, this approach appears reasonable for obtaining MWD from GPC chromatograms. The treatment presented here was applied to polystyrene, but it is believed that it will also be useful for other polymers since the chromatograms of monodisperse polymers appear to be independent of polymer type. Thus the basic set obtained with polystyrene should also be applicable to other polymers.

Literature Cited

1. Pickett, H. E., Cantow, M. J. R., Johnson, J. F., *J. Polym. Sci., Part C* (1968) **21,** 67.
2. Tung, L. H., *J. Appl. Polym. Sci.* (1966) **10,** 1271.
3. Smith, W. N., *J. Appl. Polym. Sci.* (1967) **11,** 639.
4. Hess, M., Kratz, R. F., *J. Polym. Sci., Part A-2* (1966) **4,** 731.

5. Tung, L. H., Runyon, J. R., *J. Appl. Polym. Sci.* (1969) **13,** 2397.
6. Tung, L. H., Moore, J. C., Knight, G. W., *J. Appl. Polym. Sci.* (1966) **10,** 1261.
7. Duerksen, J. H., Hamielec, A. E., *J. Appl. Polym. Sci.* (1968) **12,** 2225.

RECEIVED January 17, 1972.

16

Average Degree of Polymerization of Cellulose by GPC without Viscosity Measurements

J. I. WADSWORTH, L. SEGAL, and J. D. TIMPA

Southern Regional Research Laboratory, Southern Marketing and Nutrition Research Division, Agricultural Research Service, U. S. Department of Agriculture, New Orleans, La. 70179

In an earlier procedure applying universal calibration, viscosities of the four most concentrated fractions eluting about the peak were measured, and the intrinsic viscosities were plotted against count. The intrinsic viscosities of all the fractions were obtained by extrapolation of the plot for use in the calculations to obtain degree of polymerization (DP). In the present method the DP of each fraction is obtained from the relationship MW = (coil size/K)$^{1/1+a}$ derived from Benoit's concept and the Mark-Houwink equation. Results from the new procedure are in excellent agreement with those obtained independently on cotton by others. Anomalies in results obtained previously on some samples disappear while marked improvement is noted for others. The determination is speeded up greatly by computer processing of data, and experimental error is reduced.

The use of gel permeation chromatography (GPC) in studies of cellulose and cellulose derivatives has expanded greatly within the past few years. In common with similar studies of other polymers the matter of calibration of the GPC instrument and of conversion of GPC data into average molecular weight or average degree of polymerization (*DP*) of the polymer is of considerable importance.

The application of universal calibration in an investigation of the chain-length distributions in various celluloses by GPC has already been described in an earlier publication (*1*). In the procedure average DP values are obtained in a manner unlike that normally used in GPC. There is a totally different mathematical procedure, and measurements are made of the intrinsic viscosities of the four most concentrated fractions eluting around the peak. The intrinsic viscosities of these fractions are plotted against half-count, from which the intrinsic viscosities of all of the other, more dilute fractions are obtained by a least-square linear extrapolation of a log count *vs.* log viscosity plot. The *DP* values of the fractions, necessary in the calculations of average *DP*, are obtained by first dividing coil size (Benoit's parameter, $[\eta]M$ where $[\eta]$ is intrinsic viscosity and M is molecular weight) by intrinsic viscosity and then dividing the resulting molecular weight by the proper unit monomer weight of the nitrated anhydroglucose unit. Coil size is taken from the polystyrene calibration curve, coil size *vs.* count, at each half-count in the chromatogram. Meas-urements of areas under the curve on the chromatogram are made for obtaining the concentrations of the eluted fractions, also needed in the calculations.

Goedhart and Opschoor (*2*) developed a procedure that is basically imilar. The conversion of GPC data to molecular weight, however, is ccomplished by another mathematical approach which is unrelated to he preceding while intrinsic viscosities of the eluted fractions are meas-ured by an automatic capillary-tube viscometer coupled to the discharge eg of the syphon of the GPC instrument. These authors comment that measurement of the viscosity of very dilute polymer solutions requires igh accuracy and careful control in order to avoid anomalous results. his measurement when done manually is a tedious and time-consuming peration requiring a fair degree of skill.

The present paper describes a procedure by which the average *DP* alues can be obtained by GPC without making measurements of intrinsic iscosity, thereby drastically reducing the time necessary for GPC char-acterization of samples as well as increasing the accuracy and reproduci-ility of average *DP* values as compared with those obtained by other xperimental methods.

xperimental

The GPC instrumentation and GPC procedures used here are the ame as described earlier (*1*) but with slight alterations. The Kroeger AK-1 pads furnished with the pressure filter were replaced by Millipore -μ Mitex membranes to reduce changes in solution concentrations arising om adsorption of the solute on the fibrous material of the pad. When etention of eluted fractions was desired, two shallow aluminum pans lled with tetrahydrofuran (THF) were placed in the syphon box, and en the door was closed and latched. The saturation of the enclosed air pace by THF vapor minimized evaporation of the solution during drop rmation, drop fall, and retention of fraction in the syphon. The possi-ility of diluting the sample solution when charging the sample loop was eliminated by first removing the THF held in the loop and then drying the loop by passing air through it.

Several celluloses are included in this study. One is a purified, partially hydrolyzed cotton made available through the courtesy of K. A. Kuiken, The Buckeye Cellulose Corp. The *DP* data furnished with this cotton are to be found in Table I; the low values are the consequences of acid hydrolysis to which the cotton was subjected prior to characterization. This cotton was nitrated for GPC and handled in the manner already described (*1*). The data obtained in the previous study (*1*) of celluloses I, II, III, and IV are used as well as the data from a study of crosslinked cottons (*3*).

Table I. Degrees of Polymerization of Cellulose Samples Obtained by Conventional Methods and by the New GPC Procedure

Method	$\overline{DP}_w$	$\overline{DP}_n$	$\overline{DP}_w/\overline{DP}_n$
Conventional methods	1570[a]	725[b]	2.17
Gel permeation chromatography	1560	680	2.30

[a] By viscometry in Cadoxen.
[b] By osmometry of the nitrate in butyl acetate.

Calculations of the average *DP* values were carried out by means of the same scheme given earlier (*1*). Because the scheme is markedly different from the usual one for handling GPC data, it is repeated here in Table II for convenient reference. The required weight data of column 2 are obtained directly from the area measurement about the half-count in the chromatogram as described in the preceding paper (*1*). The *DP* (column 3), however, is calculated differently. The intrinsic viscosities necessary for calculating molecular weights of the fractions are not measured. These molecular weights are obtained by means of the relationship

$$M = \left(\frac{\text{coil size}}{K}\right)^{\frac{1}{1+a}} \quad (1)$$

which is derived from the concept of Benoit *et al.* (*4*) and the Mark-Houwink equation, $[\eta] = KM^a$. For cellulose trinitrate in THF, $K = 8.3 \times 10^{-4}$ and $a = 0.76$ for $DP \geqslant 1000$; for $DP \leqslant 1000$, $K = 0.075 \times 10^{-4}$ and $a = 1.14$ (*5*). To obtain M for a given elution volume, the value of the coil size corresponding to that elution volume is obtained from the universal calibration curve. Then M is calculated from the above equation using the appropriate values for coil size, *K*, and *a*. *DP* then follows as already mentioned.

All of the calculations were done by computer. The program, which is general and can be applied to polymers other than cellulose nitrate is available on request.

Results and Discussion

The earlier procedure for converting GPC data to average *DP* values was an advance over that of the original based on extended molecul

Table II. Calculation of Weight- and Number-Average Degrees of Polymerization from Fractionation Data[a,b]

Fraction	*Weight* (W) *in Polymer, grams*	DP	W/DP	W × DP
1	10	500	0.0200	5,000
2	20	1,000	0.0200	20,000
3	25	2,000	0.0125	50,000
4	25	2,500	0.0100	62,500
5	10	3,000	0.0033	30,000
6	10	4,000	0.0025	40,000
Total	100		0.0683	207,500

[a] Data in columns 1, 2, and 3 are strictly arbitrary.
[b] $\overline{DP}_w = \Sigma(W \times DP)/\Sigma W = 207{,}500/100 = 2{,}075$; $\overline{DP}_n = \Sigma W/\Sigma(W/DP) = 100/0.0683 = 1{,}464$; $\overline{DP}_w/\overline{DP}_n = 1.42$.

chain length in that it was founded on more acceptable principles and incorporated automation of data acquisition as well as data processing by computer. The results obtained were more in the range of those from conventional methods. Inherent in the method as developed, however, is the measurement of the viscosities of very dilute polymer solutions—*i.e.*, the eluted fractions. The possibility of eliminating this measurement appeared when consideration was given to Benoit's coil-size parameter $[\eta]M$ for universal calibration and to the Mark-Houwink equation $[\eta] = KM^a$ relating intrinsic viscosity of a polymer solution to the molecular weight of the dissolved polymer. It can easily be shown that the relationship expressed in Equation 1, where K and a are the Mark-Houwink constants for the particular system, is derived from the above equations (*6*). The molecular weight M of the fraction is obtained from the calibration curve by means of this equation.

In another approach towards eliminating the viscosity measurements, the parameter KM^{1+a}, related to the above, has been reported (*7*). It is claimed that this could be substituted for the coil-size parameter $[\eta]M$. A more complex function relating M to elution volume where no viscosity measurements are made has been proposed by Coll and co-workers (*8*). This function, derived from the Mark-Houwink equation and the equation of Ptitsyn and Eizner (*9*), lacks the simplicity of that described above.

Application of the present procedure to the characterized cotton supplied by The Buckeye Cellulose Corp. gives the results shown in the lower part of Table I. The agreement among the data is particularly good, even though the value for $\overline{DP}_n$ is somewhat lower than that determined by osmometry (this raises slightly the polymolecularity ratio $\overline{DP}_w/\overline{DP}_n$). This behavior of $\overline{DP}_n$ and the ratio has been noted previously (*1*) and is still in accord with reports of others (*10*). The polymolecularity ratios for the cellulose I, II, III, and IV samples reported in the preceding paper

Table III. Comparison of Results

Nitrated Sample	*Viscosity-Average* DP *Using Cadoxen*[a]	*Weight-Average* DP *Present Procedure*	*Weight-Average* DP *Prior Procedure*[a]
Cellulose I	4990	4140	5190
Cellulose II	3760	3380	4520
Cellulose III	3780	3370	4795
Cellulose IV	2330	2490	3390

[a] Data of Segal, Timpa, and Wadsworth (*1*).

(*1*) are unusually large. Recalculation of the data by the new procedure gives the results tabulated in Table III. The ratios are now more of the order usually obtained for material of this sort. Also, the values for *DP* are shifted; those for $\overline{DP}_w$ are lowered to the range of the Cadoxen viscosity-*DP*'s, while those for $\overline{DP}_n$ are raised.

In the preceding publication (*1*), five replications for the cellulose II sample were tabulated in Figure 3 to show that variations in $\overline{DP}_n$ were noted when the related $\overline{DP}_w$ data were in good agreement. The polymolecularity ratios for the *DP* values fluctuate widely as a consequence of this. These results are compared in Table IV with those now obtained from the same data. The improvement in the values is quite marked as the statistical analysis of the data indicates.

In recent work with cotton crosslinked with formaldehyde (*3*), it was noted that (a) $\overline{DP}_w$ remained constant with treatment time when decreasing values were anticipated from concurrent viscosity measurements, and (b) $\overline{DP}_n$ became quite low, whereby very high values appeared for the polymolecularity ratios. These values are compared in Table V with those resulting from application of the new procedure. Now $\overline{DP}_w$ decreases with treatment time as expected, while $\overline{DP}_n$ and polymolecularity ratio assume more consistent values. The inclusion in this table of the viscosity-average *DP* data presents an interesting picture. The agreement between $\overline{DP}_w$ by the present procedure and $\overline{DP}_v$ for the initial cotton is quite good; it is satisfactory for the next two samples. However, with increasing time of treatment $\overline{DP}_v$ and $\overline{DP}_n$ decrease greatly, indicating that much chain cleavage has taken place. The smaller decrease in $\overline{DP}_w$ is marked.

The lesser change here in $\overline{DP}_w$ illustrates the power of GPC in studies of changes in molecular weight brought about by various means. Factors are brought out that are not detected by other techniques. The reasons for the lesser change in $\overline{DP}_w$ in the above data become apparent when the chromatograms for these samples (*3*) are re-examined. Thus, as the combined formaldehyde in the sample increases, the chromatogram begins to rise at lower count than does that of the initial cotton, and reaches

Obtained According to Procedure Used

Number-Average DP		*Ratio* $\overline{DP}_w/\overline{DP}_n$	
Present Procedure	*Prior Procedure*[a]	*Present Procedure*	*Prior Procedure*[a]
2180	1580	1.90	3.35
1640	1040	2.07	4.67
1680	1140	2.02	4.21
1000	490	2.48	7.07

Table IV. Statistical Comparison of Results Obtained by the Present and Prior Procedures with Nitrated Cellulose II

Aliquant	*Weight-Average DP*		*Number-Average DP*		*Ratio* $\overline{DP}_w/\overline{DP}_n$	
	Present Procedure	*Prior Procedure*[a]	*Present Procedure*	*Prior Procedure*[a]	*Present Procedure*	*Prior Procedure*[a]
41 A	3500	4525	1840	1370	1.90	3.30
42 A	3460	4655	1614	980	2.15	4.76
43 A	3480	4470	1850	1390	1.88	3.21
44 A	3190	4520	1440	770	2.22	5.84
45 A	3250	4430	1430	710	2.27	6.24
Mean	3380	4520	1630	1040	2.08	4.67
Std. dev.	140	85	210	323	0.18	1.40
Coeff. of var., %	4.2	1.9	12.6	30.9	8.7	30.0

[a] Data of Segal, Timpa, and Wadsworth (*1*).

appreciable levels by the time the latter chromatogram does start up. This is evidence of nonhydrolyzed formaldehyde linkages that effectively increase molecular chain length, producing a small amount of polymer of higher molecular weight than initially present. Small amounts of such material will influence $\overline{DP}_w$ as this parameter is weight sensitive. The presence of such material is not even hinted at in the viscosity measurements, which may explain the occasional lack of agreement between viscosity and GPC data that has been observed.

The marked improvement in results presently obtained lies in the elimination of sources of error now recognized as being present in the prior procedure. A major source of error was the extrapolation from the log-log plot of half-count *vs.* intrinsic viscosity for the four eluates which lay about the peak of the chromatogram. It is this extrapolation that supplied the necessary intrinsic viscosities of the remaining more

Table V. Results Obtained According to Procedure Used

Formaldehyde Content of Sample, %	*Minutes of Treatment*	*Viscosity-Average* DP[a]	*Weight-Average* DP	
			Present Procedure	*Prior Procedure*[b]
Initial cotton	0	3430	3500	4120
0.08	1	3010	3310	3900
0.09	3	2970	3120	4130
0.18	8	2620	3060	4120
0.32	20	2460	3060	4200
0.63	60	1970	3200	4170
0.72	105	1930	2840	4130
0.83	180	1600	2780	3790
0.98	360	1360	2485	3590

[a] In ethyl acetate, $\overline{G} = 1{,}200\ \text{sec}^{-1}$, constants for conversion to *DP* according to Marx-Figini and Schulz (*11*).
[b] Data of Segal and Timpa (*3*).

dilute fractions (Figure 3, Ref. *1*). As the four points making up the plot seemed to fall about a straight line, a linear least-squares regression was applied and extrapolated. Closer study and more refined analysis of these data have disclosed that this is not a simple linear relationship. Because this is so, the values of the intrinsic viscosities of the more dilute fractions become more in error as extrapolation is made further and further from the central four points. As a consequence, the values of *DP* obtained through the calibration curve deviate from the true values.

Experimental error entered the data, of course, through the manual measurement of the viscosities of the four very dilute fractions. Another error in the viscosities was introduced by contamination of each fraction by some of the fraction preceding it. Not all of the liquid in the measuring syphon could be removed as the syphon emptied; thus a small portion of the fraction was retained and added to the incoming fraction. Furthermore, the previous procedure required exact weights in each fraction, whereas now only relative weights are necessary. The relationship developed here can be extended to polymers other than cellulose. To do this, the values of *K* and *a* have to be determined for the particular polymer, dissolved in the desired GPC solvent which was used to establish the calibration curve.

Although only $\overline{DP}_w$ and $\overline{DP}_n$ are mentioned in the tables of data presented here, the computer program includes instructions for obtaining $\overline{DP}_z$ and $\overline{DP}_{z+1}$. The latter have been helpful in indicating the extent of crosslinking produced or retained in treated cotton samples under study in connection with other work underway.

with Formaldehyde-Crosslinked Cotton After Nitration

Number-Average DP		*Ratio* $\overline{DP}_w/\overline{DP}_n$	
Present Procedure	*Prior Procedure*[b]	*Present Procedure*	*Prior Procedure*[b]
1590	800	2.21	5.15
1380	710	2.41	5.49
1215	520	2.58	7.94
1100	405	2.78	10.17
1040	440	2.96	9.55
850	380	3.79	10.97
840	370	3.39	11.16
710	280	3.91	13.54
635	290	3.92	12.38

Conclusions

A relationship has been developed by means of which more valid values are obtained by GPC for the average degrees of polymerization for cellulose. This can be extended to other polymers. With automation of data acquisition and computer processing of data, a differential molecular weight distribution and complete information on *DP* of the sample are available in very short order with a low degree of error. The ready availability of narrow, well characterized polystyrene fractions for calibration makes this procedure highly attractive in view of the lack of similar standards of cellulose.

Literature Cited

1. Segal, L., Timpa, J. D., Wadsworth, J. I., *J. Polym. Sci. A-1* (1970) **8,** 3577.
2. Goedhart, D., Opschoor, A., *J. Polym. Sci. A-2* (1970) **8,** 1227.
3. Segal, L., Timpa, J. D., *Proc. 10th Cotton Util. Res. Conf.,* New Orleans, La., Apr. 29–May 1, 1970, ARS No. 72-83.
4. Benoit, H., Grubisic, Z., Rempp, P., Decker, D., Ziliox, J. G., *J. Chim. Phys.* (1966) **63,** 1507; Grubisic, Z., Rempp, P., Benoit, H., *J. Polym. Sci. B* (1967) **5,** 753; Grubisic, Z., Reibel, L., Spach, G., *C. R. Acad. Sci., Paris, Ser. C* (1967) **246,** 1690.
5. Timpa, J. D., Segal, L., *J. Polym. Sci., Pt. A-1* (1971) **9,** 2099.
6. Weiss, A. R., Cohn-Ginsberg, E., *J. Polym. Sci., B* (1969) **7,** 379.
7. Chang, M., Abstracts, 161st National Meeting of the American Chemical Society, Los Angeles, Calif., Mar. 28–Apr. 2, 1971.
8. Coll, H., Prusinowski, L. R., *J. Polym. Sci. B* (1967) **5,** 1153; Coll, H., Gilding, D., *J. Polym. Sci. A-2* (1970) **8,** 89.

9. Ptitsyn, O. B., Eizner, Y. E., *Soviet J. Technol. Phys.* (Engl. Trans.) (1960) **4**, 1020.
10. Alliet, D. F., in "International Symposium on Polymer Characterization" (*Appl. Polymer Symp. 8*), K. A. Boni and F. A. Sliemers, Eds., Interscience, New York, p. 39; Boni, K. A., Sliemers, F. A., in "International Symposium on Polymer Characterization," p. 65; Crouzet, P., Fine, F., Mangin, P., *J. Appl. Polym. Sci.* (1969) **13**, 205.
11. Marx-Figini, M., Schulz, G. V., *Makromol. Chem.* (1962) **54**, 102.

RECEIVED January 17, 1972. Mention of a commercial product does not imply endorsement by the U. S. Department of Agriculture over similar products not mentioned.

17

Gel Permeation Chromatography

VI. Molecular Weight Averages and Molecular Weight Distribution of Cellulose Nitrate

A. C. OUANO and EDWARD M. BARRALL II

IBM Research Laboratory, San Jose, Calif. 95193

A. BROIDO

Pacific Southwest Forest and Range Experiment Station, Forest Service, U. S. Department of Agriculture, Berkeley, Calif. 94701

A. C. JAVIER-SON

Statewide Air Pollution Research Center, University of California, Riverside, Calif. 92502

Cellulose samples which have undergone various stages of thermal decomposition were characterized for changes in molecular weight and molecular weight distribution using gel permeation chromatography (GPC) and viscometry. Calculation of cellulose molecular weights (as cellulose nitrate) from the chromatogram and polystyrene calibration curves using the extended chain length–retention volume relationship (Q factor) resulted in very poor agreement between GPC and viscometric molecular weight values. Molecular weight averages determined by GPC were approximately five times greater than those obtained by viscometric technique. Application of various hydrodynamic considerations completely corrected this problem. The effects of calibration standard distribution and range are also discussed.

Gel permeation chromatography using polystyrene standards and viscometry were employed to determine molecular weight distribution and molecular weight averages of cellulose. The cellulose samples were all nitrated to about 13.5% nitrogen to make them soluble in THF (GPC solvent). The agreement between the molecular weight averages determined by GPC and viscometry ranged from very poor (order of magnitude difference) to good (within 10%) depending on the model used to calculate molecular weights from GPC data.

The previously described laboratory automation system (computer controlled) for gel permeation chromatography developed in this laboratory (*1*) was particularly well suited for calculating absolute molecular weights of cellulose nitrate from GPC data and a polystyrene calibration using the universal calibration model. The laboratory automation consisted of two main parts: data-logging and data-reduction programs. The data-logging program converts the analog signal of the GPC differential refractometer detector to digital form and stores it in a disk data-storage system. The data-reduction program calls the stored GPC data and reduces it to normalized molecular weight distribution curves and molecular weight averages using polystyrene calibration and the universal calibration of extended chain length model (Q factor). The computer programs allow a minimum of interaction with the technician. The only external input the program requires is: integration limits, sample identification, and the Mark-Houwink constants. The reduced data output consists of a normalized molecular weight distribution (Cal-Comp plots) and molecular weight averages (number-, weight-, viscosity-, and Z-average molecular weights) in digital forms.

Q-*Factor (Extended Chain Length–Retention Volume Calibration)*

This method of computing molecular weight averages from GPC data assumes that molecules having equal extended chain length have equal retention volume. Hence, by computing the average molecular size from the GPC data and the polystyrene calibration curve (plot of retention volume *vs.* extended chain length) one can presumably calculate the average molecular weight by simply multiplying the computed molecular size by a Q factor (Q = molecular weight/molecular size). Segal (*2*) reported a Q factor for cellulose of 58. A comparison between the weight-average molecular weights computed from GPC ($Q = 58$) and viscometric data is shown in Table I. The very poor agreement between the two sets of data is obvious, the GPC computed molecular weights being more than a magnitude larger than viscometric results. The very large disagreement is not unexpected and can be explained satisfactorily by the present generally accepted model of GPC separation. That is, molecules are separated according to hydrodynamic volume in solution and not according to the extended chain length. Consequently, molecules with identical extended chain length but different chain stiffness in solution will not have identical retention volumes.

Table I. Molecular Weight Averages of Cellulose by Viscometric and GPC Weight Average Methods Using Extended Chain Length Model (Q Factor)

Sample	*Viscometry* DP	*GPC* DP_W	*Disagreement, %*
Microcrystalline	195	1,650	745
1	110	970	780
2	185	1,545	735
3	210	1,810	760
4	285	2,640	810
5	310	2,820	1,530
6	790	12,900	1,535
7	850	16,700	1,870
8	900	21,800	2,320
9	965	31,000	3,120

Universal Calibration

Grubisic *et al.* (*3*) showed that for many polymers a single calibration curve can be drawn through a plot of the product of intrinsic viscosity and molecular weight ($[\eta]M$) *vs.* retention volume. This relationship certainly supports the model of molecular separation based on hydrodynamic volume since $[\eta]M$ is proportional to the hydrodynamic volume of the molecule in solution. Hence, molecular weights of the two polymers (calibration standard polymer and sample) which have identical retention volume under identical GPC analytical conditions can be expressed in terms of each other by combining the Grubisic relationship:

$$[\eta]_1 M_1 = [\eta]_2 M_2 \tag{1}$$

and the Mark-Houwink relationship

$$[\eta] = KM^{\alpha} \tag{2}$$

Thus, the expression

$$M_2 = \left[\frac{K_1}{K_2}\right]^{\left(\frac{1}{\alpha_2+1}\right)} M_1^{\left(\frac{\alpha_1+1}{\alpha_2+1}\right)} \tag{3}$$

permits the computation of the molecular weight (M_2) of a cellulose sample from the known molecular weight (M_1) of the standard polystyrene samples with identical retention volume.

The Mark-Houwink constants for polystyrene (K_1 and α_1) in THF at 30°C were measured in this laboratory to be 2.89×10^{-4} and 0.65, respectively, while those of cellulose nitrate were reported by Jenkins (*4*) to be 0.606×10^{-4} and 1.014, respectively. Dawkins, Maddock, and Coupe (*5*) suggested the use of an additional parameter for Equation 3 because of the excluded volume effect.

$$M_2 = \left[\frac{\Phi_2 K_1}{\Phi_1 K_2}\right]^{\left(\frac{1}{\alpha_2+1}\right)} M_1^{\left(\frac{\alpha_1+1}{\alpha_2+1}\right)} \quad (4)$$

The excluded volume correction Φ_2/Φ_1 for polystyrene and cellulose in THF was computed to be 0.58 from the relationship (*5*), $\Phi_{(\epsilon)} = \Phi_0(1 - 2.63\epsilon + 2.86\epsilon^2)$, where $\epsilon = (2\alpha - 1)/3$.

Meyerhoff (*6*) suggested the relationship

$$M_1^{1/2}[\eta]_1^{1/3} = M_2^{1/2}[\eta]_2^{1/3} \quad (5)$$

to be more appropriate for cellulose nitrate and polystyrene with equivalent retention volume. Equation 5, when combined with Equation 2 yields

$$M_2 = \left[\frac{K_1}{K_2}\right]^{\left(\frac{1}{\alpha_2+1.5}\right)} M_1^{\left(\frac{\alpha_1+1.5}{\alpha_2+1.5}\right)} \quad (6)$$

which is analogous to Equation 3. Rudin and Hoegy (*7*) suggested further correction because of the concentration effects on the universal calibration method. Meyerhoff's relationship and Rudin's correction, however, have not yet been applied in interpreting our results.

A comparison of the weight-average molecular weights of cellulose computed from both Equations 3 and 4 and from viscometric data are shown in Table II. It is apparent that the agreement between viscometric

Table II. Viscometric and GPC Weight Average Molecular Weight Data Using the Hydrodynamic Volume (Universal Calibration Model)

Sample	*Viscometry* DP	*GPC*[a] DP_W	*Disagreement, %*	*GPC*[b] DP_W	*Disagreement,*[b] *%*
Microcrystalline	195	240	23	135	8
1	110	145	32	110	0
2	185	220	19	165	11
3	210	260	24	200	5
4	285	350	23	265	7
5	310	375	21	285	8
6	790	1,230	56	935	18
7	850	1,580	86	1,200	41
8	900	1,970	119	1,500	67
9	965	2,650	175	2,100	118

[a] Calculated using Equation 3.
[b] Calculated using Equation 4.

and GPC computer molecular weight is much better for either Equation 3 or 4 than for the Q factor values in Table I. The poorer agreement for the higher molecular weight sample is interpreted as being a result of errors incurred in extrapolating the calibration curve beyond the highest molecular weight (1.8×10^6) polystyrene calibration standard used. Figure 1 shows the effect of extrapolation of the calibration curve on the agreement between GPC and viscosity-molecular weight. It is also interesting to note that the use of excluded volume correction, Equation 4, results in consistently lower molecular weights than the viscometric results while Equation 3 consistently yields higher molecular weight relative to viscometric values.

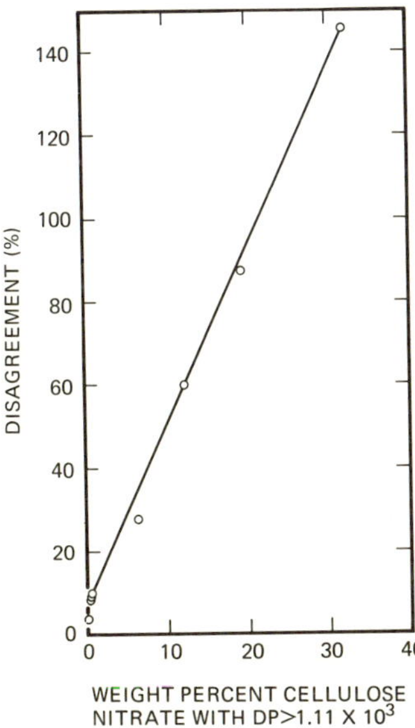

Figure 1. Disagreement between viscometric and GPC viscosity-average molecular weight vs. *weight percent of the sample component with retention volume less than the highest molecular weight calibration standard*

As evidenced in Table III, a closer agreement than those previously obtained in Tables I and II is made possible by comparing the GPC viscosity-average molecular weight (using Equation 3) and molecular weight averages obtained by viscometry. The above comparison is perhaps more valid than the previous comparisons (Table I and II) since for a polydisperse system the viscometric method of measurement determines the molecular weight average as defined by:

$$M_v = \left[\frac{\Sigma W_i M_i^{\alpha}}{\Sigma W_i}\right]^{1/\alpha} \quad (7)$$

and not the weight-average molecular weight. Since Equation 7 is in fact the relationship used in computing the viscosity-average molecular weight from the GPC chromatogram, the comparison made in Table III is justified. Note, however, that a similar computation using Equation 4 would yield poorer agreement since DP_v is less than DP_w and the values obtained from Equation 4 are already low. Some uncertainty is introduced in the value of the Mark-Houwink constants for cellulose in this work because they were obtained from $[\eta]$–M relationship of samples which had polydispersities of between 1.5 and 2.5. Since cellulose has a relatively low polydispersity ($M_w/M_n \sim 2.0$) and $\alpha \sim 1$, the values of M_w and M_n are usually within 10%. Furthermore, the literature (*8*) shows that the α values of fractionated and unfractionated cellulose are essentially the same. Hence, the above uncertainty is somewhat mitigated.

Table III. Viscometric and GPC Viscosity Average Molecular Weights Using the Universal Calibration Model

Sample	*Viscometry* DP	*GPC* DP_v[a]	*Disagreement, %*
Microcrystalline	195	200	3
1	110	115	5
2	185	185	0
3	210	230	10
4	285	310	9
5	310	340	10
6	790	1,010	28
7	850	1,360	60
8	900	1,690	88
9	965	2,370	146

[a] Calculated using Equation 3.

The results discussed in the preceding paragraphs indicate clearly that the extended chain length model (*Q* factor) is unsatisfactory for calculating cellulose molecular weight averages from the GPC retention volume distribution and polystyrene calibration curves. However, calculations based on the hydrodynamic volume of cellulose in solution give average molecular weights which agree well with results obtained by both the viscometric method and the literature values for microcrystalline cellulose (*9, 10*). The best agreement between GPC and viscometric data is obtained by comparing the viscosity-average molecular weight computed from GPC chromatograms using a model without the excluded volume effect.

Literature Cited

1. Ouano, A. C., *J. Polym. Sci.*, in print.
2. Segal, L., *J. Polym. Sci. Part C* (1968) **21**, 267.
3. Grubisic, Z. *et al.*, *J. Polym. Sci. B* (1967) **5**, 753.
4. Jenkins, R. G., Master's Thesis, The University of Waterloo.
5. Dawkins, J. V., Maddock, J. W., Coupe, D., *J. Polym. Sci. Part A-2* (1970) **8**, 1803.
6. Meyerhoff, G., *Makromol. Chem.* (1965) **89**, 282.
7. Rudin, A., Hoegy, H., *J. Polym. Sci. A-1* (1972) **10**, 217.
8. Brandrup, J., Immergut, E., Eds., "Polymer Handbook," Interscience, New York, 1967.
9. Patai, S., Halpern, Y., *Israel J. Chem.* (1970) **8**, 655.
10. Battista, O. A., Smith, P. A., *Ind. Eng. Chem.* (1962) **54**, 20.

RECEIVED January 17, 1972. Work supported in part by Grant AP00568 from Air Pollution Control Office, Environmental Protection Agency, to the University of California Statewide Air Pollution Research Center, Riverside, Calif.

18

Molecular Weight Characterization of Resole Phenol–Formaldehyde Resins

F. L. TOBIASON, CHRIS CHANDLER, and P. NEGSTAD

Pacific Lutheran University, Tacoma, Wash. 98447

F. E. SCHWARZ

Reichhold Chemicals, Inc., Tacoma, Wash. 98401

*Number-average molecular weights and intrinsic viscosities in dimethylformamide (DMF) and 1*N *sodium hydroxide have been determined for alkali-catalyzed phenol-formaldehyde resins prepared under different reaction conditions. Polyelectrolyte effects found in DMF viscosity measurements were corrected for and found to be a result of sodium ions trapped in the resin during many types of separations of the polymer from its caustic reaction solution. Corrections to vapor pressure osmometry data were made by analytically determining the* Na^+ *present by atomic absorption and specific ion electrode techniques. Number-average molecular weights were found in the range of 1300 to 10,000 for fractionated samples. Comparing the data with the Mark-Houwink curves established for linear acid-catalyzed phenol–formaldehyde (novolak) resins gives some insight into the configurational nature of resole resins in solution. The resole data fall close to but a little above the novolak data, which would indicate that the resole resins are flexible and possibly coiled in the pre-cured reaction stage.*

Base-catalyzed phenol–formaldehyde resins polymerized with a mole ratio of formaldehyde to phenol greater than one pose an interesting molecular weight characterization problem. This system is a dynamic one with active methylol end groups. Branched and crosslinked structures are formed, and in general, the separation of the resin from the reaction mixture is difficult. Figure 1 illustrates the chemical nature of a resole resin.

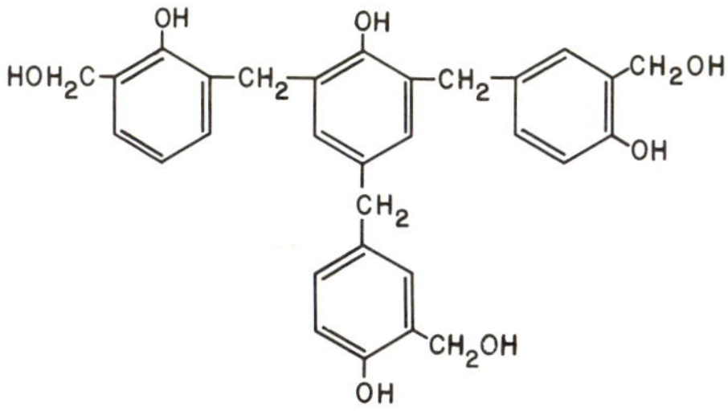

Figure 1. Representation of a branched phenol–formaldehyde resole resin

Some of these problems have been attacked (*1, 2*), but to our knowledge no satisfactory molecular weight characterization has been completed. In addition, gel permeation studies are difficult to interpret fully because of the possible range of molecular size at the same molecular weight, and the resins are generally too low in molecular weight for light scattering and ultracentrifuge studies. We have undertaken the task of characterizing several specially prepared resole resins by obtaining number-average molecular weights and intrinsic viscosity data. Comparison of these data with those obtained for linear phenol–formaldehyde novolaks (*3*) is made in an attempt to gain more information about the possible bulk structure changes and the polydispersity developing in resole resins during polymerization.

Experimental

Sample Preparation. The three primary resoles selected for inclusion in this study were synthesized in laboratory glassware in 4000 gram batches, using 99+% commercial phenol, ion-exchanged (low formic acid content) 50% aqueous formaldehyde, reagent grade 50% aqueous sodium hydroxide, and tap water. The three resins, coded A, B, and C, were allowed to react to the same apparent bulk viscosity as measured by the Gardner-Holdt bubble tube method. In addition to the same final viscosities, the resins also were synthesized from identical mole ratios: F/P = 2.0, Na/P = 0.71.

Although the above similarities among the resins might suggest little reason for their study, the reaction procedures employed were intentionally programmed to provide gross differences in resin structure and molecular weight for the three systems. Resins A and B were initiated with identical mole ratios: F/P/Na = 2.0/1.0/0.33. After an initial mixing step at a temperature below 50°C to promote polymethylol phenol formation, the polymerization was carried to completion at 85–95°C. During the polymerization a second sodium hydroxide addition of 0.38 mole was added to increase the incorporation of unreacted phenol and phenol alcohols into the polymer structure. The difference between resins A and B is merely the water content, more water being included in resin

B to allow a greater degree of polymerization to arrive at the same bulk viscosity.

The third sample, resin C, was synthesized from a different reaction procedure. The initial Na/P mole ratio was 0.5, with the formaldehyde added slowly over a 3-hour period at 95°C. The intention of this procedure was to provide conditions to maximize condensation and to minimize addition, producing maximum polymer size. The mole ratios after completion of the formaldehyde addition were F/P/Na = 2.0/1.0/0.50. A second addition of sodium hydroxide equivalent to 0.20 mole was then made, and the polymerization was completed at 85–95°C. Table I lists some of the general comparative properties for the three resins.

Table I. General Property Comparisons

Property	*Resin A*	*Resin B*	*Resin C*
Nonvolatile, %	42.0	40.0	40.0
Bulk specific gravity	1.195	1.185	1.185
pH	12.6	12.2	12.2
Sodium hydroxide, %	7.2	6.9	6.9
Viscosity (G-H), cps	600	600	600
Viscous flow (C-U), sec	49.4	61.7	70.8
Polydispersity	moderate	moderate	broad
Structure	branched	branched	linear
Cure rate	slowest	fastest	medium

The nonvolatile solids, specific gravity, pH, and Gardner-Holdt standard tube viscosity were determined by standard industrial techniques based on ASTM methods. The relative capillary flow rates (in sec) were measured using a Cannon-Ubbelohde viscometer with the flow times measured from bulk solutions diluted 1:1 with water. The polydispersity label is based on GPC scans using G50 to G200 Sephadex-packed 25-inch columns in single stages. Gravity flow (~1 ml/min) of solutions through a UV monitor set at 285 mμ was used. The solvents were water and some dilute NaCl or NaOH solutions. The qualitative polydispersity comparison is based on examining one resin relative to the other. The structure index is based on IR absorbance band ratios of the total methylol (9.8 μ) to total aromatic absorbance (12.2–13.2 μ) and on free bromine uptake. The rapid stage of free bromine uptake by methylol in alkaline solutions of the resin was compared by taking the ratio of equivalents of bromine reacted to grams of nonvolatile solids. The larger each of these ratios, the more methylol is present in the resins, and consequently the greater the probability of branch points being present. The cure rate data were determined from various resin applications in glue mixes. The data points labeled E in the figures are some general resin samples prepared similar to B and C and studied a year earlier.

Sample Separation. After the bulk resins were prepared they were kept under refrigeration to reduce any further polymerization. To separate samples from the reaction mixture for study, 20–30 grams of bulk resin was diluted with 150 ml of distilled water and titrated slowly with 1N H_2SO_4. During the titration and resin precipitation the sample was kept in an ice bath and mixed vigorously to insure homogeneity. Many acids were tried—hydrochloric, carbonic, citric, phosphoric—but the sulfuric led to the best samples with which to work. The titration was followed to a pH of 6.5–6.7 using a pH meter and a high pH tripurpose Corning electrode. The sample was allowed to mix for at least 15 min here to be sure the pH had stabilized. Samples were then washed with an equal volume of distilled water (0°C), centrifuged, and then washed with an equal portion of methanol directly from the freezer (−12°C). These samples were then centrifuged, collected, and vacuum-dried. Since the methanol removed most of the water, drying times required were only about 3 to 4 hours. It was very important to keep the samples cold, to dry them fast, and also to refrigerate the dried samples. In fact, solutions were made for study very shortly after drying was completed. The samples obtained by this method were light tan, amorphous powders which were easily soluble to 60 grams/liter in dimethylformamide (DMF). However, if a high sodium residue remained in the samples, they would be brittle, hard, and red, with very poor solubility. Table II lists the pertinent information on sample separation and fractionation.

Table II. Sample Separation and Fractionation

Resin Code	*pH of Sample*	*Wash 1 Parts Water*	*Wash 2*
A1	6.5[a]	1	acetone
A2–1	6.5[b]	2	methanol
A2–2	(evaporated from methanol wash of A2–1)		
A3	6.40	2	—
A4	6.3	1	methanol
A5–1	9.45	5	methanol
A5–2	9.32	1	methanol
A5–3	9.28	1	methanol
A5–4	5.80	1	methanol
A5–5	(evaporated from methanol wash of 5–1 to 5.4)		
A6–1	9.93[c]	2	methanol
A6–3	9.50[c]	2	methanol
A6–4	6.02	2	methanol
B2	6.5	2	methanol
B3	6.5	2	—
B4–1	9.5	2	methanol
B4–2	9.3	2	methanol
B4–3	6.5	2	methanol
C6	6.5[a]	1	acetone
C7	6.5	2	methanol
C8	6.5	2	—
C9–1	8.35[c]	2	methanol
C9–3	8.25[c]	2	methanol
C9–4	8.25[c]	2	methanol
C9–5	6.50	2	methanol

[a] Titrated with carbonic acid.
[b] Remaining samples titrated with 1*N* H_2SO_4 unless otherwise noted.
[c] These fractions were spun and mixed with equal volume water; then the pH was lowered to 6.5 with 1*N* H_2SO_4.

Fractionation. Fractionation of the resin redissolved in DMF or acetone-methanol mixtures with a non-solvent such as water was unsatisfactory. The resin appeared to be highly bound in the solvent, and only a portion of the suspended resin could be removed even by centrifugation. Consequently, we decided to fractionate the resin by pH from the caustic reaction mixture since the polymer precipitates over a range of pH. Acid was slowly added to a bulk solution diluted with distilled water until a sufficient amount of resin appeared that would not redissolve in 20 min. This sample was separated by centrifugation, and the pH was recorded. More acid was added to the remaining solution until another sample was obtained. Generally 3 or 4 fractions could be collected. Because of interference from the Na^+ ion, each fraction collected was washed once and then titrated to pH 6.5. This worked satisfactorily except that in the neutralization process some additional low molecular weight material no doubt was precipitated, thus causing a broader molecular weight distribution. However, the overall distribution appeared to be relatively narrow in terms of pH precipitation. It should be emphasized that in this study we made no attempt to include the monomeric or simple alcohols as part of the resin system.

Table III. Comparison of Atomic Absorption and Specific Ion Measurements

Resin Code	*AA Total Sodium, %*	*SI (25°C) Free Sodium, %*
A3-1	0.50	0.25
A4-3	0.29	0.25
A5-4	0.27	0.21
A5-5	1.65	1.20
C9-1	0.78	0.56
C9-4	0.49	0.31
E-0863	1.12	1.01
E-0869	0.93	1.01

Viscosity Measurements. Viscosity measurements were made in filtered DMF and 1*N* NaOH at 25 ± 0.05°C. The DMF was dried over molecular sieve. Flow times were measured in Cannon-Ubbelohde dilution viscometers, and intrinsic viscosities were obtained from the extrapolation of η_{sp}/c from the Huggins equation to zero concentration. When an apparent polyelectrolyte effect was found in DMF, the intrinsic viscosities were determined by adding $CaCl_2$ to the DMF or by extrapolation from the higher concentration portion of the Huggins curve. Intrinsic viscosities are probably accurate to about 5%.

Molecular Weight Measurements. Number-average molecular weights were determined using a Hewlett-Packard 302B vapor pressure osmometer operating at 62°C with DMF as the solvent. Molecular sieve was used in the solvent cup to help keep the DMF dry. The higher temperature was used to help sensitivity and also to keep the solvent dry. The thermistor differential voltage output curves were recorded. It sometimes took 10 min to approach asymptotically the equilibrium output voltage.

The change in voltage was determined relative to the pure solvent output, and was quite reproducible.

Calibration was carried out using benzoic acid, raffinose, poly(tetrahydrofuran), and poly(propylene)glycol. The calibration curve was rechecked with cholesterol after all the molecular weights were determined. The polymer solutions generally yielded slight curvature when $\Delta V/c$ was plotted *vs.* c instead of the straight line found for the monomers. The resin solutions also showed a similar slight curvature. This effect did not limit consistent extrapolations to determine molecular weights. However, the curvature along with the DMF system did limit the experimental molecular weights to about 10%. The biggest problem in making the molecular weight measurements was inclusion of Na^+ during separation of the polymer samples from the reaction mixture. This arises because of the enormous influence of low molecular or atomic weight impurities on colligative properties.

Sodium Measurements. The sodium ion content was measured by two techniques: 1) A Perkin-Elmer 403 atomic absorption unit was used to obtain the total amount of sodium in a particular solution; and 2) an Orion 801 pH meter with Corning sodium-specific ion electrodes was used at 25°C to determine the "free" sodium present. An attempt to measure the free sodium content at 60°C failed because of thermal instability. It was especially important to determine if the sodium ions present were free so that correct molecular weights could be computed from the VPO data. The calibration was accomplished by dissolving small amounts of sodium benzoate in the DMF, or by using DMF-distilled water (3 + 1) mixtures containing dissolved NaOH. Both calibrations gave similar final results. Even though tap water was used in the synthesis, interference from other ions (*e.g.*, Ag^+, Li^+, or NH_4^+) was considered minimal be-

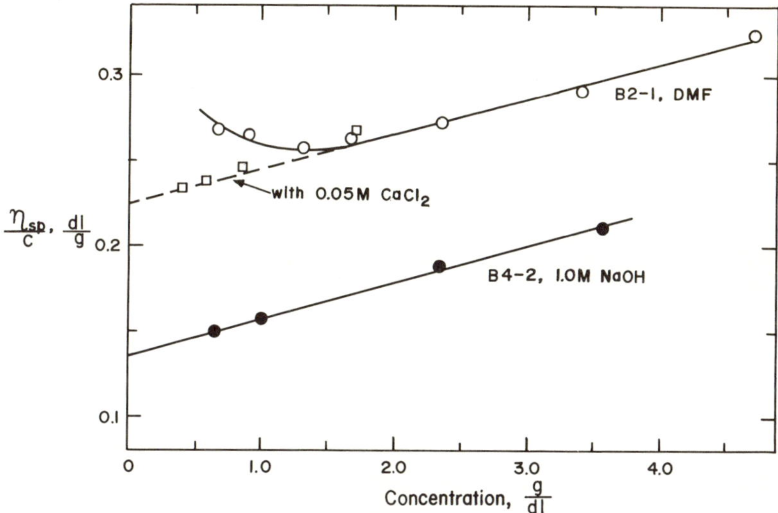

Figure 2. Viscosity data showing the polyelectrolyte behavior of resoles in DMF

Table IV. Experimental and Corrected

		M_n	
Resin Code	*Na, %*	*VPO*	0.25% *Na*
A1	0.17	2290	2290
A2–1	0.29	2110	2190
A2–2	0.26	1000	1000
A3	0.50	1700	2080
A4–3	0.29	1280	1310
A5–1	2.70	575	1450
A5–23	2.30	720	1970
A5–4	0.27	1680	1700
A5–5	1.65	1170	4010
A6–1	0.20	1670	1670
A6–3	0.19	1900	1900
A6–4	0.21	2000	2000
B2–1	0.27	2880	2950
B2–2	0.18	1250	1250
B3	0.43	2000	2370
B4–1	0.47	2000	2470
B4–2	0.34	1750	1870
B4–3	0.28	2880	2990
C6	1.50	650.	990
C7	0.23	2500	2500
C8	0.77	2400	5490
C9–1	0.78	2400	5300
C9–3	0.50	2170	2850
C9–4	0.50	2700	3807
C9–5	0.92	1810	3780
A2–1			
A4–3			
A6–3			
B4–1		viscosities in 1*N* NaOH	
B4–2			
C7–1			
C9–3			
C9–4			

[a] Calculated assuming only sodium is present.
[b] The correction to 0% sulfate is based on assumption that sodium sulfate is present. Since the sulfate was not determined quantitatively, this set of molecular weights would

cause of sample washing techniques described in the fractionation section. This is confirmed in that one measures very low sodium content in some of the samples. A comparison of results is shown in Table III. Since the sodium ions appeared to be active at room temperature, we therefore assumed that most of the sodium present was free to contribute to the VPO readings at 62°C. Corrections to the molecular weights could then be made directly from the atomic absorption data.

Molecular Weights and Viscosities

M_n			
0% Na[a]	0% SO_4[b]	$[\eta]$, $\frac{deciliter^2}{gram}$	Huggins k
2750	2820(3036)[c]	0.131	0.39
2870	2990(3442)	0.128	0.50
1125	1140(1191)	0.071	0.50
2680	2870(3684)	0.135	0.44
1520	1550(1665)	0.076	0.82
1720	2255	0.112	0.03
2510	3580	0.128	0.21
2090	2150(2350)	0.094	0.65
7140	18500	0.278	0.06
1950	1985(2110)	0.112	0.66
2250	2320(2460)	0.119	0.59
2440	2500(2720)	0.112	0.66
4330	4610(5710)	0.233	0.32
1380	1400(1450)	0.122	0.97
3180	3420(4405)	0.217	0.29
3360	3660(4980)	0.173	0.38
2350	2450(2800)	0.183	0.41
4410	4720(5930)	0.138	0.75
1110	1220(1660)[c]	0.220	0.42
3325	3460(3925)	0.198	0.17
12200	23000	0.315	0.26
12800	23000	0.220	0.42
4080	4600(7040)	0.190	0.17
6490	7800	0.194	0.17
6420	9350	0.188	0.11
		0.104	0.95
		0.059	1.05
		0.095	1.11
viscosities in 1N NaOH		0.135	1.22
		0.169	0.82
		0.200	0.95
		0.152	1.20
		0.187	0.71

have the greatest error. The first number in the column assumes the sulfate per cent weight is ½ that of the sodium. The number in parentheses is for a full correction.
[c] Acetate ion was present in separation.

Results

The nature of the viscosity determinations is illustrated in Figure 2. The upper curve shows the typical polyelectrolyte effect found for many of the resins in DMF. Adding $CaCl_2$ to the DMF causes a saturation of charge and induces a relaxation of the resin structure, resulting in normal viscosity behavior. In the NaOH solution the resins acted normally. The size of the polyelectrolyte effect and the nature of Huggins constant k seemed to be related to the amount of Na^+ trapped in the resin and carried into the DMF solution. Corrections to the number-average molecular weights from the sodium present were made by applying the equation $\overline{M}_n = \Sigma w_i / \Sigma n_i$, where w_i is the weight of the fraction and n_i is the number of moles. The amount of sodium used in the correction is that taken from the atomic absorption data. In order to have charge balance, some SO_4^{2-} ion could have been carried into solution. A final correction

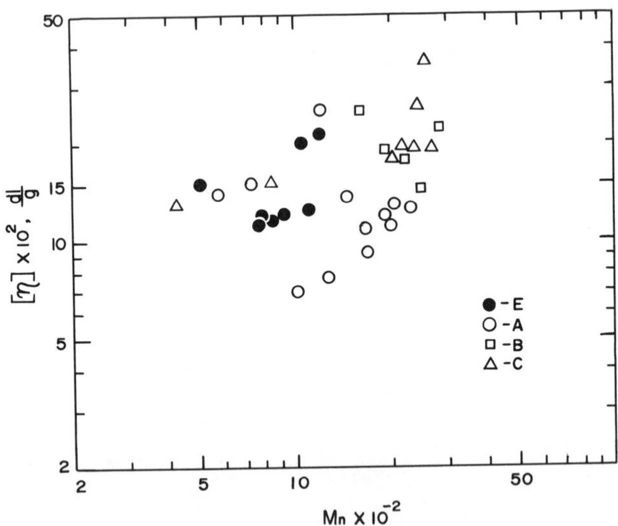

Figure 3. Viscosity (DMF) and uncorrected molecular weight data for resole resins

was made for sulfate (which was detected analytically) based on the sodium analytical work. However, since sodium sulfate has very low solubility in DMF, the amount of sulfate present is probably small. Table IV gives the sodium content and corrected and uncorrected molecular weight data. The molecular weights are corrected to 0.25 and 0% in Na. The 0.25% level was picked since a number of samples had approximately this amount. The best molecular weights are probably somewhere between the 0% Na values and the unbracketed 0% SO_4^{2-} values. The values for molecular weights that had a high sodium content are most likely to be in error.

Figure 3 illustrates the scatter in the molecular weight data before corrections. Figure 4 shows the 0% Na data on log M *vs.* log [η] plot compared with the DMF novolak curve. The very high molecular weight data in the tables were not included on the graphs because of the large possible error present. Figure 5 displays the molecular weight data cor-

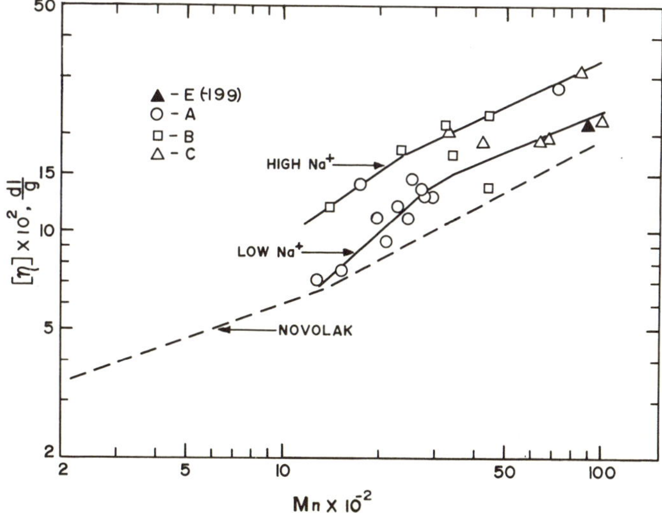

Figure 4. Mark-Houwink data with molecular weights corrected to 0% sodium

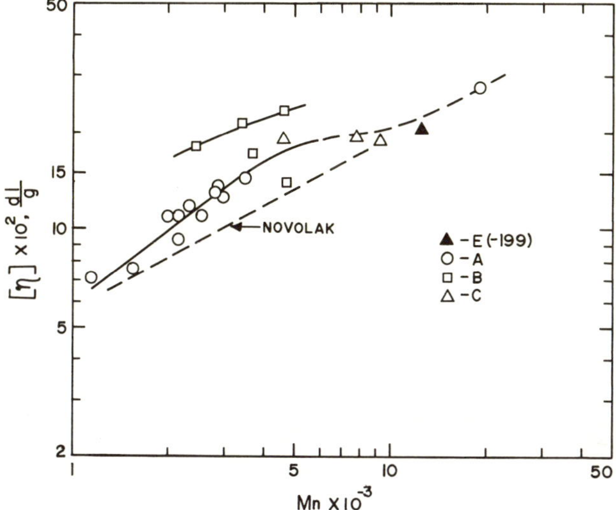

Figure 5. Mark-Houwink data with molecular weights corrected to 0% sodium and corrected assuming 25% of potential sulfate ion was present

rected for the sodium and sulfate ions and compared with the novolak data. The plotted data represent what are thought to be the best molecular weights, that is, those corrected for the smaller weight per cent sulfate. Figure 6 shows the viscosities and molecular weights in 1*N* NaOH compared with the novolak in NaOH. Sufficient data were available to correct fully only one E sample, and the result was favorable.

To investigate whether the Na^+ is trapped by the resin or is carried through in a specific interaction, several model compounds were studied in NaOH solution. The decrease in activity was measured with a specific-ion electrode, and the results are recorded in Table V. A relatively small decrease in sodium ion activity in *o*-hydroxybenzyl alcohol was noted.

Discussion

The interference of sodium in these measurements indicates an important problem for this residue can have great effects on IR, GPC, and ultracentrifuge studies. Polyelectrolyte effects can influence GPC properties drastically because of changes in the hydrodynamic volume of the resin. Our GPC labeling in Table I could possibly be influenced by this polyelectrolyte effect. This was checked by analyzing samples in water solutions with salts added. The same relative polydispersity results were obtained even though the chromatograms were changed. From our measurements it would appear that the sodium is not chelated or bound directly by phenol groups but must basically be trapped by the resin rings. Examination of the results in Table V indicates that *o*-hydroxybenzyl alcohol does not tie up much Na^+; a reaction mechanism using this has been proposed (2).

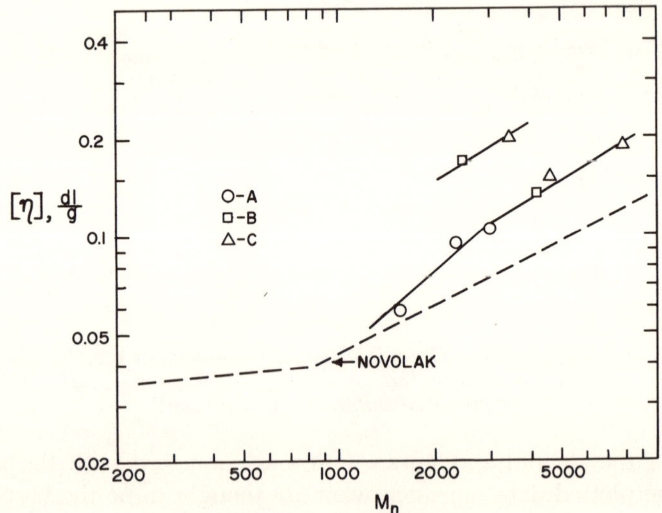

Figure 6. Mark-Houwink data of the resoles with viscosities determined in 1N NaOH

The resins A, B, and C all apparently occur in relatively narrow molecular weight ranges if one neglects the very low molecular weight water-soluble components. The ranges are: A (2500), B (4000), and C (8000). The initial polymerizations were designed to produce relative molecular weight differences of this type.

The differences in viscosity between these three resins should indicate changes in branching or stiffness along the resin molecular chain. A possible way to arrive at this insight is to compare the viscosity results directly with the linear flexible ortho-para phenol–formaldehyde novolak viscosity data in the same solvent. A possible interpretation of the DMF results displayed in Figures 4 and 5 is that the rising slope of resin A to B is induced by building branch points (more like crosslinks) into the resin chain. The evaluation of Mark-Houwink parameters is not considered valid for this study because of the nature of the fractions and system. However, on a relative basis, the observation that the Mark-Houwink slope through the A samples in Figure 5 is equal to 0.75 would be indicative that the structures are not rigid. The fact that resin C decreases as it does toward the novolak curve would seem to indicate greater linearity and flexibility built into the structure. Usually the viscosity of a simple branched polymer will be lower than that of the linear one. However, if stiffness approaching that of crosslinking were added to the polymer chain, the viscosity would certainly increase—as for resin B. The range in viscosity for resin B in a narrow molecular weight range would indicate that it has greater molecular size variance as a result of branching than the other two resins. This is probably why B is predominant in the high sodium curve. In sodium hydroxide the same general trend is found except that the C resin does not fall as close to the novolak curve. This could reflect the better viscosity measurements in this case, that is, measurements unperturbed by the polyelectrolyte effect.

Table V. Sodium Ion Activity with Model Compounds (25°C)

Sample	*Phenol-OH, M*	*Readings, mv*	*% Phenol Tied with Na*
1.0*N* NaOH	—	51.3±0.5	0
OHBA [a, b]	0.272	51.0	4
OHBA [b, c]	0.272	51.0	4
Phenol [b]	—	50.2	—
Bisphenol [b]	0.583	46.8	24
0.6*N* NaOH	—	37.4	—

[a] OHBA: *o*-hydroxybenzyl alcohol.
[b] In 1.0*N* NaOH.
[c] Alcohol was heated for 15 min at 80°C.

Although distribution curves could not be meaningfully constructed, in most cases 90% of the starting weight of resin was recovered from the precipitations. Molecular size rather than molecular weight seemed to control the fractionations. From the weights and molecular weight observations it appeared that for pH fractionation the molecular weight distribution was relatively narrow with a small amount of high and low weight material.

Literature Cited

1. Whitehouse, A. A. K., Pritchett, E. G. K., Barnett, G., "Phenolic Resins," American Elsevier Publishing Co., 1968.
2. Martin, R. W., "The Chemistry of Phenolic Resins," John Wiley and Sons, Inc., 1956.
3. Tobiason, F. L., Chandler, C., Schwarz, F., *Macromolecules* (1972) **5**, 321.

RECEIVED January 17, 1972. Work supported in part by the National Science Foundation Undergraduate Research Participant Program.

19

Approximate Solutions of Chemical Separation Equations with Diffusion

GEORGE H. WEISS

Physical Sciences Laboratory, Division of Computer Research and Technology, National Institutes of Health, Bethesda, Md. 20014

MENACHEM DISHON

Department of Applied Mathematics, Weizmann Institute of Science, Rehovot, Israel

Many problems in the analysis of biochemical separation techniques require the solution of a Fickian equation of the form

$$\frac{\partial c}{\partial \tau} = \varepsilon \frac{\partial}{\partial x}\left(f(x)\frac{\partial c}{\partial x}\right) - \frac{\partial}{\partial x}(g(x)c)$$

where ϵ *is proportional to the diffusion constant. This paper presents an approximate solution to this equation valid for small* ϵ. *For many separation systems* ϵ *is* 10^{-3} *or less, while* $f(x)$ *and* $g(x)$ *are of the order of 1. Applications are given to velocity centrifugation experiments, with pressure dependent sedimentation, and to pore gradient electrophoresis.*

Many problems related to biochemical separation systems require solution to a transport equation of the form

$$\frac{\partial c}{\partial \tau} = \varepsilon \frac{\partial}{\partial x}\left(f(x)\frac{\partial c}{\partial x}\right) - \frac{\partial}{\partial x}(g(x)c) \qquad (1)$$

in which c is solute concentration, $g(x)$ is the transport term, and $\epsilon f(x)$ is the diffusion term. Some examples of systems which fall into this category are:

(1) Pressure dependent systems in velocity sedimentation (*1, 2, 3, 4*).
(2) Gel pore electrophoresis (*5, 6*).
(3) Analytical gradient chromatography (*7, 8*).

Undoubtedly other examples can be found. In the study of the systems cited above, a common property is that the (dimensionless) parameter ϵ is very small compared with the (dimensionless) terms $f(x)$ and $g(x)$ which can be chosen to be of the order of 1. In all of the examples cited ϵ is 2.5×10^{-2} or less. In this paper we summarize a singular perturbation technique for the solution of Equation 1, which is to be solved subject to an initial condition $c(x,0)$, and on the assumption that boundary effects (in x) can be ignored.

To see why it is impossible to obtain a solution to Equation 1 by ordinary perturbation theory, consider the simple system

$$\frac{\partial c}{\partial \tau} = \varepsilon \frac{\partial^2 c}{\partial x^2} \tag{2}$$

subject to an initial condition $c(x,0) = \delta(x)$, a delta function. If we set $\epsilon = 0$, then the solution is $c(x,\tau) = \delta(x)$. On the other hand, the solution to Equation 2 is known to be:

$$c(x, \tau) = \frac{1}{\sqrt{4\pi\varepsilon\tau}} \exp\left(-\frac{x^2}{4\varepsilon\tau}\right) \tag{3}$$

that is, spreading occurs because of diffusion. The solution (Equation 3) cannot be found by expanding the variables in Equation 2 in a power series in ϵ. In particular if we consider a point $x \neq 0$, the concentration is identically equal to 0 when $\epsilon = 0$, but it can be much greater than 0 for any nonzero ϵ.

To derive an approximate solution we start by transforming the space variable x to one that moves by the convective transport mechanism—*i.e.*, a coordinate that remains fixed with the solute motion when diffusion can be neglected. This coordinate will be denoted by ξ and is

$$\xi = \int_0^x \frac{du}{g(u)} - \tau \tag{4}$$

If we define a function $H(z)$ to be the solution to

$$\int_0^{H(z)} \frac{du}{g(u)} = z \tag{5}$$

then a molecule initially at position ξ_0 (or $x_0 = H(\xi_0)$) will be transported to position $\xi_0 + \tau$ at time τ in the absence of diffusion. If we also define a new dependent variable $\psi(x,\tau)$ by

$$\psi(x, \tau) = g(x)c(x, \tau) \tag{6}$$

then Equation 1 can be transformed to

$$\frac{\partial \psi}{\partial \tau} = \varepsilon \frac{\partial}{\partial \xi}\left[\frac{F(\xi + \tau)}{G(\xi + \tau)} \frac{\partial}{\partial \xi}\left(\frac{\psi}{G(\xi + \tau)}\right)\right] \tag{7}$$

where

$$F(u) = f(H(u)),\ G(u) = g(H(u)) \tag{8}$$

Equation 7 is exact, no approximation having been made.

At this point we can introduce the perturbation procedure based on the fact that ϵ is small. Let us assume that the initial condition is $c(x,0) = c_0\delta(x)$, where $\delta(x)$ is a delta function (—*i.e.*, an initial pulse injection). The principal consequence of a small ϵ is the fact that the bandwidth of the diffusion broadened peak is narrow, going to zero as ϵ vanishes. Thus, we make the approximation that only the region near $\xi = 0$ in the terms $F(\xi + \tau)$ and $G(\xi + \tau)$ is of interest. The resulting value, denoted by $\psi_0(\xi,\tau)$, therefore satisfies the equation

$$\frac{\partial \psi_0}{\partial \tau} = \varepsilon \frac{F(\tau)}{G^2(\tau)} \frac{\partial^2 \psi_0}{\partial \xi^2} \tag{9}$$

subject to the initial condition $\psi_0(x,0) = c_0 g(x)\delta(x)$. This is a diffusion equation with a time-dependent diffusion constant. It can be reduced to a more familiar form by introducing a new dimensionless time $\Delta(\tau)$ by

$$\Delta(\tau) = \int_0^\tau \frac{F(u)}{G^2(u)}\, du \tag{10}$$

in which case Equation 9 becomes

$$\frac{\partial \psi_0}{\partial \Delta} = \varepsilon \frac{\partial^2 \psi_0}{\partial \xi^2} \tag{11}$$

The solution to this equation is straightforward and can be written

$$\psi_0(\xi, \Delta) = \frac{C_0}{\sqrt{4\pi\varepsilon\Delta(\tau)}} \exp\left(-\frac{\xi^2}{4\varepsilon\Delta(\tau)}\right) \tag{12}$$

If we can assume that $g(0) = 1$ (this can always be done without loss of generality), then the final expression for $c(x,\tau)$ becomes

$$c(x, \tau) = \frac{C_0}{g(x)\sqrt{4\pi\varepsilon\Delta(\tau)}} \exp\left(-\frac{\xi^2}{4\varepsilon\Delta(\tau)}\right) \tag{13}$$

where ξ is given in Equation 4. Notice that $c(x,\tau)$ is not generally symmetric in x because ξ is equal to a function of x that is not necessarily symmetric. The maximum peak concentration is very closely given by

$$\frac{c_{max}}{c_0} = \frac{1}{g(H(\tau))\sqrt{4\pi\varepsilon\Delta(\tau)}} \tag{14}$$

This relation enables one to determine ϵ (or equivalently, the diffusion constant) experimentally.

To put the preceding analysis into a more applied framework, let us consider the peak broadening in pore gradient electrophoresis (*5, 6, 9*). For this problem, let D and D_0 represent the diffusion coefficient in the gel and in the absence of a gel, and M and M_0 the respective mobilities. For many gels these variables experimentally satisfy Equation 5:

$$D/D_0 = M/M_0 = \exp(-x/L) \tag{15}$$

for a linear gel gradient, where x is distance and L is an experimentally measured parameter. Let V be the voltage gradient. Then the dimensionless parameters τ and ϵ that characterize the Fick equation are

$$\tau = M_0Vt/L,\ \varepsilon = D_0/(M_0VL) \tag{16}$$

In addition we will let $z = x/L$ be a dimensionless distance. The transport equation for this system is

$$\frac{\partial c}{\partial \tau} = \varepsilon \frac{\partial}{\partial z}\left(e^{-z}\frac{\partial c}{\partial z}\right) - \frac{\partial}{\partial z}(e^{-z}c) \tag{17}$$

An exact solution to this equation subject to an initial pulse of amount c_0 can be found (*9*), and the following expression can be found from it:

$$\frac{c(x,\tau)}{c_0} = \frac{1}{L\tau(2\pi\varepsilon)^{1/2}}(1+\lambda^2)^{-1/4}\left(\frac{\lambda}{\sqrt{1+\lambda^2}-1}\right)^{1+\frac{1}{\varepsilon}} \times \exp\left[\frac{1}{\varepsilon}\left(\sqrt{1+\lambda^2}+\frac{z}{2}-\frac{1+e^z}{2}\right)\right] \tag{18}$$

in which $\lambda = (z/\tau)\exp(z/2)$. Although this expression is very close to being exact, it is difficult to work with. Since ϵ is small, we can use the theory developed in earlier paragraphs. The functions required are easily found to be

$$\begin{aligned} \xi &= e^z - 1 - \tau \\ H(y) &= \ln(1+y) \\ \Delta(\tau) &= \tau(1+\tau/2) \end{aligned} \tag{19}$$

A comparison of the results of the exact calculation and the approximate one is shown in Figure 1 for the values $\epsilon = 5 \times 10^{-3}$, $\tau = 5$. The error is fairly small, and the relative error is generally less than 5%. There is no great advantage to using the perturbation theory for the example just discussed because an exact solution is available. However, a gel for which the diffusion coefficients and mobilities are

$$D/D_0 = \exp(-\alpha x/L), \; M/M_0 = \exp(-x/L) \tag{20}$$

for example, requires an approximate theory because an exact one is not available.

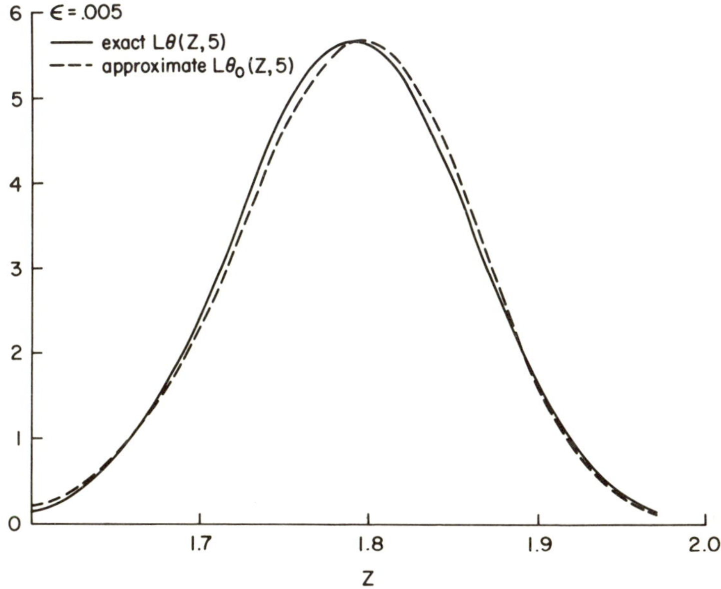

Figure 1. A comparison of the exact and approximate concentration profiles ($\theta \equiv c/c_0$) (Equations 13 and 18) for pore gradient electrophoresis, for $\varepsilon = 5 \times 10^{-3}$, $\tau = 5$

Extensive application of the theory has been made in treating pressure effects on velocity centrifugation. If ω denotes angular speed, s_p the sedimentation coefficient at pressure p, r_a the radius of the meniscus, $x = (r/r_a)^2$, the usual representation of pressure effects leads to the relation (*10*),

$$s_p/s_0 = 1 + m(x - 1) \tag{21}$$

where m is a parameter to be determined experimentally, that contains the solvent compressibility. The parameters ϵ and τ for this model are

$$\varepsilon = 2D/(s_0\omega^2 r_a^2), \ \tau = 2\omega^2 s_0 t \tag{22}$$

and the Lamm equation is

$$\frac{\partial c}{\partial \tau} = \varepsilon \frac{\partial}{\partial x}\left(x \frac{\partial c}{\partial x}\right) - \frac{\partial}{\partial x}[x(1 + mx - m)c] \tag{23}$$

This equation is to be solved subject to the initial condition

$$c(x, 0) = c_0 H_0(x - 1) \text{ where } \begin{matrix} H_0(y) = 1 \ y>0 \\ H_0(y) = 0 \ y<0 \end{matrix} \tag{24}$$

The approximate solution for the concentration for the model of Equation 23 is (*11*),

$$\frac{c(x, \tau)}{c_0} = e^{-(1+m)\tau}\left(\frac{1 + m}{1 + m - mx + mxe^{-(1+m)\tau}}\right)^2 \phi\left(\frac{\xi}{\sqrt{2\varepsilon\Delta(\tau)}}\right) \tag{25}$$

in which

$$\xi = \frac{1}{1 + m} \ln \frac{x}{1 + m - mx} - \tau$$

$$\phi\ (x) = (2\pi)^{-1/2} \int_{-\infty}^{x} \exp\ (-\ u^2/2)dn \tag{26}$$

$$\Delta(\tau) = \frac{1}{(1 + m)^4}\Big[1 - e^{-(1+m)\tau} + 3m(1 + m)\tau$$

$$+ 3m^2(e^{(1+m)\tau} - 1) + \frac{m^3}{2}(e^{2(1+m)\tau} - 1)\Big]$$

The relative error in using Equation 25 is less than 0.005 over most of the concentration profile (*11*) for parameters that are characteristic of several polymer systems (*12*) while the error in the concentration gradient is generally less than 0.02 for these same parameters.

An iterative method for developing higher order approximations to the solution to Equation 1 can be devised by using some of the ideas of singular perturbation theory (*15*). To display the systematics of the procedure let us rewrite Equation 7 as

$$\frac{\partial \psi}{\partial \tau} = \varepsilon A(\xi + \tau)\frac{\partial^2 \psi}{\partial \xi^2} + \varepsilon B(\xi + \tau)\frac{\partial \psi}{\partial \xi} + \varepsilon C(\xi + \tau)\ \psi \tag{27}$$

in which

$$A(u) = F(u)/G^2(u)$$

$$B(u) = F'(u)/G^2(u) - 2F(u)G'(u)/G^3(u) \tag{28}$$
$$C(u) = 3F(u)G'(u)/G^4(u) - F(u)G''(u)/G^3(u) - F'(u)G'(u)/G^3(u)$$

where the prime denotes differentiation with respect to u. In Equation 27 we now define a new space variable $\rho = \xi/\sqrt{\epsilon}$ and introduce an expansion for ψ in powers of $\sqrt{\epsilon}$

$$\psi = \psi_0 + \varepsilon^{1/2}\psi_1 + \varepsilon\psi_2 + \varepsilon^{3/2}\psi_3 + \ldots \tag{29}$$

in which we will assume that $\psi_n(\rho,0) = 0$ for $n \neq 0$—that is, the initial value for ψ_0 is the same as that for ψ. When we make the indicated substitutions and collect the terms, multiplying successive powers of $\epsilon^{1/2}$, we find:

$$\varepsilon^0: \quad \frac{\partial\psi_0}{\partial\tau} = A(\tau)\frac{\partial^2\psi_0}{\partial\rho^2}$$
$$\varepsilon^{1/2}: \quad \frac{\partial\psi_1}{\partial\tau} - A(\tau)\frac{\partial^2\psi_1}{\partial\rho^2} = \rho A'(\tau)\frac{\partial^2\psi_0}{\partial\rho^2} + B(\tau)\frac{\partial\psi_0}{\partial\rho}$$
$$\varepsilon: \quad \frac{\partial\psi_2}{\partial\tau} - A(\tau)\frac{\partial^2\psi_2}{\partial\rho^2} = \frac{1}{2}\rho^2 A''(\tau)\frac{\partial^2\psi_0}{\partial\rho^2} + \rho B'(\tau)\frac{\partial\psi_0}{\partial\rho} \tag{30}$$
$$+ c(\tau)\psi_0 + \rho A'(\tau)\frac{\partial^2\psi_1}{\partial\rho^2} + B(\tau)\frac{\partial\psi_1}{\partial\rho}$$

and so forth.

The successive equations in Equation 30 can be represented in the general form

$$\frac{\partial\psi_0}{\partial\tau} = A(\tau)\frac{\partial^2\psi_0}{\partial\rho^2}$$
$$\frac{\partial\psi_n}{\partial\tau} - A(\tau)\frac{\partial^2\psi_n}{\partial\rho^2} = U_n(\rho, \tau) \tag{31}$$

where the function $U_n(\rho,\tau)$ is calculated in terms of $\psi_0, \psi_1, \ldots, \psi_{n-1}$, and can be considered known at any stage of the calculation. Equation 31 can be solved by means of a Fourier transformation, the final result being

$$\psi_n(\rho_1\tau) = \frac{1}{\sqrt{4\pi}}\int_0^\tau \frac{d\tau'}{\sqrt{\Delta(\tau) - \Delta(\tau')}}\int_{-\infty}^{\infty} U_n(\rho', \tau') \times$$
$$\exp\left\{-\frac{(\rho - \rho')^2}{4\,[\Delta(\tau) - \Delta(\tau')]}\right\} d\rho' \tag{32}$$

For an initial pulse distribution we have, from Equation 12 the result

$$\psi_0(\rho_1\tau) = \frac{1}{\sqrt{4\pi\varepsilon\Delta(\tau)}} \exp\left(-\frac{\rho^2}{4\Delta(\tau)}\right) \tag{33}$$

If we insert this into the equation for ψ_1 that appears in Equation 30, we find, after some lengthy calculation, that

$$\varepsilon^{1/2}\psi_1 = \exp\left(-\frac{\rho^2}{4\Delta(\tau)}\right)\left\{\frac{3\rho}{(16\pi)^{1/2}\Delta^{5/2}(\tau)}\int_0^\tau A'(\tau')[\Delta(\tau) - \Delta(\tau')]^2 d\tau' + \frac{\rho^3}{(64\pi)^{1/2}\Delta^{7/2}(\tau)}\int_0^\tau A'(\tau')\Delta(\tau')d\tau' - \frac{\rho}{(16\pi)^{1/2}\Delta^{3/2}(\tau)} \times \times \int_0^\tau [A'(\tau') + B(\tau')]d\tau' \right. \tag{34}$$

showing no correction at the peak maximum, but there is some shift in the tails of the concentration profile. Detailed calculations can also be made for the initial condition $c(x,0) = H(x - 1)$ where $H(u) = 0$, $u < 0$, $H(u) = 1$, $u > 0$. This case is appropriate to the problem of ultracentrifugation and leads to a result different from that given in Equation 34. Further details of this correction procedure will be given in a forthcoming publication, but for most applications at the present stage of separation technology the simple approximation given by ψ_0 should suffice for whatever information is required.

It would be of considerable interest to extend the technique just presented to problems involving nonlinear equations because there are many situations in ultracentrifugation where nonideality is a dominant feature. Furthermore, it is known (*4, 14*) that even for two-component systems with nonideality the theory for estimating the sedimentation constant based on a diffusion-free ($\epsilon = 0$) approximation can lead to systematic error. Therefore, the development of an approximate procedure for nonlinear equations would be useful for further progress in analytical separation methods.

Literature Cited

1. Mosimann, H., Signer, R., *Helv. Chim. Acta* (1944) **27**, 1123.
2. Fujita, H., *J. Amer. Chem. Soc.* (1956) **78**, 3598.
3. Dishon, M., Weiss, G. H., Yphantis, D. A., *J. Polym. Sci.* A-2 (1970) **8**, 2163.
4. Dishon, M., Stroot, M. T., Weiss, G. H., Yphantis, D. A., *J. Polym. Sci.* A-2 (1971) **9**, 939.
5. Rodbard, D., Chrambach, A., *Proc. Nat. Acad. Sci.* (1970) **65**, 970.
6. Rodbard, D., Kapadia, G., Chrambach, A., *Anal. Biochim.* (1971) **40**, 135.
7. Ackers, G. K., *Advan. Protein Chem.* (1969) **24**, 343.

8. Halvorson, H. R., Ackers, G. K., *J. Polym. Sci. A-2* (1971) **9**, 245.
9. Weiss, G. H., Rodbard, D., *Sep. Sci.* (1972) **7**, 217.
10. Fujita, H., "Mathematical Theory of Sedimentation Analysis," p. 131, Academic, New York, 1962.
11. Weiss, G. H., Dishon, M., *Biopolymers* (1970) **9**, 865.
12. Billick, I. H., *J. Phys. Chem.* (1962) **66**, 565, 1941.
13. Morse, P. M., Feshbach, H., "Methods of Theoretical Physics," p. 857, McGraw-Hill, New York.
14. Dishon, M., Weiss, G. H., Yphantis, D. A., *Biopolymers* (1967) **5**, 691.
15. Cole, J. D., "Perturbation Methods in Applied Mathematics," Ginn, Blaisdell, San Francisco, 1968.

RECEIVED January 17, 1972.

20

A New Way of Determining Molecular Weight Distribution, Including Low Molecular Weight, from Equilibrium Sedimentation

MATATIAHU GEHATIA and D. R. WIFF

Air Force Materials Laboratory, Wright-Patterson Air Force Base, Ohio 45433
University of Dayton Research Institute, Dayton, Ohio 45424

Formulas leading to determination of molecular weight distribution (MWD) from equilibrium sedimentation have been subjected to mathematical analysis. It has been shown that these formulas are a particular case of the Fredholm integral equation of the first kind, which is an improperly posed problem in the Hadamard sense. To cope with this problem the Tikhonov method of regularization has been implemented, and new computation-oriented equations have been derived. Mathematical analysis shows that very precise results can be obtained in case of a narrow molecular weight range. Therefore, the method suggested can be successfully used to determine MWD of a low molecular weight polymer. For a wide MWD or if the distribution is multimodal and very assymmetrical, this method becomes more cumbersome.

Molecular weight determination of a monodisperse macromolecular system from equilibrium sedimentation was devised by Svedberg and Fahraeus in 1925 (*1*). They applied the following formula

$$M = \frac{2RT}{(1 - V\rho)\omega^2} \frac{d \ln c}{d\, r^2} \quad (1)$$

Here M is molecular weight, R is the gas constant, T is the absolute temperature in degrees Kelvin, V is the partial specific volume of the solute, ρ is the density of solution, ω is the angular velocity, r is the distance from the center of rotation, and c is the concentration measured at r. Under ideal conditions Equation 1 can be easily integrated. The integral form will be presented in the following by adopting the Fujita formalism which won a wide acceptance (*2*)

$$\frac{c}{c_0} = \frac{\lambda M}{1 - e^{-\lambda M}} e^{-\lambda M \xi} \tag{2}$$

where c_0 is the initial concentration, $\lambda = (1 - V\rho)\omega^2(r_b^2 - r_a^2)/2RT$, $\xi = (r_b^2 - r^2)/(r_b^2 - r_a^2)$, r_a and r_b are the values of r at meniscus and bottom, respectively.

Almost simultaneously with the first attempt to determine molecular weight from equilibrium sedimentation, Rinde tried to widen this method to include determination of the molecular weight distribution (MWD) of a polydisperse system (*3*). Unfortunately, this attempt proved to be more complicated and did not result in establishment of a reliable routine. Since the appearance of Rinde's dissertation in 1928, many investigators have tried to determine MWD. Most of these efforts, however, did not provide a successful comprehensive technique (*4–16*). This objective has been accomplished only in a few cases under very limited conditions, such as in case of a Gaussian or near Gaussian MWD, in which only characterizing parameters had to be determined. Scholte (*17, 18*) determined MWD by performing an experimental procedure based on several equilibrium experiments.

If it is assumed that the lack of success in developing a comprehensive method leading to MWD determination (a single experiment at a single rotor velocity) stems from mathematical difficulties, the expressions relating MWD to concentrations or concentration gradients must be more closely analyzed. There are methods other than the classical equilibrium sedimentation which can be used to obtain a molecular weight distribution. Two such methods are "equilibrium density gradient" and "sedimentation velocity." The former method has been developed for aqueous solutions where the density gradient is achieved primarily by use of cesium chloride. In the study of synthetic polymers, which dissolve only in a limited number of organic solvents, the experimental conditions required by this method have not been sufficiently developed for all cases. The latter method, that of sedimentation velocity, could be applied to fractionated synthetic polymers for molecular weight determination. However, the determination of MWD by applying this method may be complicated by various factors. For example, if the molecular weight is not high enough ($M > 100{,}000$), the correction for diffusion becomes cumbersome. The MWD of spherical molecules large enough to neglect diffusion was investigated by Gralen and Lagermalm (*19*). Methods to correct the resulting curves by introducing the contribution of diffusion

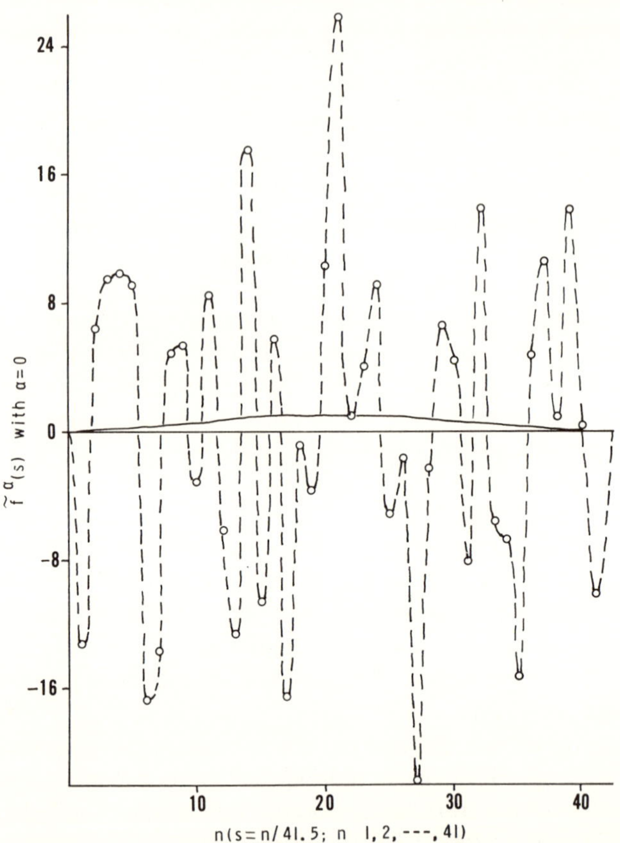

Figure 1. MWD calculated from Equation 10. The solid line represents the originally assumed distribution f(M), and the dashed line represents the curve $\overline{f(M)}$ resulting from the inverse operation.

are discussed by Hengstenberg (*20*). According to this analysis such corrections may be complicated. However, when the material under investigation is of sufficiently high molecular weight, the influence of diffusion can be neglected, and the distribution of the sedimentation constant of the unfractionated material may be directly obtained (*21*). In any case, to infer MWD from velocity sedimentation experiments, auxiliary measurements must be made to correlate the molecular weight of each fraction with its appropriate sedimentation constant. Since each method has its own limitations, this paper deals only with the technique of obtaining a MWD from an equilibrium sedimentation experiment performed at a single angular velocity.

The Mathematical Analysis of Formulas Relating MWD to Concentrations or Concentration Gradients

MWD can be expressed by several equivalent formulas derived from the theory of equilibrium sedimentation at ideal conditions. In the following the well-known Fujita formalism (*see* Equation 2) will be used. This expression related the density of MWD, f(M), to concentrations or concentration gradients, respectively:

$$\frac{c(\xi)}{c_0} = \int_0^\infty \left[\frac{\lambda M}{1 - e^{-\lambda M}} e^{-\lambda M \xi} \right] \mathrm{f}(M) dM, \tag{3}$$

$$- \frac{dc(\xi)}{c_0 d\xi} = \int_0^\infty \left[\frac{\lambda^2 M^2}{1 - e^{-\lambda M}} e^{-\lambda M \xi} \right] \mathrm{f}(M) dM \tag{4}$$

The left sides of Equations 3 and 4 depending only on ξ will be denoted as u(ξ). The expressions in brackets under the integral sign which are known for every ξ and M will be called kernels and denoted as K(ξ, M). In this way Equations 3 and 4 reduce to a single formula:

$$\mathrm{u}(\xi) = \int_0^\infty \mathrm{K}(\xi, M)\mathrm{f}(M) dM = \int_0^{M_{\max}} \mathrm{K}(\xi, M)\mathrm{f}(\mathrm{M}) dM \tag{5}$$

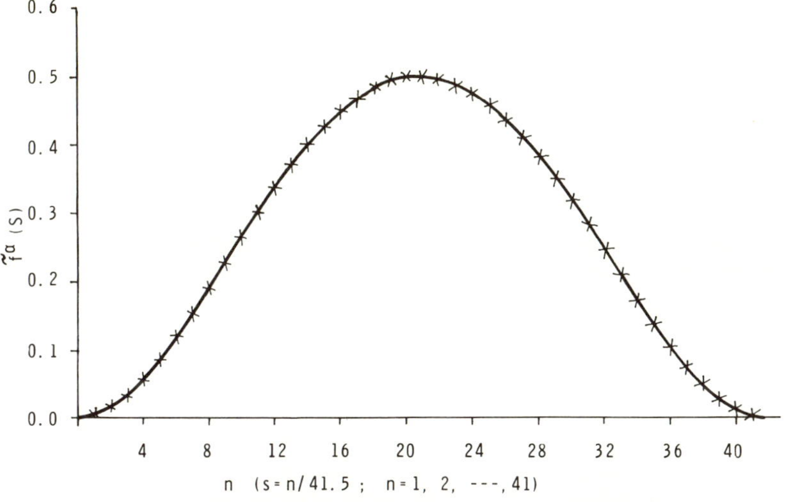

Figure 2. Unimodal MWD using regularization without linear programming. The solid line is the originally assumed distribution, and the crosses are the computed distribution values.

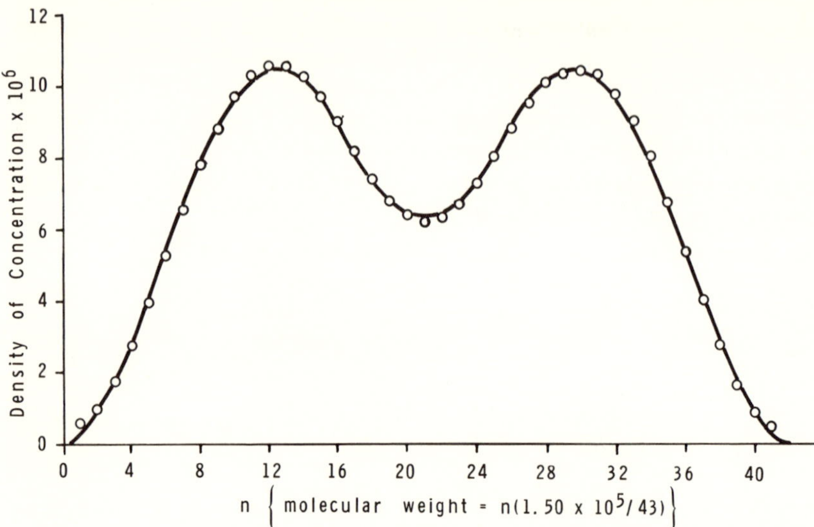

Figure 3. Symmetrical bimodal MWD using regularization with linear programming. The solid line represents the initial distribution, and the circles represent the computed distribution.

which is the well-known Fredholm integral equation of the first kind. The upper integration limit (∞) has been replaced by M_{max} since mathematically it becomes important later, and beyond M_{max} $f(M) = 0$.

Since $K(\xi, M)$ is always known, only two types of different computations may result from Equation 5: computation of $u(\xi)$ if $f(M)$ is known, defined as the "straight operation," and computation of $f(M)$ if $u(\xi)$ is known, defined as the "inverse operation." While the straight operation is a simple integration always leading to reliable results, the inverse operation is a more complicated procedure, and its results are less reliable. It has been shown that the inverse operator of Equation 5 is unstable. A small change in $u(\xi)$ may cause a very large change and sometimes even uncontrollable oscillations of $f(M)$ (22). Hadamard defined this type of mathematical operation an "improperly posed problem" (IPP) and excluded it from mathematical applications (23). Unfortunately, Equations 3 and 4 are mathematically IPP's, and therefore, their direct application did not result in a comprehensive and reliable method to determine MWD.

To demonstrate the instability of Equations 3 and 4, a symmetrical unimodal distribution $f(M)$ was arbitrarily assumed, and the corresponding $u(\xi) = c/c_0$ was determined by applying a straight operation on Equation 3. This newly obtained $u(\xi)$ function was now used as an input to evaluate back $f(M)$, by applying an inverse operation on the same Equation 3. The parameter λ was taken from a real experiment conducted in the laboratory. The inverse operation was accomplished with the aid of variational calculus which minimizes the following functional:

$$N[\overline{f(M)}, u(\xi)] = \int_0^1 \Big[\int_0^{M_{max}} K(\xi, M)\overline{f(M)}dM - u(\xi) \Big]^2 d\xi \tag{6}$$

By applying the Euler equation, this expression becomes:

$$b(M) = \int_0^{M_{max}} F(M, x)f(x)dx \tag{7}$$

where

$$F(M, x) = \int_0^1 K(\xi, M)K(\xi, x)d\xi \tag{7a}$$

and

$$b(M) = \int_0^1 K(\xi, M)u(\xi)d\xi \tag{7b}$$

Here x and M denote molecular weight. x denotes a molecular weight variable within a single equation, while M is a fixed molecular weight for this equation. By making M, x, and $f(M)$ discrete, and by introducing a constant interval $h = x_i - x_{i-1} = \Delta M$, Equation 7 becomes:

$$b(M_j) = \sum_{i=l}^{I} \sigma_i F(M_j, x_i) f_i h \tag{8}$$

Here the σ_i's are the Simpson integration coefficients. Equation 8 represents a set of linear equations:

$$\mathbf{b} = A\mathbf{f} \tag{9}$$

Here $\mathbf{b} = (b(M_1), b(M_2), \ldots b(M_I))$ and $\mathbf{f} = (f_1, f_2, \ldots f_I)$ are vectors and $A = \{\alpha_{ji}\}$ is a matrix, where $a_{ji} = \sigma_i F(M_j, x_i)h$. Thus the determination of MWD reduces to computation of the vector $\mathbf{f}$ with the aid of the inverse matrix $Q = A^{-1}$:

$$Q\mathbf{b} = \mathbf{f} \tag{10}$$

This procedure was applied in conjunction with the above-mentioned unimodal distribution. The computation resulted in a "noisy" $f(M)$ curve which had no resemblance to the originally assumed f(M) curve (Figure 1).

Approximate Solution of Certain IPP's

Since many vital relationships in science and technology are IPP's, there was a need for obtaining a stable and reliable solution. Tikhonov (*22, 24–39*) and Phillips (*40*) suggested a method leading to an approximate solution for this general type of problem. They introduced a regularizing function, which if added to the functional expressed by Equation 6 causes damping of oscillations induced by an unstable inverse operator. Functions which can be computed this way are called regularizable functions. Since f(M) expressed by Equations 3 and 4 proved to be a regularizable function at least in the most typical cases, the Tikhonov–Phillips method was applied to transform these theory-oriented equations (*see* Equations 3 and 4) into new computation-oriented expressions. This task can be accomplished by substituting a new functional for the functional given in Equation 6.

$$M^{\alpha}[\mathrm{f}(M);\, \mathrm{u}(\xi)] = N[\mathrm{f}(\mathrm{M});\, \mathrm{u}(\xi)] + \alpha\,\Omega\,[\mathrm{f}(M)] \qquad (11)$$

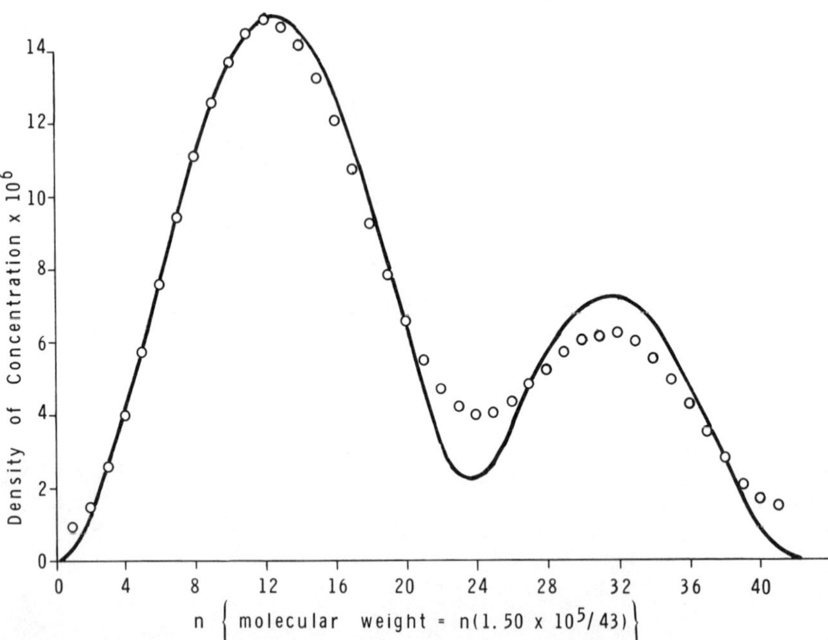

Figure 4. Asymmetrical bimodal MWD using regularization with linear programming. The solid line represents the initial distribution, and the circles represent the computed distribution.

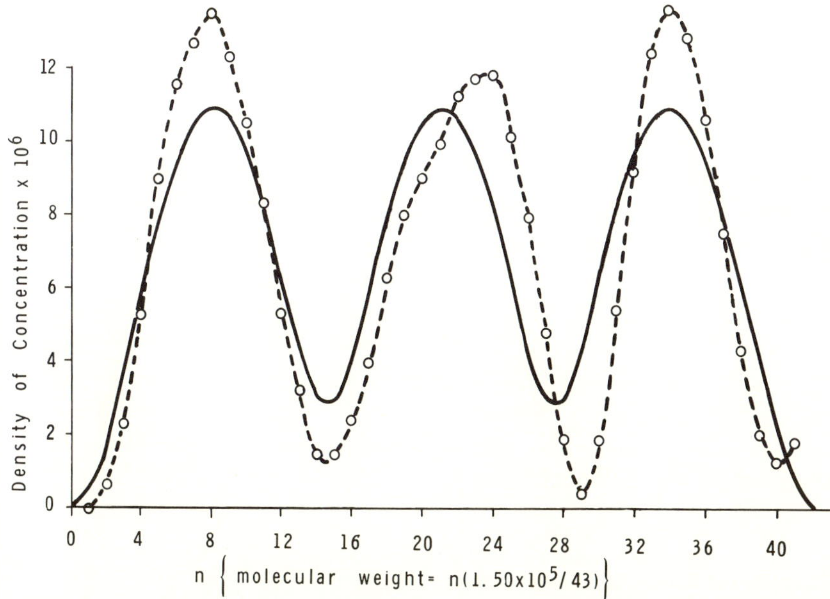

Figure 5. Symmetrical trimodal MWD using regularization with linear programming. The solid line represents the initial distribution, and the circles (dashed curve) represent computed distribution.

where

$$\alpha\,\Omega = \int_0^{M_{\max}} \left\{ \sum_{i=0}^{n+1} \alpha_n \left(\frac{d^i \mathrm{f}}{dM^i} \right)^2 \right\} dM \tag{12}$$

α_n's are adjustable numerical parameters. The expression defined by Equation 12 is the regularizing function.

Equation 11 can be transformed by applying variational calculus analogously to the transformation of Equation 6. This procedure results in the following expression:

$$\mathrm{b}(M_j) = \sum_{i=1}^{I} \sigma_i F(M_j, x_i) \mathrm{f}_i h - R_j \tag{13}$$

where

$$R_j = \sum_{n=1}^{N} (-1)^{n+1} \alpha_n \sum_{k=1}^{2n+1} (-1)^{k-1} \binom{2n}{k-1} h^{-2n} \mathrm{f}_{j+n-k+1} \tag{14}$$

The following boundary conditions were applied: for $i < 1$,

$$f_{-i} = f_{i+1} \tag{15a}$$

and for $i > I$,

$$f_{I+i} = f_{I-i+1} \tag{15b}$$

First, one of the terms composing the sum given in Equation 14 was chosen, and then the numerical values of the corresponding α_n parameters were arbitrarily assumed. Under such conditions a set of linear equations defined by Equation 13 leads to determination of a corresponding vector $\overline{\mathbf{f}}_{\{\alpha_1, \alpha_2, \ldots \alpha_n\}} = \overline{\mathbf{f}}_{\{\alpha_n\}}$. This in turn leads to the determination of the corresponding $\overline{\mathbf{u}}_{\{\alpha_n\}}(\xi)$ by implementing Equation 3 or 4. In general the function $\overline{\mathbf{u}}_{\{\alpha_n\}}$ determined in this way is different from the function $\mathbf{u}(\xi)$ provided by the experiment. Such a difference can be expressed by the following norm:

$$||\overline{\mathbf{u}} - \mathbf{u}||_{\{\alpha n\}} = \left\{ \int_0^1 [\overline{\mathbf{u}}_{\{\alpha n\}}(\xi) - \mathbf{u}(\xi)]^2 \right\}^{1/2} d\xi \tag{16}$$

By applying a high-speed digital computer, a search for different sets $\{\alpha_n\}$ of parameters α_n can be conducted to minimize the norm $||\overline{\mathbf{u}} - \mathbf{u}||_{\{\alpha_n\}}$. Finally, a special $\overline{\mathbf{f}}_{\{\alpha_n\}}$ will be obtained which corresponds to the $\inf\{\min||\overline{\mathbf{u}} - \mathbf{u}||_{\{\alpha_n\}}\}$. It is obvious that this special value is the closest approximation of the "true" distribution $f(M)$.

The method of regularization gave excellent results in the case of a unimodal distribution (*41*). A symmetrical arbitrarily assumed distribution, which previously had led to a "noisy" curve if no regularization was applied, gave very good agreement if determined by implementing Equation 13 (*see* Figure 2). In the case of a bimodal asymmetrical distribution, the use of regularization proved not to be adequate and this method had to be incorporated into linear programming.

Regularization and Linear Programming

Even though regularization smooths the "high frequency" noise, a certain amount of "low frequency" noise still exists in the $\overline{\mathbf{f}}_{\{\alpha_n\}}$. This is especially noticeable when the true distribution $f(M)$ has more structure and/or a narrow peak somewhere in the distribution. In these cases negative values of $\overline{\mathbf{f}}_{\{\alpha_n\}}$ will exist, and such a region is commonly referred to as an "overshoot." From purely physical reasoning the negative portion of $\overline{\mathbf{f}}_{\{\alpha_n\}}$ is meaningless. A technique which would automatically admit only positive values for $\overline{\mathbf{f}}_{\{\alpha_n\}}$ would be helpful. One such technique

Table I. Parameters Characterizing the Conducted Experiments

Sample	Concentration, grams of polymer/100 grams of Solution	Relative % Concentration	Rotor Speed, rpm: Velocity Sed	Rotor Speed, rpm: Equil Sed	Equil Sedimentation: Starting Velocity, rpm	Equil Sedimentation: Duration, hr
A	0.496	56.9	40,000	16,000	16,000	48
B	0.434	34.1	40,000	10,000	10,000	74
C	0.528	9.0	40,000	4,800	5,000	72
P	0.449	—	—	10,000	10,000	104

[a] Sum of concentrations = 100.0.

is linear programming. Scholte (*17, 18*) has applied linear programming in his method, but when we tried to apply the same for low molecular weights ($0 \leqslant M \leqslant 150{,}000$) erratic fluctuations resulted for $M < 10{,}000$. If, however, linear programming is incorporated into the regularization scheme discussed in the previous section, the erratic fluctuations at low molecular weights no longer exist (*42*). The regularization part ensures an analytical molecular weight distribution, and the linear programming eliminates the negative values of the distribution and simultaneously contributes further smoothening. For specific details regarding linear programming the reader is referred to the books by Cass (*43*) or Hadley (*44*). An excellent review article by Rabinowitz (*45*) may also be of considerable help.

In the present case two algorithms for the best linear approximation on a discrete set were considered, one dealing with the L_1 norm and the other with the L_∞ norm (or Chebyshev approximation). The L_1 norm approximation consistently gave better results in comparison with the assumed initial molecular weight distributions. Therefore, the application of the L_∞ approximation was stopped, and all that follows relates to the L_1 approximation. In matrix notation Equation 13 may be expressed as:

$$\mathbf{Q}^\alpha \mathbf{f}^\alpha = \mathbf{b} \tag{17}$$

Here the superscript α is retained to emphasize that for each value of the parameter α there exists a corresponding set of equations. If one admits that error exists in $\mathbf{b}(M)$ (*see* Equation 7a) when using the experimental values $u(\xi_i)$, then

$$\mathbf{Q}^\alpha \mathbf{f}^\alpha = \mathbf{b} + \varepsilon \tag{18}$$

where ϵ denotes both positive and negative error. Letting

$$\varepsilon_i = \delta_i - \beta_i \tag{19}$$

where $\delta_i \geqslant 0$ and $\beta_i \geqslant 0$, the linear programming formulation with the L_1 approximation and regularization follows.

The objective function which is to be minimized is

$$\sum_{i=J+1}^{3J} \mathrm{f}_j^{\alpha} \tag{20}$$

$$\begin{bmatrix} q^{\alpha}{}_{11} & q^{\alpha}{}_{12} & \cdots & q^{\alpha}{}_{1J} & 1 & -1 & 0 & 0 & \cdots & 0 & 0 \\ q^{\alpha}{}_{21} & q^{\alpha}{}_{22} & \cdots & q^{\alpha}{}_{2J} & 0 & 0 & 1 & -1 & \cdots & 0 & 0 \\ \cdot & \cdot & & \cdot & \cdot & \cdot & \cdot & \cdot & & \cdot & \cdot \\ \cdot & \cdot & & \cdot & \cdot & \cdot & \cdot & \cdot & & \cdot & \cdot \\ \cdot & \cdot & & \cdot & \cdot & \cdot & \cdot & \cdot & & \cdot & \cdot \\ q^{\alpha}{}_{J1} & q^{\alpha}{}_{J2} & \cdots & q^{\alpha}{}_{JJ} & 0 & 0 & 0 & 0 & \cdots & 1 & -1 \end{bmatrix} \begin{bmatrix} \mathbf{f}_1^{\alpha} \\ \mathbf{f}_2^{\alpha} \\ \cdot \\ \cdot \\ \cdot \\ \mathbf{f}^{\alpha}{}_J \\ \mathbf{f}^{\alpha}{}_{J+1} \\ \cdot \\ \cdot \\ \cdot \\ \cdot \\ \mathbf{f}^{\alpha}{}_{3J} \end{bmatrix} = \begin{bmatrix} b_1 \\ b_2 \\ \cdot \\ \cdot \\ \cdot \\ b_J \end{bmatrix} \tag{20a}$$

subject to the constraints where $\mathbf{f}_j^{\alpha} \geqslant 0$ and $j = 1, 2, \ldots, 3J$. The quantities $\mathbf{f}_j^{\alpha}$ for $J + 1 \leqslant j \leqslant 3J$ are called "slack variables" and correspond to the δ's and β's in Equation 19. For each value of the parameter α, there will correspond a set $\{\mathbf{f}_j^{\alpha}\}$. The "best fit" criterion used was $\inf\{\min||\bar{\mathbf{u}}_i - \mathbf{u}_i||_{\{\alpha_n\}}\}$, where $||\bar{\mathbf{u}}_i - \mathbf{u}_i||_{\{\alpha_n\}}$ has been defined by Equation 16. In this case excellent results were obtained for a unimodal and a symmetrical bimodal MWD (*see* Figure 3). Good results were obtained when an asymmetrical bimodal and a trimodal distribution were mathematically assumed as the initial distributions (Figures 4 and 5).

Up to this point we have considered only those evaluations where the following procedure was invoked. (1) An initial mathematical functional form distribution $\mathrm{f}(M)$ was assumed. (2) Analogous experimental data $\mathbf{u}(\xi)$ were numerically computed. (3) A distribution $\overline{\mathrm{f}(M)}$ was inferred using Equations 20 and 21. (4) Finally, the computed $\bar{\mathbf{u}}(\xi)$ and the observed $\mathbf{u}(\xi)$ were compared. In what follows a true experimental justification of the validity of this technique is discussed.

Verification of the Newly Developed Method

To substantiate the credibility of the newly developed method, a polydisperse system with $\mathrm{f}(M)$ known in advance was subjected to an equilibrium sedimentation experiment. Then the new computation-oriented equations were implemented to infer $\overline{f(M)}$ from the experimental data. Finally, this resulting $\overline{f(M)}$ was compared with the *a priori* f(M).

To perform this verification three narrow fractions of linear polystyrene denoted as A, B, and C were investigated to determine their molecular weights. This task was accomplished by applying velocity as well as equilibrium sedimentation. In addition, these three individual fractions were combined according to a known weight ratio into a new polydisperse sample denoted as P. Sample P was also subjected to equilibrium sedimentation. All samples were investigated in cyclohexane at 35°C (near the Θ temperature); *see* Table I.

Table II. Tabulation of Parameters Determined

Sample	*Velocity Sedimentation*			*Equilibrium Sedimentation*			*Viscosity*
	s *Sed.*	D × *10*[1], *cm*[2]*/sec*	*Mol Wt*	M "*mono-disperse*" [a]	M_w [b]	M_w [c]	M [d]
A	1.33	17.50	6,300	5,600	5,700	e	5,270
B	2.73	5.50	41,100	36,700	38,000	42,400	29,100
C	4.97	3.16	130,700	146,500	148,000	156,100	114,200

[a] By Equation 22.
[b] By the new method for individual samples.
[c] By the new method from the P curve.
[d] Estimated by the Mellon Institute.
[e] Molecular weight intervals (about 5200) were too large for obtaining reliable results.

Sedimentation velocity experiments were performed using a 12-mm Kel-F double-sector capillary synthetic boundary cell. Each equilibrium experiment was performed with aid of two separate 12-mm, 4° single-sector aluminum cells, one containing solvent and the other solution. In all cases the schlieren angle was 65°.

Processing Velocity Sedimentation Data

The sedimentation constant s has been determined by applying the moving boundary method (*46*)

$$\log (r_{\max}) = \text{constant } \lambda + 0.4343\omega^2 s t_{\exp} \qquad (22)$$

Here $r_{\max}$ is the radial distance to the moving boundary and $t_{\exp}$ is an experimental time relative to some time t_0. The diffusion constant D was determined from knowledge of the highest $(\partial c/\partial r)$ coordinate measured at different times and by applying the following relationship:

$$\left(\frac{\partial c}{\partial r}\right)_{\max} = H = \frac{c_0 e^{-2\beta t}}{\sqrt{4\pi D \bar{\alpha} t}} \tag{23}$$

where $\beta = s\omega^2$ and $\bar{\alpha} = (1 - e^{2\beta t})/2\beta t$. Since

$$\bar{\alpha} = 1 + \beta t + 2/3\beta^2 t^2 + \ldots \approx e^{\beta t} \tag{24}$$

one can derive the following approximate expression:

$$H^2 \approx c_0^2 e^{-5\beta t}/4\pi D t \tag{25}$$

Therefore,

$$\left(\frac{c_0}{H}\right)^2 \approx 4\pi D t + \ldots \tag{26}$$

and

$$\left(\frac{H}{c_0}\right)^2 \approx \frac{\theta}{4\pi D} + \ldots \tag{27}$$

where

$$\theta = \frac{1}{t} e^{-5\beta t}$$

There is no way to measure the "absolute" time appearing in Equations 26 and 27. The most feasible way to circumvent this problem is to introduce an elapsed time t_0 and an experimental time t^*, *i.e.*,

$$t = t_0 + t^* \tag{28}$$

By plotting $(c_0/H)^2$ *vs.* t^* and extrapolating $(c_0/H)^2$ to zero, one can obtain an approximate value of t_0 which makes the time correction possible. In addition, since the diffusion constant, D, determined from Equation 26 is not always reliable, the connected time can be substituted into Equation 27 resulting in a better determination of D from a plot of $(H/c_0)^2$ *vs.* θ.

Finally the molecular weight can be determined from the well-known Svedberg equation (*1*)

$$M = \frac{RT}{1 - V\rho} \frac{s}{D} \tag{29}$$

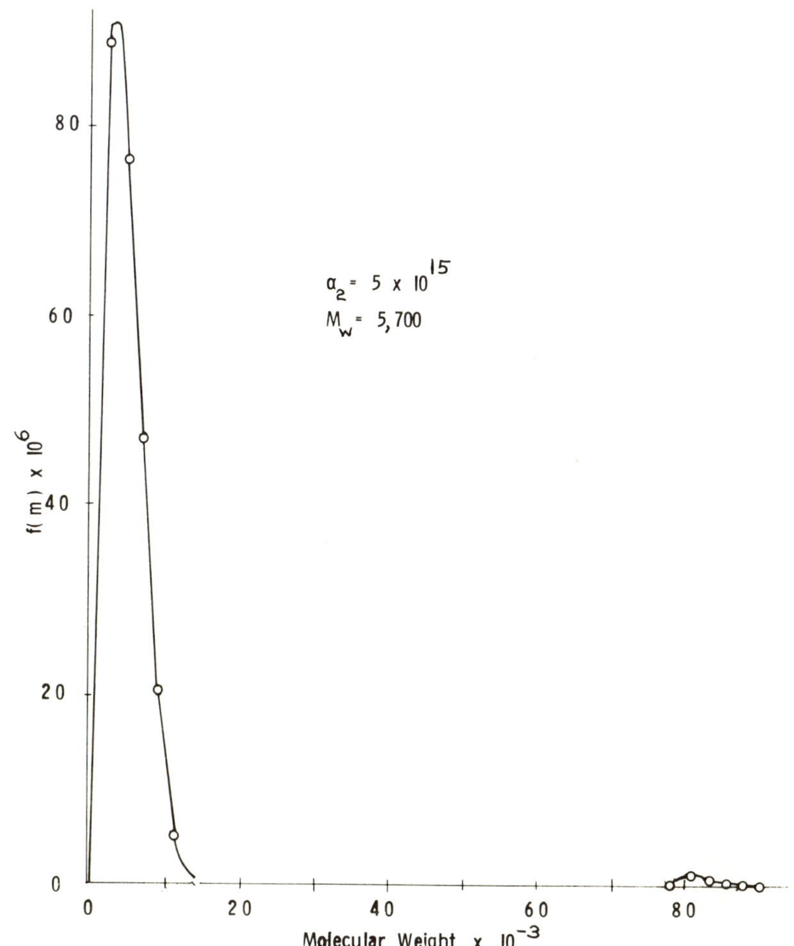

Figure 6. MWD for sample A with $\Delta M = 2381$ *and* $\int f(M)dM = 0.569$

Processing Equilibrium Sedimentation Data

For a monodisperse fraction one can derive from Equation 4 the following formula

$$\log\left(-\frac{dc(\xi)}{d\xi}\right) = \text{constant} + 0.4343\lambda M\xi \tag{30}$$

which leads to the determination of molecular weight or an average molecular weight in case of a polydisperse system. In addition, experimental results for fractions A, B, and C were also analyzed by the newly suggested method. Thus the molecular weight distribution of each fraction has been individually determined, and from these distributions the weight averages, $<M_w>$, were mathematically evaluated. Finally, these distributions were superimposed so as to compare the resulting distribution with the known $f(M)$ function.

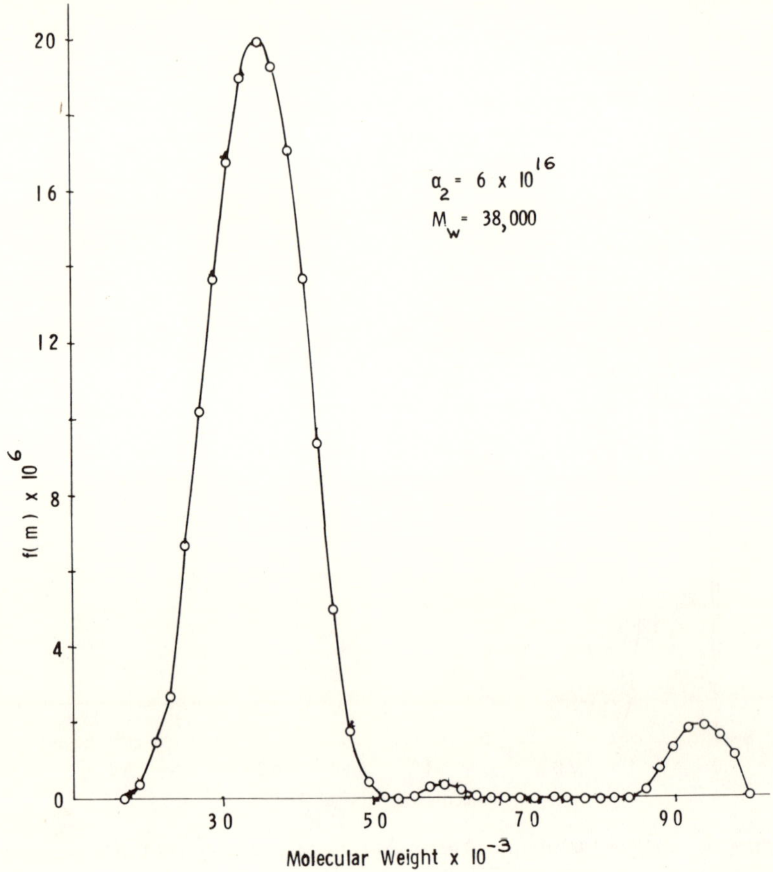

Figure 7. MWD for sample B with $\Delta M = 2024$ *and* $\int f(M)dM = 0.341$

The equilibrium sedimentation experiment of sample P was analyzed only by applying the computation-oriented expressions. The resulting $\overline{f(M)}$ was compared with the previously obtained $f(M)$. From the curve of $f(M)$ the average $<M_w>$ values were computed and compared with the $<M_w>$ values computed in the same way from individual distribution curves of fractions A, B, and C. These $<M_w>$ values were also compared with M values obtained in a conventional way from equilibrium sedimentation by applying Equation 2, and with M values determined from velocity sedimentation experiment, Equation 30 (*see* Table II).

Results

In all calculations the buoyancy factor $(1 - V\rho)$ at 35°C was taken as 0.31. The computation-oriented expression Equation 13 was applied in order to obtain the results shown in Figures 6–9 for sample A, with $0 \leqslant M \leqslant 100{,}000$ and $\alpha_2 = 6 \times 10^{17}$ (Figure 6); sample B, with $15{,}000 \leqslant M \leqslant 100{,}000$ and $\alpha_2 = 2 \times 10^{16}$ (Figure 7); sample C, with $100{,}000 \leqslant M \leqslant 180{,}000$ and $\alpha_2 = 2 \times 10^{16}$ (Figure 8); and sample P, with $0 \leqslant M \leqslant 180{,}000$ and $\alpha_2 = 1.5 \times 10^{16}$ (Figure 9). The integrals $\int f(M)dM$ of individual samples A, B, and C were adjusted so as to correspond with the relative concentrations, characterizing the composition of the polydisperse sample P (*see* Table I). It is obvious that for sample P $\int f(M)dM = 1$. The weight-average molecular weight was computed by $\langle M_w \rangle = \int Mf(M)dM \int f(M)dM$. In all cases a 41-

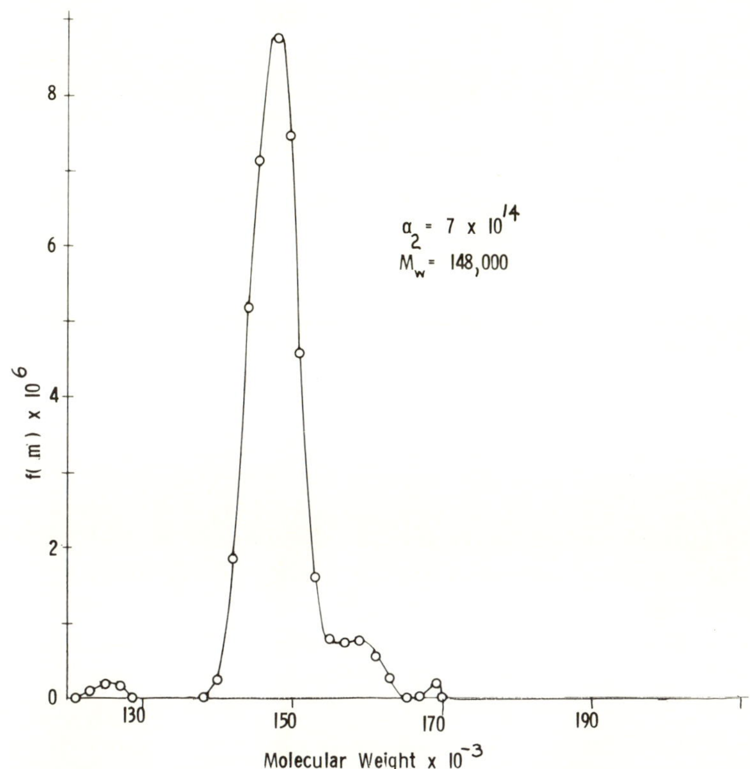

Figure 8. MWD for sample C with $\Delta M = 1910$ *and* $\int f(M)dM = 0.090$

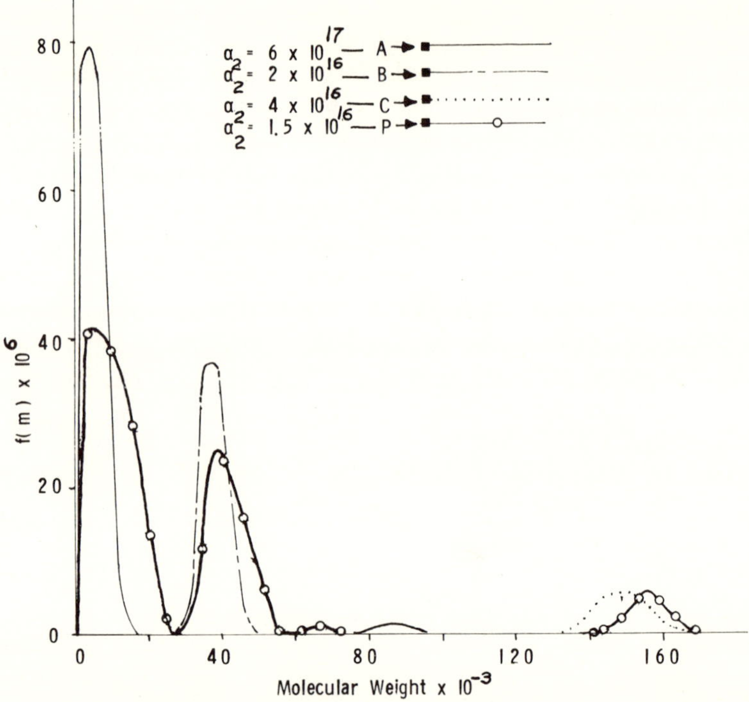

Figure 9. MWD for sample P with ΔM = 5120 and $\int f(M)dM =$ 1.000. The solid thin line is the distribution of sample A, the dashed line represents sample B, and the dotted line represents sample C.

point mesh was applied. It limited the precision of the resulting function f(M) which was computed in steps corresponding to the intervals ΔM equal to 2381, 2024, 1910, and 5120, respectively. The parameters determined in all the caluculations are tabulated in Table II.

Conclusion

In general, the results obtained by the different methods discussed above are in good agrrement thereby proving the applicability of the proposed computation-oriented equation for MWD determination. The precision of the results obtained by this method depends, however, on the molecular weight range, $M_{max} - M_{min}$, where $M_{min} \leqslant M \leqslant M_{max}$. Every discrete step in determining the corresponding f(M) value depends upon the interval

$$\Delta M = (M_{max} - M_{min})/N \tag{31}$$

where N is the order of the matrix (or mesh) applied.

If molecular weight range is very wide, one is forced by the size of the matrices, *i.e.*, the allowable digital computer storage available, to settle for less resolution by having a larger molecular weight interval. On the other hand, the narrower the range the smaller is the interval ΔM and the better the computation precision. One obvious advantage to this method is that it seems especially precise for low molecular weights where the range is $M_{max} \geqslant M \leqslant 0 = M_{min}$, and M_{max} is not a large number.

Another limitation of this method is that it tends to present the most compact distribution. That is, the structure of a multimodal curve is suppressed unless one is willing to invest the time to tediously parametrize for a better "fit" to the experimental data. The eventual limiting factor will be the experimental error itself.

Literature Cited

1. Svedberg, T., Fahraeus, R., *J. Amer. Chem. Soc.* (1926) **48,** 430.
2. Fujita, H., "Mathematical Theory of Sedimentation Analyses," Academic Press, New York, 1962.
3. Rinde, H., "The Distribution of the Sizes of Particles in Gold Sols," Dissertation, Upsala, Sweden, 1928.
4. Schulz, G. V., *Z. Phys. Chem., Abt. B* (1939) **43,** 25.
5. Kramer, E. O., In "The Ultracentrifuge," p. 350, Svedberg, T., Pedersen, K. O., Eds., Oxford University Press, London, 1940.
6. Wales, M., Thesis, University of Wisconsin, 1947.
7. Wales, M., *J. Phys. Colloid Chem.* (1948) **52,** 235.
8. Wales, M., Williams, J. W., Thompson, J. O., Ewart, R. H., *J. Phys. Chem.* (1948) **52,** 983.
9. Wesselan, H., *Macromol. Chem.* (1956) **20,** 111.
10. Gardner, D. G., Gardner, T. C., Lausch, G., Meinke, W. W., *J. Chem. Phys.* (1959) **31,** 978.
11. Donnelly, T. H., *J. Phys. Chem.* (1960) **64,** 1830.
12. Blair, J. E., In "Analysis and Fractionation of Polymers," *J. Polym. Sci., Part C* (1965) **8,** 287.
13. Sundelof, L. O., *Ark. Kem.* (1965) **25,** 1.
14. Donnelly, T. H., *J. Phys. Chem.* (1966) **70,** 1862.
15. Provencher, S. W., *J. Chem. Phys.* (1967) **46,** 3229.
16. Sunderlof, L. O., *Ark. Kem.* (1967) **29,** 279.
17. Scholte, Th. G., *J. Polym. Sci., Part A-2* (1968) **6,** 91.
18. Scholte, Th. G., *J. Polym. Sci., Part A-2* (1968) **6,** 111.
19. Gralen, N., Lagermalm, G., "Festskrift Tillagnad," Hedvall, J. A., Ed., Goteborg, 1948.
20. Hengstenberg, J., "Die Physik der Hochpolymeren," Vol. 2, "Das Makromolekuel in Loesungen," p. 491, Stuart, H. A., Ed., Springer Verlag, Berlin, 1953.
21. Gehatia, M. *et al.*, *Kolloid-Z., Z. Polym.* (1960) **169,** 162.
22. Tikhonov, A. N., *Dokl. Akad. Nauk SSSR* (1943) **39,** 195.
23. Hadamard, J. S., "Lectures on Cauchy's Problem," p. 4, Dover, New York, 1952.
24. Tikhonov, A. N., *Dokl. Akad. Nauk SSSR* (1963) **1155,** 501.
25. Tikhonov, A. N., *Dokl. Akad. Nauk SSSR* (1963) **153,** 49.
26. Tikhonov, A. N., *Dokl. Akad. Nauk SSSR* (1964) **156,** 268.
27. Tikhonov, A. N., *Dokl. Akad. Nauk SSSR* (1965) **156,** 6.

28. Tikhonov, A. N., *Dokl. Akad. Nauk SSSR* (1965) **161,** 5.
29. Tikhonov, A. N., *Dokl. Akad. Nauk SSSR* (1965) **162,** 763.
30. Tikhonov, A. N., *Dokl. Akad. Nauk SSSR* (1965) **163,** 591.
31. Tikhonov, A. N., *Dokl. Akad. Nauk SSSR* (1965) **164,** 507.
32. Tikhonov, A. N., *Zh. Vych. Mat. Mat. Fiz.* (1965) **5,** 718.
33. Tikhonov, A. N., *Zh. Vych. Mat. Mat. Fiz.* (1966) **6,** 81.
34. Tikhonov, A. N., "Computational Methods and Programming. Collective Works of Computation Center of Moscow University," No. 1, VIII, Moscow University Press, 1967.
35. Tikhonov, A. N., Samarskii, A. A., *Zh. Vych. Mat. Mat. Fiz.* (1961) **1,** 5.
36. Tikhonov, A. N., Glasko, B. V., *Zh. Vych. Mat. Mat. Fiz.* (1964) **44,** 564.
37. Tikhonov, A. N., Glasko, B. V., *Zh. Vych. Mat. Mat. Fiz.* (1965) **5,** 463.
38. Tikhonov, A. N., Arsenid, V. Ya., Vladimirov, L. A., Droshenko, G. G., Lumor, L. A., *Izv. Akad. Nauk SSSR, Ser. Fiz.* (1965) **29,** 815.
39. Tikhonov, A. N., Gatkin, V. Ya., *Zh. Vych. Mat. Mat. Fiz.* (1967) **7,** 2.
40. Phillips, D. L., *J. Ass. Compt. Mach.* (1962) **9,** 1.
41. Gehatia, M., Wiff, D. R., *J. Polym. Sci., Part A-2* (1970) **8,** 2039.
42. Wiff, D. R., Gehatia, M., *J. Macromol. Sci.-Phys. B* (1972) **6,** 287.
43. Cass, S. I., "Linear Programming, Methods and Applications," McGraw-Hill, New York, 1964.
44. Hadley, G., "Linear Programming," Addison-Wesley, Reading, Mass., 1960.
45. Rabinowitz, P., *SIAM Rev.* (1968) **10,** 121.
46. Gehatia, M. T., *Kolloid-Z., Z. Polym.* (1959) **167,** 1.

RECEIVED January 17, 1972.

21

Molecular Weight Distributions from Sedimentation Equilibrium Experiments

E. T. ADAMS, Jr. and PETER J. WAN
Texas A & M University, College Station, Texas 77843

DONALD A. SOUCEK
Illinois Institute of Technology, Chicago, Ill. 60616

GRANT H. BARLOW
Molecular Biology Department, Abbott Laboratories, Inc., North Chicago Ill. 60064

Since 1928 it has been realized that molecular weight distributions (MWD's) of polymeric solutes could be obtained from sedimentation equilibrium experiments. Two currently used methods developed by Donnelly and by Scholte to obtain MWD's require ideal, dilute solutions and the use of ultracentrifuge cells with sector-shaped centerpieces. These restrictions are removed in this paper. Equations analogous to those of Donnelly and Scholte are presented for non-sector-shaped centerpieces. A method of correcting for nonideal behavior is also presented along with a method of application of this nonideal correction to the measurement of MWD's by Donnelly's or by Scholte's methods, using centerpieces of any shape.

In 1924 Svedberg and Rinde (*1*) published the first paper on the ultracentrifuge. Four years later Rinde (*2*) advocated the idea of obtaining a molecular weight distribution (MWD) from sedimentation equilibrium experiments. He tried to obtain the distribution of radii from colloidal gold sols. Subsequent methods proposed for obtaining MWD's from sedimentation equilibrium experiments were largely based on Rinde's (*2*) pioneering work. There have been a number of attempts to obtain MWD's from sedimentation equilibrium experiments, using the concentration (c) or concentration gradient (dc/dr) distribution at sedimentation equilibrium. In some cases (*3–5*) specific models of the MWD, such as the most probable distribution, were used. Wales and his associates

(*6–9*) attempted to avoid specific models for the MWD; they also tried to include nonideal behavior. These methods, as well as a recent one by Sundelöf (*10*) based on the Fourier convolution theorem, depended on data (c or dc/dr *vs.* r) obtained from one sedimentation equilibrium experiment; sometimes the experimental error caused negative weight fractions to be obtained for some of the polymeric components.

This then was the state of the art until two separate and elegant breakthroughs, the first by Donnelly (*11, 12*) and the second by Scholte (*13, 14*) showed new ways to solve the problem. Donnelly pointed out that the concentration distribution of the polymeric solutes at sedimentation equilibrium was in the form of a Laplace transform. He showed methods for obtaining analytical expressions for the Laplace transform. Once these expressions were available, one could obtain the MWD from the inverse of the Laplace transform. While this method is theoretically rigorous, it appears to be applicable only to unimodal distributions. Donnelly's (*11, 12*) method requires only one sedimentation equilibrium experiment. Refinements to the Laplace transform method and also to the Fourier convolution theorem method have been reported by Provencher and Gobush (*15, 16*). Scholte's (*13, 14*) method uses sedimentation equilibrium experiments at different speeds on the same solution to obtain MWD's. He showed that his method could be applied to bimodal and trimodal distributions. Both Donnelly's (*11, 12*) and Scholte's (*13, 14*) methods were restricted to ideal Θ solutions and to ultracentrifuge cells with sector-shaped centerpieces. We show here how both of these restrictions can be removed. The ideal case is developed first, followed by a method for obtaining MWD's under nonideal conditions from sedimentation equilibrium experiments using cells with sector- or non-sector-shaped centerpieces.

Ideal Solutions

Basic Sedimentation Equilibrium Equation. Sedimentation equilibrium experiments are performed at constant temperature. The condition for sedimentation equilibrium is that the total molar potential, $\bar{\mu}_i$, for all components i be constant everywhere in the solution column of the ultracentrifuge cell. Mathematically this can be expressed as

$$\bar{\mu}_i = \mu_i + \Phi_i = \mu_i - \frac{M_i\omega^2 r^2}{2} = \text{constant} \tag{1}$$

$$\omega = \frac{2\pi \text{RPM}}{60} = \text{angular velocity}$$

Here μ_i is the molar chemical potential of component i, $\Phi_i = -M_i\omega^2r^2/2$

its molar centrifugal potential, M_i the molecular weight of component i, and r the distance (from the center of rotation) in the solution column of the ultracentrifuge cell. The quantity r is restricted to values between and including r_m, the air-solution meniscus radial position, and r_b, the radial position of the cell bottom. For simplicity we shall assume that the solution is incompressible. It is not necessary to make this assumption, but the treatment is easier to follow if we do so. Excellent treatment of compressible solutions will be found in the monograph by Fujita (*17*) or the review by Casassa and Eisenberg (*18*). At constant temperature and for incompressible systems, one notes that the following relations apply for a solution containing q polymeric solutes.

$$\mu_i = f(c_1, c_2, \ldots, c_q, P, T) \tag{2}$$

$$(\partial \mu_i/\partial P)_{T,\, c} = M_i \bar{v}_i \tag{3}$$

The chemical potential of component i can be expressed as

$$\mu_i = \mu^\circ_i + RT \ln y_i c_i \tag{4}$$

Here c_i is the concentration of component i in grams per liter, and y_i is the activity coefficient of component i on this concentration scale. The quantity μ_i° is the standard state chemical potential of component i and is a function of temperature only. The standard state of solute component i is chosen so that $\ln y_i \to 1$ as $c_i \to 0$. In Equation 4, R is the universal gas constant, 8.314×10^7 ergs/(deg mole), and T is the absolute temperature. The quantity $\ln y_i$ is a function of the concentration of all q solutes; thus

$$\ln y_i = f(c_1, c_2, \ldots, c_q) \tag{5}$$

We will denote the solvent by the subscript 0. Usually $\ln y_i$ is expressed as a Maclaurin series in which only the first term is retained—*i.e.*,

$$\ln y_i = \sum_j \left(\frac{\partial \ln y_i}{\partial \delta c_j}\right)^\circ_{T,c_{k\neq j}} c_j + \ldots = M_i \sum_j B_{ij} c_j + \ldots \tag{6}$$

Here $(\partial \ln y_i/\partial c_j)^\circ_{T,c_{k\neq j}}$ is the value of the derivative at infinite dilution of the solutes. For ideal solutions $\ln y_i$ is taken to be zero at all concentrations. In this case Equation 1 can be expressed as

$$d\mu_i = M_i \omega^2 d(r^2)/2$$

but

$$d\mu_i = \left(\frac{\partial \mu_i}{\partial c_i}\right) dc_i + \left(\frac{\partial \mu_i}{\partial P}\right) dP$$

and

$$dP = \omega^2 \rho d(r^2)/2$$

so that

$$\frac{RT}{c_i} dc_i = \frac{M_i(1 - \bar{v}_i \rho)\omega^2 d(r^2)}{2} \tag{7}$$

or

$$\frac{d \ln c_i}{d(r^2)} = \frac{M_i(1 - \bar{v}_i \rho)\omega^2}{2RT} = A_i M_i \tag{7a}$$

This then is our basic, working equation in sedimentation equilibrium experiments. To obtain the weight-average molecular weight, M_w, it is necessary to assume that the specific refractive index increments, Ψ_i, of the polymeric solutes are equal, *i.e.*, $\Psi_i = \Psi_j = \Psi$; it is also necessary to assume that the partial specific volumes, $\bar{v}_i$, of the polymeric solutes are equal. The most common optical systems available on modern ultracentrifuges are the Rayleigh and the schlieren optical systems, which give information proportional to the refractive index difference, $n - n_0$, between the solution and the solvent (Rayleigh optics) or to the derivative $d(n - n_0)/dr$ (schlieren optics) at each radial position r. Some ultracentrifuges are also equipped with an absorption optical system which measure ultraviolet absorption at 248, 254, and 265 nm. This was one of the earliest optical systems used on the ultracentrifuge (*19*); it is used extensively for studies on viruses or nucleic acids. Here the optical density or absorbance is proportional to concentration. A photoelectric scanner has been developed for the absorption optical system (*20–22*), and with the use of a monochromator one can do studies in the wavelength range of 220–560 nm. For refractometric optics (Rayleigh and/or schlieren optics) one notes that

$$n - n_0 = \sum_{i=1}^{q} (n_i - n_0) = \sum_{i=1}^{q} \Psi'_i c_i \tag{8}$$

When the refractive index increments, Ψ_i, of the solutes are equal, then Equation 8 can be written as

$$n - n_0 = \Psi' c \tag{8a}$$

The Rayleigh interference fringes result because of a difference in optical path between solution and solvent. The number of fringes, which is the number of wavelengths difference in the optical path between solution and solvent, is denoted by J and is defined by

$$J = \frac{h}{\lambda}(n - n_0) = \frac{h}{\lambda}\Psi c \tag{9}$$

Here λ is the wavelength of the light being used (546 nm for the green line of mercury), and h is the thickness of the centerpiece (12 mm is most common).

Equation 7a can be rewritten as

$$\frac{dc_i}{d(r^2)} = Ac_iM_i \tag{10}$$

where $A = (1 - \bar{v}\rho)\omega^2/2RT$. The summation of Equation 10 over all q solutes leads to

$$\frac{dc}{d(r^2)} = A\sum_{i=1}^{q} c_iM_i = AcM_{wr} \tag{11}$$

Here $M_{wr} = \Sigma c_{ir}M_i/c_r$ is the weight-average molecular weight at any radial position r between and including r_m and r_b. Equation 11 only gives M_{wr} if $\bar{v}_i = \bar{v}_j = \bar{v}$ and $\Psi_i = \Psi_j = \Psi$. When this condition is not met, then one measures a quantity called M_{1r} (*23*). For this situation Equation 11 becomes

$$\frac{d(n - n_0)}{d(r^2)} = \sum_i A_i\Psi_i c_{ir}M_i \tag{12}$$

$$= \overline{M}_{r1}\sum_i A_i\Psi_i c_{ir}$$

The quantity $\overline{M}_{1r}$ is defined by

$$\overline{M}_{1r} = \sum_i A_i\Psi_i c_iM_i/\sum_i A_ic_i\Psi_i \tag{12a}$$

Here $A_i = (1 - \bar{v}_i\rho)\omega^2/2RT$.

To simplify matters we will restrict the discussion to situations where $\Psi_i = \Psi_j = \Psi$ and $\bar{v}_i = \bar{v}_j = \bar{v}$. We will also write c_{ir} as c_i and c_r as c; the initial concentrations will be denoted by c_{0i} or c_0.

Concentration and Concentration-Gradient Distributions. Equation 10 can be written for convenience in dimensionless quantities using methods initiated by Fujita (*17*); thus, Equation 10 becomes

$$\frac{d\ln c_i}{d\xi} = -\Lambda M_i \tag{13}$$

where

$$\Lambda = (1 - \bar{v}\rho_0)\omega^2(r_b^2 - r_m^2)/2RT \tag{14}$$
$$\simeq (1 - \bar{v}\rho)\omega^2(r_b^2 - r_m^2)/2RT$$

and

$$\xi = \frac{r_b^2 - r^2}{r_b^2 - r_m^2} \tag{15}$$

In Equation 14 it has been assumed that the density of the solution, ρ, is approximately equal to the density of solvent, ρ_0; this approximation improves as the solution becomes more dilute. Note that ξ, defined by Equation 15, varies from 0 at r_b to 1 at r_m. When the variables in Equation 13 are separated, the equation can be integrated between $\xi = 0$ and $\xi = \xi$ to give

$$\ln\,[c_i(\xi)/c_i(\xi = 0)] = -\Lambda M_i\xi \tag{16}$$

Conversion to the exponential form gives

$$c_i(\xi) = c_i(\xi = 0)\exp(-\Lambda M_i\xi) \tag{17}$$

So far the equations that have been developed are independent of the shape of the cell. In order to make Equation 17 more useful, it is necessary to relate $c_i(\xi = 0)$ to the initial concentration of component i, c_{0i}. To do this one applies conservation of mass, which states that in a closed system (the solution column in the ultracentrifuge cell) the total amount of solute is constant at all times. Since mass equals concentration times volume, it is necessary to know the volume of the centerpiece, and this depends on the shape of the centerpiece. Two types of centerpieces are currently used. One is a sector-shaped centerpiece, and the other is a six-channel, equilibrium [or Yphantis (*24*)] centerpiece. Top views of both centerpieces are shown in Figure 1. In either centerpiece, one side is reserved for solvent and the other for solution.

For the sector-shaped centerpiece the conservation of mass equation can be expressed as (for θ in radians)

$$\theta h\int_{r_m}^{r_b} c_i d(r^2) = \theta h c_{0i}(r_b^2 - r_m^2) \tag{18}$$

or

$$\int_0^1 c_i d\xi = c_{0i} \tag{18a}$$

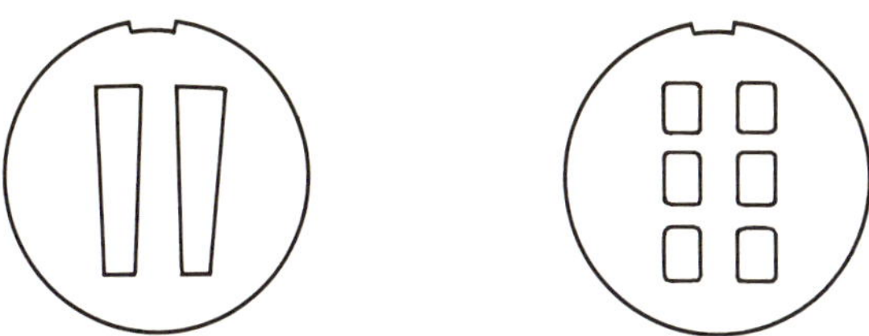

Figure 1. Top view of double-sector (left) and Yphantis (right) centerpieces

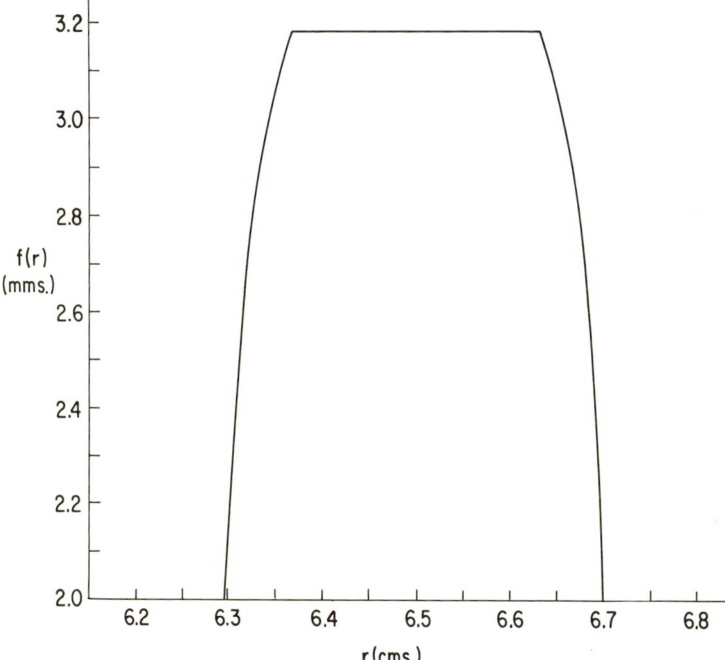

Figure 2. Plot of the cross-sectional width f(r) vs. *radial distance* r *for an Yphantis centerpiece*

Here θ is the sector angle (2.5° for a double-sector centerpiece), and h is the cell thickness which can vary from 12 mm (most common) to 30 mm. Since θ and h are constants, they drop out. Note that for a sector-shaped cell $dv = \theta h d(r^2)/2$. One can also write the conservation of mass equation for the total concentration; this is done by summing Equation 18a over all q solutes. Thus

$$c_0 = \sum_i c_{0i} = \int_0^1 c d\xi \tag{19}$$

The total concentration at the meniscus, $c(\xi = 1)$, can be evaluated from

$$c_0 - c(\xi = 1) = \int_0^1 [c(\xi) - c(\xi = 1)]d\xi \tag{20}$$

For the Yphantis cell one notes $dv = hf(r)dr$, where $f(r)$ is the cross-sectional width. This will vary with r since the holes are curved at each end. A typical plot of $f(r)$ *vs.* r is shown in Figure 2. For component i the conservation of mass equation is

$$h\int_{r_m}^{r_b} c_i f(r)dr = hc_{0i}\int_{r_m}^{r_b} f(r)dr \tag{21}$$

or

$$\int_0^1 c_i dx = c_{0i} \tag{21a}$$

Here $dx = f(r)dr \Big/ \int_{r_m}^{r_b} f(r)dr$. The conservation of mass equation for the total concentration is

$$c_0 = \int_0^1 c dx \tag{22}$$

It is obtained by summing Equation 21a over all q solutes. One can obtain c_m, the meniscus concentration, from

$$c_0 - c_{r_m} = \int_0^1 (c - c_{r_m})dx \tag{23}$$

The conservation of mass equations are used with refractometric optics. The absorption optical system will give absorbance (optical density) which is directly proportional to concentration. Unless conservation of mass equations are used, one can only obtain concentration differences from refractometric optics. The Rayleigh optical system will give information proportional to $c_r - c_{r_m}$; thus Equation 20 or 23 would have to be used to obtain c_{r_m}. Note also that the initial concentration c_0 is needed. This must be measured by differential refractometry, by boundary-forming experiments, or from ultraviolet light absorption.

For sector-shaped centerpieces the substitution of Equation 17 into Equation 18a leads to

$$c_i(\xi = 0) = \frac{\Lambda M_i c_{0i}}{1 - \exp(-\Lambda M_i)} \tag{24}$$

When Equation 24 is substituted into Equation 17, the result is

$$c_i(\xi) = \frac{\Lambda M_i c_{0i} \exp(-\Lambda M_i \xi)}{1 - \exp(-\Lambda M_i)} \tag{25}$$

The summation of Equation 25 over all i solute components, followed by division by c_0, leads to

$$\theta(\xi) = \frac{c(\xi)}{c_0} = \sum_i \frac{\Lambda M_i f_i \exp(-\Lambda M_i \xi)}{1 - \exp(-\Lambda M_i)} \tag{26}$$

Here $f_i = c_{0i}/c_0$ is the weight fraction of component i. Now differentiate Equation 26 with respect to ξ to obtain

$$-\frac{1}{c_0}\frac{dc(\xi)}{d\xi} = \frac{-d\theta(\xi)}{d\xi} = U(\Lambda, \xi) = \sum_i \frac{\Lambda^2 M_i^2 f_i \exp(-\Lambda M_i \xi)}{1 - \exp(-\Lambda M_i)} \tag{27}$$

Equations 26 and 27 are the equations needed for obtaining MWD's when sector-shaped centerpieces are used.

The corresponding equations for the Yphantis centerpiece are

$$c_i(\xi = 0) = \frac{c_{0i}}{\int_0^1 e^{-\Lambda M_{\xi i}} dx} \tag{28}$$

$$c_i(\xi) = \frac{c_{0i} \exp(-\Lambda M_i \xi)}{\int_0^1 [\exp(-\Lambda M_i \xi)] dx} \tag{29}$$

$$\hat{\theta}(\xi) = \frac{c(\xi)}{c_0} = \sum_i \frac{f_i \exp(-\Lambda M_i \xi)}{\int_0^1 [\exp(-\Lambda M_i \xi)] dx} \tag{30}$$

$$-\frac{1}{c_0}\frac{dc(\xi)}{d\xi} = -\frac{d\theta(\xi)}{d\xi} = V(\Lambda, \xi) = \sum_i \frac{\Lambda f_i M_i \exp(-\Lambda M_i \xi)}{\int_0^1 [\exp(-\Lambda M_i \xi)] dx} \tag{31}$$

Here we have used Equations 17 and 21a.

Donnelly's Method for MWD's. If one assumes a continuous distribution of molecular weights, then one can write Equation 26 as

$$\theta(\xi) = \int_0^\infty \frac{\Lambda M f(M) \exp(-\Lambda M \xi) dM}{1 - \exp(-\Lambda M)} \tag{32}$$

for sector-shaped centerpieces. Here the summation over all solute components (from $i = 1$ to q) has been replaced by an integral from 0 to ∞, and the weight fraction of i, f_i, has been replaced by $f(M)dM$. The quantity $f(M)$ is the differential MWD, and $f(M)dM$ is the weight fraction of solute having a molecular weight between M and $M + dM$. Now make the following substitutions.

$$t = \Lambda M \tag{33}$$

$$\varphi(t) = \frac{\Lambda M f(M)}{1 - \exp(-\Lambda M)} = \frac{tg(t)}{1 - e^{-t}} \tag{33a}$$

$$\xi = s \tag{33b}$$

$$dM = dt/\Lambda \tag{33c}$$

Then one can write Equation 32 as

$$\Lambda\theta(\xi) = \int_0^\infty \varphi(t)e^{-st}dt \tag{34}$$

$$= L\{\varphi(t)\}$$

In other words, the quantity $\Lambda\theta(\xi)$ is the Laplace transform of $\varphi(t)$, the Laplace transform being denoted by $L\{\varphi(t)\}$. Donnelly (*11, 12*) showed how one could use the quantity $(d \ln c/d(r^2))$ to obtain an analytical expression for the Laplace transform. This is now illustrated for a non-sector-shaped centerpiece.

When the Yphantis centerpiece is used, the analog of Equation 32 is

$$\hat{\theta}(\xi) = \int_0^\infty \frac{f(M) \exp(-\Lambda M\xi)dM}{\int_0^1 [\exp(-\Lambda M\xi)]dx} \tag{35}$$

Now let

$$\gamma(t) = \frac{f(M)}{\int_0^1 [\exp(-\Lambda M\xi)]dx} \tag{36}$$

When Equations 33, 33b, 33c, and 36 are used, one obtains

$$\Lambda\hat{\theta}(\xi) = \int_0^\infty \gamma(t)e^{-st}dt \tag{36a}$$

$$= L\{\gamma(t)\}$$

Here $\Lambda\theta(\xi)$ is the Laplace transform of $\gamma(t)$.

An analytical expression for the Laplace transform can be obtained from values of $(d \ln c/d(r^2))$ at various radial positions. Note that

$$\int_{r_m}^{r} \frac{d \ln c}{d(r^2)} d(r^2) = \ln c_r/c_{r_m} \tag{37}$$

and

$$\exp \int_{r_m}^{r} \frac{d \ln c}{d(r^2)} d(r^2) = c_r/c_{r_m} = c(\xi)/c(\xi = 1) \tag{38}$$

However, what we need is $\Lambda\theta(\xi) = \Lambda c(\xi)/c_0$. Therefore, Equation 38 is multiplied by $1/K = \Lambda c(\xi = 1)/c_0$ to give

$$\Lambda\hat{\theta}(\xi) = L\{\gamma(t)\} = (1/K) \exp \int_{r_m}^{r} \frac{d \ln c}{d(r^2)} d(r^2) \tag{39}$$

Define

$$F(n, u) = 1 \Big/ \left(\frac{d \ln c}{d(r^2)} \right) \tag{40}$$

and

$$u = \frac{r^2 - r_m^2}{r_b^2 - r_m^2} = 1 - \xi = 1 - s \tag{41}$$

If $F(n, u)$ can be expressed as a function in u, this can be converted to a function in s since $u = 1 - \xi = 1 - s$. Laplace transforms are algebraic expressions in s, so one can use a table of Laplace transforms to obtain the inverse of the Laplace transform.

Suppose that when $F(n, u)$ is plotted against u, a straight line is obtained. The equation for this straight line can be written as

$$F(n, u) = P - Qu \tag{42}$$

Here P is the intercept at $u = 0$ and Q is the slope. Note that

$$du = \frac{d(r^2)}{r_b^2 - r_m^2}$$

or

$$d(r^2) = (r_b^2 - r_m^2)du = bdu \tag{43}$$

Using Equations 42 and 43

$$\int_0^u \frac{bdu}{F(n, u)} = \int_0^u \frac{bdu}{P - Qu}$$

$$= \frac{b}{Q} \ln \frac{P}{P - Qu} \tag{44}$$

$$= \frac{b}{Q} \ln \frac{P}{Q\left(\frac{P}{Q} - 1 + s\right)}$$

When Equation 44 is substituted into Equation 39, the result is

$$\Lambda\hat{\theta}(\xi) = L\{\gamma(t)\} = \frac{1}{K}\left[\frac{P}{Q\left(\frac{P}{Q} - 1 + s\right)}\right]^{b/Q} = \frac{(P/Q)^{b/Q}}{K}\left[\frac{1}{s+a}\right]^{n} \quad (45)$$

Here $a = (P/Q) - 1$ and $n = b/Q$.

$$\int_0^\infty t^{n-1}e^{-(s+a)t}dt = \frac{\Gamma(n)}{(s+a)^n} \quad (46)$$

Here $\Gamma(n)$ is the gamma function of n. Thus it follows that

$$L\{\gamma(t)\} = \int_0^\infty \gamma(t)e^{-st}dt \quad (47)$$

$$= \int_0^\infty \frac{(P/Q)^{b/Q}}{K} \frac{t^{(b/Q)-1}}{\Gamma(b/Q)} e^{-[(P/Q)-1]t}e^{-st}dt$$

and

$$\gamma(t) = \frac{(P/Q)^{b/Q}}{K} \frac{t^{[(b/Q)-1]}}{\Gamma(b/Q)} e^{-[(P/Q)-1]t} \quad (48)$$

Finally

$$f(M) = \{\int_0^1 [\exp(-\Lambda M\xi)]dx\}\gamma(t) \quad (49)$$

For a sector-shaped centerpiece one would obtain

$$f(M) = \left[\frac{1 - \exp(-\Lambda M)}{\Lambda M}\right]\varphi(t) \quad (50)$$

If $F(n, u) = P - Qu$, then the expression for $\gamma(t)$ or $\varphi(t)$ is given by Equation 48. In other words, $\gamma(t)$ is the same as $\varphi(t)$ for the same choice of $F(n, u)$. However, to use Donnelly's (*11, 12*) method the total solute concentration, c, must be known at any radial position for the evaluation of $d \ln c/d(r^2)$, and c is obtained from the appropriate conservation of mass equation when Rayleigh or schlieren optics are used.

Table I gives values of $\gamma(t)$ or $\varphi(t)$ in the third column for the three types of equations that are obtained from a plot of $F(n, u)$ *vs.* u. The equations for $F(n, u)$ are listed in column 1. From these values of $F(n, u)$ the Laplace transforms which are listed in column 2 are obtained. The appropriate expression for $f(M)$ can be obtained from values of $\gamma(t)$ or $\varphi(t)$ by the use of Equation 49 or 50.

Scholte's Method. Scholte's (*13, 14*) method requires sedimentation equilibrium experiments at different speeds on the same solution. When sedimentation equilibrium has been attained at one speed, the Rayleigh and schlieren data are recorded, then the speed is changed. When sedimentation equilibrium is attained at the new speed, the Rayleigh and schlieren data are recorded again. This procedure is repeated at each speed. At lower speeds the concentration distribution of the lower molecular weight solutes is relatively slight, whereas at higher speeds the higher molecular weight solutes are relegated to the bottom of the cell, and the lower molecular weight solutes are redistributed throughout the cell. In a way this simulates fractionation of the sample.

EVALUATION OF THE WEIGHT FRACTION OF SOLUTE. *Sector-Shaped Centerpieces.* For sector-shaped centerpieces Scholte (*13, 14*) starts his analysis with Equation 27 which he writes as

$$U(\Lambda, \xi)_j = \sum_i f_i \frac{\Lambda_j^2 M_i^2 \exp(-\Lambda_j M_i \xi_j)}{1 - \exp(-\Lambda_j M_i)} + \delta_j = \sum_i f_i K_{ij} + \delta_j \qquad (27)$$

Here

$$K_{ij} = \frac{\Lambda_j^2 M_i^2 \exp(-\Lambda_j M_i \xi_j)}{1 - \exp(-\Lambda_j M_i)} \qquad (51)$$

and δ_j is a term expressing experimental error. Since f_i depends on M_i, and M_i is independent of any speed chosen, the subscript i is used here.

Table I. MWD's by Donnelly's Method[a]

F(n, u)	f(s)	γ(t) *or* φ(t)
P	$\dfrac{e^{b/P}e^{-bs/P}}{K}$	$\left(\dfrac{e^{b/P}}{K}\right)\delta\left(t - \dfrac{b}{P}\right)$
$P - Qu$	$\dfrac{1}{K}\left[\dfrac{P}{Q\left(\dfrac{P}{Q} - 1 + s\right)}\right]^{b/Q}$	$\dfrac{\left(\dfrac{P}{Q}\right)^{b/Q} t^{[(b/Q)-1]} e^{-[(P/Q)-1]t}}{K\Gamma(b/Q)}$
$\dfrac{P - Qu}{1 + R(P - Qu)}$	$\dfrac{\left(\dfrac{P}{Q}\right)^{b/Q} e^{bR}e^{-bRs}}{K\left(\dfrac{P}{Q} - 1 + s\right)^{b/Q}}$	$\dfrac{\left(\dfrac{P}{Q}\right)^{b/Q} e^{-(bRP/Q)} (t-bR)^{[(b/Q)-1]} e^{-[(P/Q)-1]t}}{K\Gamma(b/Q)}$

[a] Note that (1) P, Q, and R are constants; (2) $b = r_b^2 - r_m^2$; (3) $s = \xi = 1 - u$; (4) $\delta[t - (b/P)]$ is the Dirac delta function of $t - (b/P)$; (5) $1/K = \Lambda c(\xi = 1)/c_0$; (6) $t = \Lambda M$; and (7) $\Gamma(b/Q)$ is the gamma function of b/Q.

The values of Λ and ξ do depend on the speed, so the subscript j is used to indicate this. Usually ξ is read at intervals of one-fourth starting at $\xi = 1$ (the air–solution meniscus) and going to $\xi = 0$ (the cell bottom). At higher speeds it may not be possible to obtain values of $U(\Lambda, \xi)_j$ (or $V(\Lambda, \xi)_j$—*see* Equation 31) at $\xi = 1/4$ or $\xi = 0$. To use Scholte's method one assumes a discrete range of molecular weights (M_i) which will bracket the molecular weight range of the sample. By experience it was found that the interval between molecular weights should be logarithmic, and that the lowest interval was about 2. Thus the choice of molecular weights might be $M_1 = 2000$, $M_2 = 4000$, $M_3 = 8000$, and so on until a molecular weight greater than the molecular weight of the largest polymeric component is obtained. The experiments should be designed so that there are more values of $U(\Lambda, \xi)_j$ than there are choices of M_i. The first choice of molecular weights is known as the first series. Once the first series has been chosen, one can use the $U(\Lambda, \xi)_j$ values to write a set of linear equation as follows

$$\begin{aligned} U(\Lambda, \xi)_1 &= f_1K_{11} + f_2K_{21} + f_3K_{31} + \ldots + \delta_1 \\ U(\Lambda, \xi)_2 &= f_1K_{12} + f_2K_{22} + f_3K_{32} + \ldots + \delta_2 \\ &\vdots \\ U(\Lambda, \xi)_n &= f_1K_{1n} + f_2K_{2n} + f_3K_{3n} + \ldots + \delta_n \end{aligned} \tag{52}$$

This set of linear equations is solved for f_i by a linear programming technique. It is perfectly set up for linear programming, since (1) we have a set of linear equations containing no negative quantities; (2) the sum of all weight fractions must be 1, *i.e.*, $\sum_i f_i = 1$; (3) any f_i must satisfy the relation $0 \leqslant f_i \leqslant 1$; and (4) the f_i values are sought for which $\Sigma|\delta_j|$ is a minimum and which also satisfy condition 2.

The set of f_i obtained in the first series is not unique. One could choose another set of M_i and solve for f_i, and this is what Scholte does. If the data are good, a second series in M_i is obtained from the first series in M_i by multiplying each M_i of the first series by $2^{1/4}$, *i.e.*, M_i(second series) $= M_i$(first series) $\times 2^{1/4}$. The $U(\Lambda, \xi)_j$ values are then used to obtain f_i for the second series. A third series, in which M_i(third series) $= M_i$(first series) $\times 2^{1/2}$, and a fourth series, in which M_i(fourth series) $= M_i$(first series) $\times 2^{3/4}$, of molecular weights are chosen, and the f_i's are obtained for each series from the $U(\Lambda, \xi)_j$ values. Now note that

$$\sum_i f_i = 1 \quad \text{for any one series} \tag{53}$$

$$\sum_i f_i = 4 \quad \text{for all series} \tag{54}$$

or

$$\sum_i f_i/4 = 1 \quad \text{for all series} \tag{54a}$$

Thus, Equation 54a is also a solution. If one wished to plot the f_i *vs.* M_i in order to display a differential MWD, the use of Equation 54a would give a better or smoother plot since more points are included than would be included in a plot based on data obtained from one series only.

Non-Sector-Shaped Centerpieces. Although Scholte (*13, 14*) set up his method originally for equations applicable to sector-shaped centerpieces, this restriction can be disposed of readily. For non-sector-shaped centerpieces start with Equation 31 written as

$$V(\Lambda, \xi)_j = \sum_i f_i \frac{\Lambda_j M_i \exp(-\Lambda_j M_i \xi_j)}{\int_0^1 [\exp(-\Lambda_j M_i \xi_j)] dx} + \delta_j \tag{31}$$

$$= \sum_i f_i H_{ij} + \delta_j$$

Here

$$H_{ij} = \frac{\Lambda_j M_i \exp(-\Lambda_j M_i \xi_j)}{\int_0^1 [\exp(-\Lambda_j M_i \xi_j)] dx} \tag{55}$$

and δ_j has the same meaning as before. (It is a term expressing experimental error.) One can then proceed to choose the various series in M_i. $V(\Lambda, \xi)_j$ can be expressed as a set of linear equations as follows

$$\begin{aligned} V(\Lambda, \xi)_1 &= f_1 H_{11} + f_2 H_{21} + f_3 H_{31} + \ldots + \delta_1 \\ V(\Lambda, \xi)_2 &= f_1 H_{12} + f_2 H_{22} + f_3 H_{32} + \ldots + \delta_2 \\ &\vdots \\ V(\Lambda, \xi)_n &= f_1 H_{1n} + f_2 H_{2n} + f_3 H_{3n} + \ldots + \delta_n \end{aligned} \tag{56}$$

This set of equations can be solved for f_i by linear programming methods since these equations meet the same criteria that were set up for the sector-shaped centerpiece case. Again one can obtain the f_i for all four series once the series in M_i have been chosen. For data that are not too precise, Scholte suggests using three series in M_i instead of four series. Here $2^{1/3}$ is used for the second series, and $2^{2/3}$ for the third series (*13, 14*).

HOW TO OBTAIN A CONTINUOUS MWD. Once the f_i's have been obtained from Equation 52 or 56, one can obtain a plot of the differential MWD using Scholte's procedures (*13, 14*) by following the steps below.

$$\sum_i f_i = 1 = \int_0^\infty f(M)dM \tag{57}$$

Now

$$\int_0^\infty f(M)dM = \int_0^\infty Mf(M)\frac{dM}{M} = 1 \tag{58}$$

Then use the rectangular rule (or trapezoidal rule if the values of $Mf(M)$ at the extremes of the interval are zero) to write

$$\int_0^\infty Mf(M)\frac{dM}{M} \simeq \Delta(\ln M)\sum_{i=1}^{n}[Mf(M)]_i = 1 \tag{59}$$

Here n is the number of intervals used in the numerical integration. If the interval between successive molecular weights is $2^{1/4}$, then

$$\Delta(\ln M) = \tfrac{1}{4}\ln 2 = \frac{0.693}{4} \tag{60}$$

and

$$\sum_{i=1}^{n}[Mf(M)]_i = \frac{4}{0.693} \tag{61}$$

Thus it follows (*cf.* Equation 54) that

$$\sum_i \frac{f_i(\text{all series})}{0.693} = \frac{4}{0.693} \tag{62}$$

So, $[Mf(M)]_i$ is given by $f_i/0.693$, where f_i is one of the f_i's in the four molecular weight series. One then plots $Mf(M)$ (or $f_i/0.693$) *vs.* ln M. Figure 3, which is redrawn from Scholte's work (*13, 14*), shows such a plot of $Mf(M)$ *vs.* ln M for polyethylene in biphenyl at 123.2°C. A sector-shaped centerpiece was used in these experiments.

Nonideal Solutions

Estimation of the Ideal Values for $d \ln c/d(r^2)$ and $dc/d\xi$. For the nonideal case we can use Equations 1–4 and 6 to obtain the basic sedimentation equilibrium equation for component i. In the Fujita notation (*17*) this equation is

$$-\Lambda M_i c_i = \frac{dc_i}{d\xi} + c_i M_i \sum_k B_{ik}'\frac{dc_k}{d\xi} \tag{63}$$

Here

$$B_{ik}' = B_{ik} + \frac{\bar{v}}{1000M_k} \tag{64}$$

In this treatment concentrations c or c_i are in grams per liter, and Λ, defined by Equation 14, is used in place of Fujita's λ. Following the procedures of Albright and Williams (*25*) and Osterhoudt and Williams (*26*), we assume that the $dc_k/d\xi$ in the nonideal term (the term containing B_{ik}') can be written as

$$\frac{dc_k}{d\xi} \simeq -\Lambda c_k M_k \tag{65}$$

Thus Equation 63 becomes

$$-\Lambda M_i c_i = \frac{dc_i}{d\xi} - \Lambda c_i M_i \sum_k B_{ik}' c_k M_k \tag{66}$$

Now

$$\left(\frac{\sum\limits_k B_{ik}' c_k M_k}{\sum\limits_k c_k M_k}\right) \sum_k c_k M_k = \langle B_{ik}' \rangle \sum_k c_k M_k = \langle B_{ik}' \rangle c M_{wr} \tag{67}$$

$\langle B_{ik}' \rangle$ is the z-average value of B_{ik}' at any radial position r for $r_m \leqslant r \leqslant r_b$. The substitution of the right-hand side of Equation 67 into Equation 66 leads to

$$-\Lambda M_i c_i (1 - \langle B_{ik}' \rangle c M_{wr}) = \frac{dc_i}{d\xi} \tag{68}$$

The summation over all i components followed by rearrangement leads to

$$\frac{dc}{d\xi} = -\frac{\Lambda c M_{wr}}{1 + \langle B_{ik}' \rangle c M_{wr}} \tag{69}$$

This equation can be converted to

$$\frac{1}{A_0} \frac{d \ln c}{d(r^2)} = \frac{M_{wr}}{1 + \langle B_{ik}' \rangle c M_{wr}} = M_{wr_{\text{app}}} \tag{70}$$

Since

$$\Lambda = (1 - \bar{v}\rho_0)\omega^2(r_b^2 - r_m^2)/2RT \tag{71}$$

$$A_0 = (1 - \bar{v}\rho_0)\omega^2/2RT \tag{72}$$

and

$$d(r^2) = -(r_b^2 - r_m^2)d\xi \tag{73}$$

In Equation 70 $M_{wr_{\text{app}}}$ is the apparent weight-average molecular weight at any radial position r, where $r_m \leqslant r \leqslant r_b$. Now Equation 70 can be written

$$\frac{1}{M_{wr_{\text{app}}}} = \frac{1}{M_{wr}} + \langle B_{ik}' \rangle c \tag{74}$$

Equation 74 tells us that when $\langle B_{ik}' \rangle$ is positive and greater than zero, the effect of the nonideal term, $\langle B_{ik}' \rangle$, will be to offset the effect of decreases in values of $1/M_{wr}$ with increasing r. Negative values of $\langle B_{ik}' \rangle$ are usually interpreted as polymer–polymer interactions (*27, 28*) which are usually interpreted as self-associations. More negative values of $\langle B_{ik}' \rangle$ could lead to phase separation (*28*). The discussion that follows will be restricted to situations where $\langle B_{ik}' \rangle$ is positive.

The problem at hand with Equation 74 is how to deal with the quantity $\langle B_{ik}' \rangle$, and this complication is what has stymied sedimentation equilibrium work with nonideal solutions containing polymeric solutes. This problem has been discussed by Goldberg (*29*), Wales (*6*), and Fujita (*17*). One way to overcome the problem is to assume that

$$\langle B_{ik}' \rangle \simeq B_{\text{LS}} \tag{75}$$

Here B_{LS} is the second virial coefficient of the polymeric solute in the original solution before ultracentrifugation. B_{LS} is a quantity which can be obtained in light-scattering experiments (*17, 25, 30*) or in Archibald experiments (*31*), provided it is calculated from a plot of $1/M_{w_{\text{app}}}{}^{\circ}$ *vs.* c. Here $1/M_{w_{\text{app}}}{}^{\circ}$ is obtained from values of $M_{w_{\text{app}}}$ (at r_m or r_b) that have been extrapolated to zero time. The reason for using Equation 75 is that it leads to a simple method of estimating the MWD in nonideal solutions.

An alternative method for obtaining $\langle B_{ik}' \rangle$ is to assume in Equation 6 that the B_{ik}'s are the same for each polymeric component. According to Fujita (*32*) this is not a bad assumption for a solution containing synthetic polymers. When the B_{ik}'s are equal then one can write Equation 74 as

$$\frac{1}{M_{wr_{\text{app}}}} = \frac{1}{M_{wr}}\left(1 + \frac{\bar{v}c}{1000}\right) + Bc \tag{76}$$

Here $B = B_{ik}$. It then follows that

$$\frac{1}{M_{wr}} = \left(\frac{1}{M_{wr_{\text{app}}}} - Bc\right) \Big/ \left(1 + \frac{\bar{v}c}{1000}\right) \tag{77}$$
$$= 1 \Big/ \left[\frac{1}{A}\left(\frac{d \ln c}{d(r^2)}\right)_{\text{ideal}}\right]$$

The application of Equation 75 leads to a similar equation, namely

$$\frac{1}{M_{wr}} = \frac{1}{M_{wr_{\text{app}}}} - B_{\text{LS}}c \tag{78}$$
$$= 1 \Big/ \left[\frac{1}{A}\left(\frac{d \ln c}{d(r^2)}\right)_{\text{id}}\right]$$

where the subscript id means ideal.

The quantity $(d \ln c/d(r^2))_{\text{id}}$, which is available from Equation 77 or 78, can be used to calculate $(c_r/c_{r_m})_{\text{id}}$ since

$$\int_{r_m}^{r} \left(\frac{d \ln c}{d(r^2)}\right)_{\text{id}} d(r^2) = \ln (c_r/c_{r_m})_{\text{id}} \tag{79}$$

These values can be used with the appropriate conservation of mass equation to obtain $(c_{r_m})_{\text{id}}$. Once this is known, the ideal concentration distribution can be obtained. For sector-shaped centerpieces the conservation of mass equation is written as

$$\int_0^1 \left[\frac{c(\xi)}{c(\xi = 1)}\right]_{\text{id}} d\xi = \frac{c_0}{c(\xi = 1)_{\text{id}}} \tag{80}$$

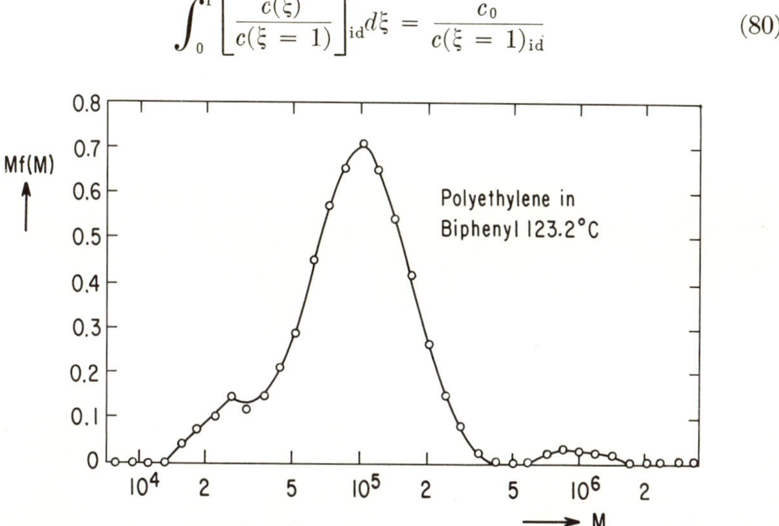

Figure 3. Molecular weight distribution of polyethylene in biphenyl

For non-sector-shaped centerpieces the corresponding equation is

$$\int_0^1 \left[\frac{c(\xi)}{c(\xi = 1)}\right]_{\text{id}} dx = \frac{c_0}{c(\xi = 1)_{\text{id}}} \tag{81}$$

In order to use Donnelly's (*11, 12*) method, we must convert the observed values of c_r and $d \ln c/d(r^2)$ in nonideal solutions to ideal values. Then we can use the ideal value of $d \ln c/d(r^2)$ to calculate MWD's as has been described previously here and elsewhere (*11, 12*). One must also remember to use Equation 49 for non-sector-shaped centerpieces.

In an earlier attempt to analyze nonideal MWD's (*25, 33*), we had started out with Equation 14 of Albright and Williams. This amounts to replacing $\Sigma_k B_{ik}' c_k M_k$ in Equation 66 by $B_{\text{LS}} c_0 M_w^\circ$. Instead of obtaining Equation 70, this leads to

$$\frac{d \ln c}{d(r^2)} = A_0 M_{wr_{\text{app}}}' = \frac{A_0 M_{wr}}{(1 + B_{\text{LS}} c_0 M_w^\circ)} \tag{82}$$

The term $1 + B_{\text{LS}} c_0 M_w^\circ$ in Equation 82 is a constant, whereas the term $1 + \langle B_{ik}' \rangle c M_{wr}$ in Equation 70 varies with r from r_m to r_b. Note when $\langle B_{ik}' \rangle = B_{\text{LS}}$, the denominator in Equation 70 becomes $1 + B_{\text{LS}} c M_{wr}$ which varies with r from r_m to r_b and more closely mirrors the true situation. Thus the treatment reported here, which is based on Equations 75–78, appears to be a better approach to the problem.

For Scholte's method one needs $(dc/d\xi)_{\text{id}}$. When $B_{ik} = B$, then Equation 6 becomes $\ln y_i = M_i B c$; thus Equations 70–72 can be used to obtain

$$\frac{1}{\left(\frac{dc}{d(r^2)}\right)} - \frac{B}{A} = \frac{1}{AcM_{wr}} = \frac{1}{\left(\frac{dc}{d(r^2)}\right)_{\text{id}}} \tag{83}$$

Then we use $(dc/d(r^2))_{\text{id}}$ to calculate the values of $U(\Lambda, \xi)_j$ or $V(\Lambda, \xi)_j$ needed for Equation 52 or 56 and proceed with the methods described earlier.

How to Evaluate B_{LS}. The final item to dispose of is how to evaluate B_{LS}. Methods for evaluating B_{LS} using sector-shaped centerpieces will be outlined first.

SECTOR-SHAPED CENTERPIECES. *Method 1.* Do the Archibald experiment and calculate $M_{w_{\text{app}},t}$ at r_m and r_b using standard procedures (*32, 34*). Here $M_{w_{\text{app}},t}$ is the value of the apparent weight-average molecular weight at r_m or r_b. Extrapolate these values of $M_{w_{\text{app}},t}$ to zero time to get $M_{w_{\text{app}}}^\circ$; details for this extrapolation have been given by Fujita *et al.* (*32*).

$$M_{w_{app}}{}^{\circ} = \lim_{t \to 0} M_{w_{app},t} \tag{84}$$

The quantity $M_{w_{app}}{}^{\circ}$ represents the value of $M_{w_{app}}$ for the original solution whose initial concentration is c_0. Now note that (*32*)

$$\frac{1}{M_{w_{app}}{}^{\circ}} = \frac{1}{M_w} + B_{LS}c_0 \tag{85}$$

Thus B_{LS} can be obtained from the limiting slope of a plot of $1/M_{w_{app}}{}^{\circ}$ *vs.* c_0. The Archibald method is restricted to sector-shaped centerpieces.

Method 2. Follow the procedure of Albright and Williams (*25*) and do sedimentation equilibrium experiments on the same solution at different speeds. Then calculate $M_{w_{app}}$(cell) from

$$M_{w_{app}}(\text{cell}) = \frac{c_{r_b} - c_{r_m}}{\Lambda c_0} \tag{86}$$

A plot of $M_{w_{app}}$(cell) values *vs.* Λ and extrapolation to $\Lambda = 0$, which is true at zero speed, gives $M_{w_{app}}{}^{\circ}$. Alternatively, the limiting slope of a plot of $(c_{r_b} - c_{r_m})/c_0$ *vs.* Λ gives $M_{w_{app}}{}^{\circ}$. Now B_{LS} is obtained from Equation 85 and the limiting slope of a plot of $1/M_{w_{app}}{}^{\circ}$ *vs.* c_0. One could also use M_w(cell vol) (*35, 36*), where

$$M_{w_{app}}(\text{cell vol}) = \frac{\ln (c_{r_b}/c_{r_m})}{\Lambda} \tag{87}$$

since

$$\lim_{\Lambda \to 0} M_{w_{app}}(\text{cell vol}) = M_{w_{app}}{}^{\circ}$$

Method 3. For sector-shaped centerpieces, Fujita showed that (*30*) if Λ was the same for each initial concentration c_0, then

$$\frac{1}{M_{w_{app}}(\text{cell})} = \frac{1}{M_w} + B'c_0 \tag{89}$$

where

$$B' \simeq B_{LS}\left(1 + \frac{\Lambda^2 M_z}{12}\right) \tag{90}$$

Thus a plot of $1/M_{w_{app}}$(cell) *vs.* c_0, could yield B' from the limiting slope, and B_{LS} could be estimated from Equation 90, provided one had an estimate of M_z, the z-average molecular weight. This method was restricted to lower speeds (to avoid solute redistribution problems) and to solutions that were not too nonideal. More recently, Deonier and Williams (*37*) showed that

$$\bar{c} \simeq c_0\left(1 + \frac{\Lambda^2 M_z}{12}\right) \tag{91}$$

where

$$\bar{c} = (c_{r_b} + c_{r_m})/2$$

Thus one could plot $1/M_{w_{app}}(\text{cell})$ *vs.* $\bar{c}$ since Equation 85 could be rewritten as

$$\frac{1}{M_{w_{aap}}(\text{cell})} \simeq \frac{1}{M_w} + B_{LS}\bar{c} \tag{92}$$

In carrying out MWD determinations by Scholte's (*13, 14*) method, one has the required information available to use any or all of the three methods discussed here for the evaluation of B_{LS}. When all the B_{ik}'s in Equation 6 are equal, then B_{LS} turns out to be $B_{LS} = B + \bar{v}/1000M_w°$. Here $B = B_{ik}$.

NON-SECTOR-SHAPED CENTERPIECES. For non-sector-shaped centerpieces one is restricted at present to Method 2. Here the equation for $M_{w_{app}}(\text{cell})$ is given by

$$-\frac{1}{\Lambda c_0}\int_0^1 \frac{dc}{d\xi}\,dx = M_{w_{app}}(\text{cell}) \tag{93}$$

$$= M_w\left[1 - \frac{1}{c_0 M_w}\int_0^1 \langle B_{ik}'\rangle_r (cM_{wr})^2 dx + \ldots\right]$$

At zero speed the concentrations of the polymeric components become uniform, so that

$$cM_{wr} \rightarrow c_0 M_w(\text{cell}) = c_0 M_w°$$

and

$$\langle B_{ik}'\rangle_r \rightarrow B_{LS}$$

Thus

$$M_{w_{app}}° = \lim_{\substack{w\rightarrow 0\\(\Lambda\rightarrow 0)}} M_{w_{app}}(\text{cell}) \tag{94}$$

$$= M_w°[1 - B_{LS}M_w°c_0 + \ldots]$$

and

$$\frac{1}{M_{w_{app}}°} = \frac{1}{M_w°} + B_{LS}c_0 \tag{95}$$

Here $M_w{}^\circ = M_w(\text{cell})$ is the weight-average molecular weight of the original sample. Thus a plot of $1/M_{w_{\text{app}}}{}^\circ$ *vs.* c_0 has a slope of B_{LS}.

Eventually it may be possible to develop an analog of Method 3 for a non-sector-shaped cell. Here one lets

$$I = \int_0^1 <B_{ik}'>_r (cM_{wr})^2 dx \tag{96}$$

Now expand I in a Maclaurin series in Λ; thus

$$I = I_0 + \left(\frac{dI}{d\Lambda}\right)_{\Lambda=0} \Lambda + \ldots \tag{97}$$

$$= I_0 \left[1 + \left(\frac{d \ln I}{d\Lambda}\right)_{\Lambda=0} \Lambda + \ldots\right]$$

The quantity $I_0 = B_{\text{LS}} c_0 M_w{}^\circ$ is the value of I at zero speed. The substitution of Equation 97 into Equation 93 leads to

$$\frac{1}{M_{w_{\text{app}}}} = \frac{1}{M_w} + B_{\text{LS}} c_0 \left[1 + \left(\frac{d \ln I}{d\Lambda}\right)_{\Lambda=0} \Lambda + \ldots\right]$$

Thus at $\Lambda = 0$ Equation 95 is definitely obtained, which can be used for the evaluation of B_{LS} as described. Whether the quantity $c_0 \, [1 + (d \ln I/d\Lambda)_{\Lambda=0}\Lambda + \ldots] \simeq \bar{c}$ or not is not known yet.

Concluding Remarks

There are three points to emphasize. First, the expressions for the concentration or concentration gradient distribution for non-sector-shaped centerpieces can be applied to other methods for obtaining MWD's, such as the Fourier convolution theorem method (*10, 15, 16*), or to more recent methods developed by Gehatia and Wiff (*38–40*). The second point is that the method for the nonideal correction is general. Since these corrections are applied to the basic sedimentation equilibrium equation, the treatment is universal. The corrected sedimentation equilibrium equation (see Equation 78 or 83) forms the basis for any treatment of MWD's. Third, the Laplace transform method described here and elsewhere (*11, 12*) is not restricted to the three examples presented here. For those cases where the plots of $F(n, u)$ *vs.* u will not fit the three cases described in Table I, it should still be possible to obtain an analytical expression for $F(n, u)$ which is different from those in Table I. This expression for $F(n, u)$ could then be used to obtain an equation in s using procedures described in the text (*see* Equations 39 and 44). Equation 39 would then be used to obtain the desired Laplace transform.

The application of the complex inversion formula (*41, 42*) may lead to the inverse Laplace transform from which the MWD is obtained.

Literature Cited

1. Svedberg, T., Rinde, H., *J. Amer. Chem. Soc.* (1924) **46,** 2677.
2. Rinde, H., Ph.D. Dissertation, University of Uppsala, 1928.
3. Lansing, W. D., Kraemer, E. O., *J. Amer. Chem. Soc.* (1935) **57,** 1369.
4. Williams, J. W., Saunders, W. M., Cicirelli, J. S., *J. Phys. Chem.* (1954) **58,** 774.
5. Williams, J. W., Saunders, W. M., *J. Phys. Chem.* (1954) **58,** 854.
6. Wales, M., *J. Phys. Colloid Chem.* (1948) **52,** 235.
7. Wales, M., *J. Phys. Colloid Chem.* (1951) **55,** 282.
8. Wales, M., Adler, F. T., Van Holde, K. E., *J. Phys. Colloid Chem.* (1951) **55,** 145.
9. Wales, M., Williams, J. W., Thompson, J. O., Ewart, B. H., *J. Phys. Chem.* (1948) **52,** 983.
10. Sundelöf, L. O., *Ark. Kem.* (1968) **29,** 297.
11. Donnelly, T. H., *J. Phys. Chem.* (1966) **70,** 1862.
12. Donnelly, T. H., *Ann. N. Y. Acad. Sci.* (1969) **164,** 147.
13. Scholte, Th. G., *J. Polym. Sci., Part A-2* (1968) **6,** 111.
14. Scholte, Th. G., *Ann. N. Y. Acad. Sci.* (1969) **164,** 156.
15. Provencher, S. W., Gobush, W., in "Characterization of Macromolecular Structure," p. 143, Publication 1573, National Academy of Sciences, Washington, D. C., 1968.
16. Provencher, S. W., *J. Chem. Phys.* (1967) **46,** 3229.
17. Fujita, H., "Mathematical Theory of Sedimentation Analysis," Chapter V, Academic Press, New York, 1962.
18. Casassa, E. F., Eisenberg, H., *Advan. Protein Chem.* (1964) **19,** 287.
19. Svedberg, T., Pedersen, K. O., "The Ultracentrifuge," Oxford University Press, 1940.
20. Schachman, H. K., *Biochemistry* (1963) **2,** 887.
21. Schachman, H. K., in "Ultracentrifugal Analysis in Theory and Practice," p. 171, J. W. Williams, Ed., Academic Press, New York, 1963.
22. Chervenka, C. H., *Fractions* (1971) No. 1.
23. Van Holde, K. E., Baldwin, R. L., *J. Phys. Chem.* (1968) **62,** 734.
24. Yphantis, D. A., *Biochemistry* (1964) **3,** 297.
25. Albright, D. A., Williams, J. W., *J. Phys. Chem.* (1967) **71,** 2780.
26. Osterhoudt, H. W., Williams, J. W., *J. Phys. Chem.* (1965) **69,** 1050.
27. Tanford, C., "Physical Chemistry of Macromolecules," Chapter 4, Section 12, Wiley, New York, 1961.
28. Van Holde, K. E., "Physical Biochemistry," pp. 33–37, Prentice-Hall, Englewood Cliffs, N. J., 1971.
29. Goldberg, R. J., *J. Phys. Chem.* (1953) **57,** 194.
30. Fujita, H., *J. Phys. Chem.* (1959) **63,** 1326.
31. Fujita, H., Inagaki, H., Kotaka, T., Utimaya, H., *J. Phys. Chem.* (1962) **66,** 4.
32. Fujita, H., *J. Phys. Chem.* (1969) **73,** 1759.
33. Adams, E. T., Soucek, D. A., Barlow, G., Wan, P. J., *Polym. Prepr., Amer. Chem. Soc., Div. Polym. Chem.* (1971) **12,** 891.
34. Kegeles, G., Klainer, S. M., Salem, W. J., *J. Phys. Chem.* (1957) **61,** 1286.
35. Adams, E. T., Jr., *Proc. Nat. Acad. Sci., U. S.* (1964) **51,** 109.
36. Adams, E. T., Jr., in "Characterization of Macromolecular Structure," p. 106, Publication 1573, National Academy of Sciences, Washington, D. C., 1968.

37. Deonier, R. C., Williams, J. W., *Proc. Nat. Acad. Sci., U. S.* (1969) **64**, 828.
38. Gehatia, M., Wiff, D. R., *J. Polym. Sci., Part A-2* (1970) **8**, 7039.
39. Gehatia, M., Wiff, D. R., *J. Chem. Phys.* (1972) **57**, 1070.
40. Gehatia, M., *Polym. Prepr., Amer. Chem. Soc., Div. Polym. Chem.* (1971) **12**, 875.
41. Churchill, R. V., "Operational Mathematics," 2nd ed., Chapter 6, McGraw-Hill, New York, 1958.
42. Spiegel, M. R., "Schaum's Outline of Theory and Problems of Laplace Transforms," Chapter 7, McGraw-Hill, New York, 1965.

RECEIVED January 17, 1972. Work supported in part by grants (to E. T. A., Jr.) GM 15551 and GM 17611 from the U. S. Public Health Service, National Institutes of Health.

22

The Study of Mixed Associations by Sedimentation Equilibrium and by Light Scattering Experiments

ALLEN H. PEKAR,[1] PETER J. WAN,[2] and E. T. ADAMS, Jr.[2]

Illinois Institute of Technology, Chicago, Ill. 60616 and Texas A&M University, College Station, Tex. 77843

The analysis of mixed associations by light scattering and sedimentation equilibrium experiments has been restricted so far to ideal, dilute solutions. Also it has been necessary to assume that the refractive index increments as well as the partial specific volumes of the associating species are equal. These two restrictions are removed in this study. Using some simple assumptions, methods are reported for the analysis of ideal or nonideal mixed associations by either experimental technique. The advantages and disadvantages of these two techniques for studying mixed associations are discussed. The application of these methods to various types of mixed associations is presented.

Associations between two macromolecules, A and B, of the type

$$nA + mB \rightleftarrows A_nB_m \qquad (n,m = 1,2, \ldots), \text{ or} \tag{1}$$

$$\begin{aligned} A + B &\rightleftarrows AB \\ 2A &\rightleftarrows A_2 \end{aligned} \tag{2}$$

as well as other related associations are known as mixed associations. These associations can occur in a variety of ways, and as Equation 2 indicates both complex formation and self-association can occur simul-

[1] Present address: The Lilly Research Laboratories, Eli Lilly & Co., Indianapolis, Ind. 46206.
[2] Present address: Texas A&M University, College Station, Tex. 77843.

taneously. This behavior may occur in the formation of the trypsin–trypsin inhibitor complex (*1, 2*) or the insulin–protamine (*3*) complex. Perhaps the best studied example of a mixed association is the antigen–antibody reaction (*4, 5, 6*). Although mixed associations have been studied by a variety of ways (*7, 8, 9, 10, 11*), relatively little has been reported on the study of mixed associations by light scattering experiments (*1, 6, 12*), by sedimentation equilibrium experiments (*13, 14, 15*), or by the latter's relative, the Archibald experiment (*14, 16, 17*).

With ideal mixed associations one can utilize the redistribution of components in sedimentation equilibrium experiments to evaluate the equilibrium constant(s); this has been described extensively in previous publications (*13, 14, 15*). In nonideal systems, the analysis of mixed associations is more complicated so that one must resort to sedimentation equilibrium experiments at different speeds on the same solution. Here one calculates M_w (cell mass) or M_z (cell), the weight—(M_w) or z—(M_z) average molecular weights, over the cell at each speed. For nonideal solutions one calculates the apparent values of these quantities. These values are extrapolated to zero speed, in a manner similar to that used by Albright and Williams (*18*) with nonideal, nonassociating polymer solutions. The required calculations are done with these values at zero speed. With the Archibald experiment (*14*) one extrapolates the values of M_{wa}, the apparent weight-average molecular weight, obtained at various times at the extremes (r_m or r_b) of the solution column in the ultracentrifuge, to zero time. In both cases the analysis becomes quite similar to that used with the light scattering experiment; hence the reason for discussing these methods together.

Ideal Solutions

Preliminary Thermodynamic Relations. Here it will be assumed that we are dealing with incompressible systems; this is a very good assumption for aqueous solutions since the isothermal compressibility of water is so small. At constant temperature the equilibrium condition for any mixed association (*see* Equations 1 and 2 for example) is

$$n\mu_{\mathrm{A}} + m\mu_{\mathrm{B}} = \mu_{\mathrm{A}_n\mathrm{B}_m} \; (n,m = 0,1,2, \ldots) \tag{3}$$

Here μ_i ($i = $ A, B, or A_nB_m) is the molar chemical potential of reacting species i. Equation 3 is valid for self-associations as well since n or m is zero in that case. Under ideal (theta) solution conditions the activity coefficient y_i of each of the associating species is one, so that

$$c_{\mathrm{A}_n\mathrm{B}_m} = Kc_{\mathrm{A}}{}^{n}c_{\mathrm{B}}{}^{m} \tag{4}$$

Here $K = K_{A_nB_m}$. In the discussion that follows, we use the mixed association described by Equation 1 as an example since the methods we will develop can, in general, be applied to other mixed associations. The total concentration c (we will use grams/liter) of all the associating species is given by

$$c = c_A + c_B + Kc_A{}^n c_B{}^m = f(c_A, c_B) \tag{5}$$

From Equations 1 and 5 we note that c is a function of c_A and c_B only. Now note that we can differentiate c with respect to c_A or c_B; thus

$$(\partial c/\partial c_A)_{c_B} = 1 + nKc_A{}^{n-1}c_B{}^m, \text{ and} \tag{6}$$

$$(\partial c/\partial c_B)_{c_A} = 1 + mKc_A{}^n c_B{}^{m-1} \tag{7}$$

Also, note that

$$M_{A_nB_m} = nM_A + mM_B \tag{8}$$

For any mixed association the weight-average molecular weight ($M_w{}^{eq}$) depends on the total concentration c of the associating species and also on the initial proportion of the reactants (A and B). A solution for which $c = 10$ grams/liter could be made up of a variety of blends of A and B; each blend would have the same value of c but they would each have a different value of $M_w{}^{eq}$. This suggests that we use the following procedure to simplify matters. Let us assume concentrated stock solutions of A and B are prepared. In work with proteins (or other polyelectrolytes) this means that the concentrated stock solutions (one for A and one for B) are dialyzed against buffer, with several changes of buffer. Now the dialyzed, concentrated stock solutions of A and B are blended so that the initial concentration ratio of A to B in the blend (assuming no chemical reaction has occurred) is $c_A{}^0/c_B{}^0 = \beta$. This blend is known as the working stock; dilutions are prepared from it, using the buffer that was in dialysis equilibrium with the concentrated stock solution. Note that β is the same for all dilutions, yet c is different for each of the solutions (working stock plus dilutions). If this procedure is followed, $M_w{}^{eq}$ or its apparent value M_{wa} will vary with c since the chemical equilibrium will cause more A_nB_m to form as c increases. This effect is shown in Figure 1; here we have plotted $1/M_{wa}$ *vs.* c at constant β for four different situations: ideal, nonassociating mixture of A and B (curve 1); nonideal, nonassociating mixture of A and B (curve 2); ideal, mixed association of A and B (curve 3); nonideal, mixed association of A and B (curve 4). Clearly such a plot gives much useful information.

The quantity M_{wa} can be defined by $1/M_{wa} = (1/M_w) + Bc$, where M_w is the true weight-average molecular weight and B is the second virial coefficient. For ideal, nonassociating mixtures of macromolecules, B is zero and $1/M_{wa} = 1/M_w$. Thus a plot of $1/M_{wa}$ *vs.* c will have a slope of zero, which is illustrated in curve 1 of Figure 1. For a nonideal, nonassociating mixture of macromolecules the second virial coefficient will be greater than zero. Since M_w is a constant when there is no association, a plot of $1/M_{wa}$ *vs.* c will have an intercept of $1/M_w$ and a slope (or limiting slope) equal to B, the second virial coefficient. This is shown in curve 2 of Figure 1.

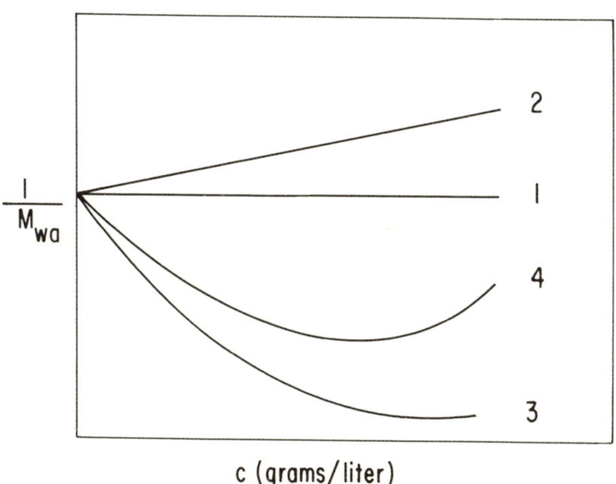

Figure 1. Curve 1 could represent an ideal polymer solution containing A and B but undergoing no association; the nonideal counterpart of this is shown in curve 2. An ideal mixed association between A and B, such as described by Equation 1 might be described by curve 3, whereas, curve 4 could represent a nonideal, mixed association.

Solutions exhibiting greater deviations from ideal behavior would give a curve instead of the straight line shown in curve 2 of Figure 1. If a mixed association occurs, then $M_w = M_w^{eq}$ will increase with c (at constant β) because of the chemical equilibrium; this means that $1/M_w^{eq}$ will decrease with increasing c as is shown in curve 3 of Figure 1. Here $M_{wa} = M_w = M_w^{eq}$. For a nonideal, mixed association with the second virial coefficient (B) greater than zero, one could encounter a minimum in the plot of $1/M_{wa}$ *vs.* c. This is because the term Bc increases with increasing c while $1/M_w^{eq}$ decreases with increasing c; such a situation is illustrated in curve 4 of Figure 1.

The quantity cM_w^{eq} is defined by

$$\begin{aligned} cM_w^{eq} &= c_A M_A + c_B M_B + K c_A{}^n c_B{}^m M_{A_n B_m} \\ &= c_A M_A (\partial c/\partial c_A)_{T,c_B} + c_B M_B (\partial c/\partial c_B)_{T,c_A} \\ &= \sum_i c_i M_i (\partial c/\partial c_i)_{T,c_j} \qquad (i,j = \text{A or B}) \end{aligned} \tag{9}$$

Equations 6–8 have been used to obtain Equation 9; Equation 9 is a general equation which can be applied to all mixed associations. The final form of Equation 9 indicates that there are only two solute components (independent variables) involved in the mixed association. When no association occurs, then $(\partial c/\partial c_i)_{T,c_j} = 1$. The superscript "eq" is used to indicate a mixed association is present. Now that the quantity cM_w^{eq} has been defined, how do we obtain M_w^{eq}, and how do we use it (or its analogs) to obtain the equilibrium constant or constants and the nonideal terms if they are present?

Evaluation of M_w^{eq}. SEDIMENTATION EQUILIBRIUM EXPERIMENTS AND THE ARCHIBALD METHOD. At constant temperature the condition for sedimentation equilibrium is that the total potential $\bar{\mu}_i$ for each associating species i (i = A, B, or A_nB_m) be constant at every radial position r in the solution column of the ultracentrifuge; thus

$$\bar{\mu}_i = \mu_i - \frac{M_i \omega^2 r^2}{2} = \text{constant} \tag{10}$$

Here μ_i is the molar chemical potential of species i, $\omega = 2\pi\text{rpm}/60$ is the angular velocity (radians/second) of the rotor, and M_i is the molecular weight of associating species i. From this relation it can be shown that (*14*)

$$\left(\frac{dc}{d(r^2)}\right)_T = \left(\frac{dc_A}{d(r^2)}\right)_T \left(\frac{\partial c}{\partial c_A}\right)_{T,c_B} + \left(\frac{dc_B}{d(r^2)}\right)_T \left(\frac{\partial c}{\partial c_B}\right)_{T,c_A} = L \sum_i c_i M_i (\partial c/\partial c_i)_{T,c_j} = LcM_w^{eq} \tag{11}$$

$$\text{Here, } L = (1 - \bar{v}\rho)\omega^2/2RT \tag{12}$$

In obtaining Equation 11 it has been assumed that the partial specific volumes $\bar{v}$ of the associating species are equal; we have also assumed that the specific refractive index increments Ψ of the associating solutes are equal. In Equation 12 R is the universal gas constant (8.314×10^7 ergs/deg-mole), ρ is the density of the solution (gram/ml), and T is the absolute temperature. Equation 11 is also valid for the Archibald experiment but *only* at r_m or r_b, the radial positions (in the solution column of the ultracentrifuge cell) of the air-solution meniscus and of the cell bottom, respectively. Archibald (*19*) was the first to point out that there is no flow at the extremes (r_m or r_b) of the ultracentrifuge cell solution column since it is a closed system; this fact could be used to show that one can obtain $M_{w,t}$ at various times t at r_m or r_b. Here one notes that

$$\frac{1}{2rcL}\frac{dc}{dr} = M_{w,t} \text{ at } r_m \text{ or } r_b \tag{13}$$

The centrifugal field causes a redistribution of the reacting species in such a way that chemical equilibrium is maintained; in order to get M_w^{eq} at c_0 it is necessary to extrapolate values of $M_{w,t}$ to zero time, *i.e.*,

$$\lim_{t \to 0} M_{w,t} = M_w^{eq} \text{ at } c_0 \tag{14}$$

To utilize Equation 13 it is necessary to know c_{r_n} or c_{r_b}; here we use the following equations, provided a plateau (a region where the concentration gradient $dc/dr = 0$) exists:

$$c_{r_m} = c_0 - \frac{1}{r_m^2}\int_{r_m}^{r_p} r^2 \frac{dc}{dr}\, dr \tag{15}$$

Here r_p is any radial position in the plateau region.

$$c_{r_b} = c_0 + \frac{1}{r_b^2}\int_{r_p}^{r_b} r^2 \frac{dc}{dr}\, dr \tag{16}$$

Equations 15 and 16 were first obtained by Kegeles and his co-workers (*16, 20, 21*) and they are valid only for cells with sector-shaped centerpieces. Because of convective disturbances in the solution column of the ultracentrifuge cell during the transient state when nonsector-shaped cells are used, it is customary to do Archibald experiments in cells with sector-shaped centerpieces which avoid this problem. For details on the Archibald method—pitfalls, extrapolation to zero time—one should consult the papers of LaBar (*17*) and Fujita *et al.* (*22*).

The quantity cM_w^{eq} can also be evaluated from sedimentation equilibrium experiments done at different speeds on the same solution. Here one gets to sedimentation equilibrium with a given solution at one speed; the necessary information (concentration and/or concentration gradient) is recorded. Then the speed is changed; once sedimentation equilibrium is attained at the second speed, c and/or dc/dr are recorded. This procedure is repeated at two or more additional speeds. Then one calculates M_w (cell mass) or M_w (cell vol) from the following equations:

$$M_{\mathrm{w}}\ (\text{cell mass}) = \frac{c_{r_b} - c_{r_m}}{\Lambda c_0} \tag{17}$$

$$M_{\mathrm{w}}\ (\text{cell vol}) = \frac{\ln[c_{r_b}/c_{r_m}]}{\Lambda} \tag{18}$$

$$\text{Here } \Lambda = L(r_b^2 - r_m^2) = \frac{(1 - \bar{v}\rho)\omega^2\,(r_b^2 - r_m^2)}{2RT} \tag{19}$$

Because of the redistribution of reacting species by the centrifugal field, one notes (*23*) that M_{w} (cell mass) and M_{w} (cell vol) do *not* represent the value of $M_{\mathrm{w}}^{\mathrm{eq}}$ at c_0, the original concentration. However, if these values are extrapolated to zero speed, they become equal to $M_{\mathrm{w}}^{\mathrm{eq}}$, *i.e.*,

$$\lim_{\Lambda \to 0} M_{\mathrm{w}}\ (\text{cell mass}) = \lim_{\Lambda \to 0} M_{\mathrm{w}}\ (\text{cell vol}) = M_{\mathrm{w}}^{\mathrm{eq}} \text{ at } c_0 \tag{20}$$

Thus one has two ways to obtain $M_{\mathrm{w}}^{\mathrm{eq}}$. In addition one can also calculate M_{z} (cell) or M_{z} (cell vol) where

$$M_{\mathrm{z}}\ (\text{cell}) = \frac{\left(\frac{1}{r}\frac{dc}{dr}\right)_{r_b} - \left(\frac{1}{r}\frac{dc}{dr}\right)_{r_m}}{2L(c_{r_b} - c_{r_m})} \tag{21}$$

$$\text{and } M_{\mathrm{z}}\ (\text{cell vol}) = \frac{\ln\left[\left(\frac{1}{r}\frac{dc}{dr}\right)_{r_b} \Big/ \left(\frac{1}{r}\frac{dc}{dr}\right)_{r_m}\right]}{\Lambda} \tag{22}$$

Here one notes that

$$\lim_{L \to 0} M_{\mathrm{z}}\ (\text{cell}) = \lim_{\Lambda \to 0} M_{\mathrm{z}}\ (\text{cell vol}) = M_{\mathrm{z}} \text{ at } c_0 \tag{23}$$

Light Scattering Experiments. This discussion is concerned only with elastic light scattering (*24*). In these experiments one measures the reduced scattering intensity (sometimes called the Rayleigh ratio) $R(\theta)$, where

$$R(\theta) = \frac{i_\theta r^2}{I_0\ (1 + \cos^2 \theta)} \tag{24}$$

Here r is the distance of the photomultiplier tube from the center of the cell, i_θ is the intensity of the scattered light at an angle θ to the path of the incident light beam, and I_0 is the intensity of the incident light. For molecules whose maximum dimension is less than $\lambda/20$, where λ is the wavelength of the light being used, $R(\theta)$ is usually independent of angle. For very large molecules a correction for the angular dependence of the light scattering must be made; this is usually done by means of a Zimm plot (*24, 25, 26, 27*). It will be assumed that angular dependence, if present, has been overcome, and we will use the symbol $R(0)$ to indicate this. One can evaluate M_w^{eq} from the equation

$$\frac{1000\Delta R(0)}{K\Psi'^2} = cM_w^{eq} \tag{25}$$

Here K is the well known light scattering constant and is defined by

$$K = 2\pi^2 n_0^2/\lambda^4 N$$

In this equation n_0 is the refractive index of the solvent, λ is the wavelength of light, and N is Avogadro's number. The quantity $\Delta R(0)$ in Equation 25 is defined by

$$\Delta R(0) = R(0)\text{solution} - R(0)\text{solvent}$$

Analysis of Mixed Associations Using M_w^{eq}. CONVENTIONAL SEDIMENTATION EQUILIBRIUM EXPERIMENTS. Usually sedimentation equilibrium experiments are run at one speed only for each concentration; these are known as conventional sedimentation equilibrium experiments. Although the choice of speed is arbitrary, practical considerations force one to choose a speed to evaluate the concentration (c) or the concentration gradient (dc/dr) near the cell bottom; otherwise blurring or defocusing of the optical images in that region at high speeds results. For this situation the basic sedimentation equilibrium equation becomes

$$\frac{d\ \ln\ c_i}{d\ (r^2)} = LM_i,\ (i = \text{A or B}) \tag{26}$$

$$L = (1 - \bar{v}\rho)\omega^2/2RT$$

for each component. This equation can be integrated between r_m and r and then converted to exponential form to give

$$\begin{aligned} c_{ir} &= c_{ir_m} e^{LM_i(r^2 - r_m^2)} \\ &= c_{ir_m} e^{\phi_{ir}} \end{aligned} \tag{27}$$

where $\phi_{ir} = LM_i(r^2 - r_m^2)$. The total concentration c_r can be expressed as

$$\begin{aligned} c_r &= c_{Ar} + c_{Br} + Kc_{Ar}^{\ n} c_{Br}^{\ m} \\ &= c_{Ar_m} e^{\phi_{Ar}} + c_{Br_m} e^{\phi_{Br}} + Kc_{Ar_m}^{\ n} c_{Br_m}^{\ m} e^{n\phi_{Ar} + m\phi_{Br}} \end{aligned} \tag{28}$$

for a mixed association described by Equation 1. One can take values of c_r at three different radial positions (not too close to each other) and obtain three simultaneous equations which can be solved by standard methods to give $c_{\mathrm{A}r_m}$, $c_{\mathrm{B}r_m}$, and $Kc_{\mathrm{A}r_m}{}^n c_{\mathrm{B}r_m}{}^m$ (*14, 15*). From these quantities one can calculate K. By comparing results obtained with different radial positions and two or more experiments, one can test for ideal behavior since K would be the same in all cases if the solution were ideal. One can also use the concentration gradient to evaluate K and test for ideality. First one notes that

$$\frac{1}{2rL}\,\frac{dc}{dr} = cM_{\mathrm{w}}{}^{\mathrm{eq}} = c_{\mathrm{A}}M_{\mathrm{A}} + c_{\mathrm{B}}M_{\mathrm{B}} + Kc_{\mathrm{A}}{}^n c_{\mathrm{B}}{}^m M_{\mathrm{A}_n\mathrm{B}_m} \tag{29}$$

Now divide Equation 29 by M_{A} and then subtract c to give

$$\frac{1}{2rLM_{\mathrm{A}}}\,\frac{dc}{dr} - c = c_{\mathrm{B}}\left(\frac{M_{\mathrm{B}}}{M_{\mathrm{A}}} - 1\right) + Kc_{\mathrm{A}}{}^n c_{\mathrm{B}}{}^m \left(\frac{M_{\mathrm{A}_n\mathrm{B}_m}}{M_{\mathrm{A}}} - 1\right) \tag{30}$$

At this point apply Equation 27 to Equation 30 and then multiply both sides by $e^{-\phi_{\mathrm{B}r}}$; this leads to

$$Y = \left\{\frac{1}{2rLM_{\mathrm{A}}}\,\frac{dc}{dr} - c\right\}e^{-\phi_{\mathrm{B}r}} \tag{31}$$

$$= c_{\mathrm{B}r_m}\left(\frac{M_{\mathrm{B}}}{M_{\mathrm{A}}} - 1\right) + Kc_{\mathrm{A}r_m}^{\;n}c_{\mathrm{B}r_m}^{\;m}\left(\frac{M_{\mathrm{A}_n\mathrm{B}_m}}{M_{\mathrm{A}}} - 1\right)e^{n\phi_{\mathrm{A}r}+(m-1)\phi_{\mathrm{B}r}}$$

A plot of Y *vs.* $e^{n\phi_{\mathrm{A}r}+(m-1)\phi_{\mathrm{B}r}}$ will have a slope equal to:

$$\text{Slope} = Kc_{\mathrm{A}r_m}^{\;n}c_{\mathrm{B}r_m}^{\;m}\left(\frac{M_{\mathrm{A}_n\mathrm{B}_m}}{M_{\mathrm{A}}} - 1\right) \tag{32}$$

The intercept at $r = r_m$ is

$$\text{Intercept} = c_{\mathrm{B}r_m}\left(\frac{M_{\mathrm{B}}}{M_{\mathrm{A}}} - 1\right) + \left(\frac{M_{\mathrm{A}_n\mathrm{B}_m}}{M_{\mathrm{A}}} - 1\right)Kc_{\mathrm{A}r_m}^{\;n}c_{\mathrm{B}r_m}^{\;m} \tag{33}$$

Since $c_{r_m} = c_{\mathrm{A}r_m} + c_{\mathrm{B}r_m} + Kc_{\mathrm{A}r_n}{}^n c_{\mathrm{B}r_m}{}^m$ one can calculate K. Upward curvature in the plot based on Equation 31 indicates the presence of higher aggregates whereas downward curvature may be caused by nonideal effects.

Archibald Experiments, Sedimentation Equilibrium Experiments at Different Speeds, and Light Scattering Experiments The analysis of a mixed association described by Equation 1 is similar for these three methods. First of all let us introduce the conservation of mass equations. For

$$nA + mB \rightleftarrows A_nB_m \qquad (n, m = 0, 1, 2, \ldots) \tag{1}$$

the initial concentrations c_i^0 (i = A or B) are related to the chemical equilibrium concentrations c_i as follows:

$$c_A^0 = c_A + \frac{nKc_A{}^nc_B{}^m}{M_{A_nB_m}} M_A \tag{34}$$

$$\text{and } c_B^0 = c_B + \frac{mKc_A{}^nc_B{}^m}{M_{A_nB_m}} M_B \tag{35}$$

Equations 34 and 35 are known as the conservation of mass equations. From Equations 34 and 35 it follows that:

$$\begin{aligned} c &= c_A^0 + c_B^0 \\ &= c_A + c_B + Kc_A{}^nc_B{}^m \end{aligned} \tag{36}$$

One can also define a quantity M_w^0, which would represent the value of M_w that a solution containing A and B in a nonreacting mixture would have. Thus,

$$\begin{aligned} cM_w^0 &= c_A^0M_A + c_B^0M_B \\ &= c_AM_A + c_BM_B + \frac{Kc_A{}^nc_B{}^m}{M_{A_nB_m}}[nM_A^2 + mM_B^2] \end{aligned} \tag{37}$$

On the other hand, cM_w^{eq} can be written as:

$$\begin{aligned} cM_w^{eq} &= c_AM_A(1 + nKc_A{}^{n-1}c_B{}^m) + c_BM_B(1 + mKc_A{}^nc_B{}^{m-1}) \\ &= c_AM_A + c_BM_B + Kc_A{}^nc_B{}^mM_{A_nB_m} \\ &= c_AM_A + c_BM_B + \frac{Kc_A{}^nc_B{}^m}{M_{A_nB_m}}[n^2M_A^2 + 2nmM_AM_B + m^2M_B^2] \end{aligned} \tag{38}$$

Then the quantity $\Delta(cM_w)$ becomes

$$\begin{aligned} \Delta(cM_w) &= cM_w^{eq} - cM_w^0 \\ &= \frac{Kc_A{}^nc_B{}^m}{M_{A_nB_m}}[(n^2 - n)M_A^2 + 2mnM_AM_B + (m^2 - m)M_B^2] \qquad (39) \\ &= \frac{Kc_A{}^nc_B{}^m}{M_{A_nB_m}} Q \end{aligned}$$

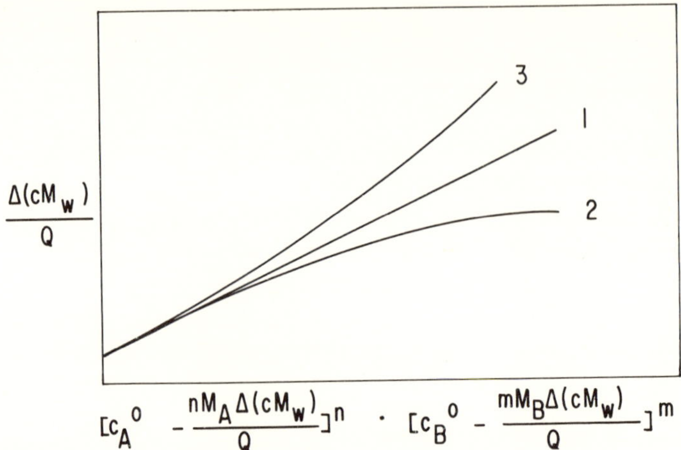

Figure 2. This plot is based on Equation 41. An ideal association described by Equation 1 could be represented by curve 1; its nonideal counterpart could be represented by curve 2. The presence of higher aggregates, such as A_nM_{m+1}, might be described by curve 3.

where $Q = (n^2 - n)M_A{}^2 + 2nmM_AM_B + (m^2 - m)M_B{}^2$. This procedure was first suggested by Steiner (*1*). From Equation 39 we obtain

$$\frac{\Delta(cM_w)}{Q} = \frac{Kc_A{}^n c_B{}^m}{M_{A_nB_m}} \tag{40}$$

We can substitute Equation 40 into the conservation of mass equations (34 and 35) to obtain new expressions for c_A and c_B. Then we can substitute these new relations for c_A and c_B back into Equation 40 to obtain

$$\frac{\Delta(cM_w)}{Q} = \frac{K}{M_{A_nB_m}}\left[c^0{}_A - \frac{nM_A\Delta(cM_w)}{Q}\right]^n \cdot \left[c^0{}_B - \frac{mM_B\Delta(cM_w)}{Q}\right]^m \tag{41}$$

A plot of $\Delta(cM_w)/Q$ *vs.* the product on the right side of Equation 41 has a slope equal to $K/M_{A_nB_m}$. A plot based on Equation 41 has the following advantages: (1) it gives a test for ideality since downward curvature may be the result of nonideal effects; (2) it provides a test for the type of association present since upward curvature may be indicative of the presence of higher aggregates ($A_{n+1}B_m$, etc.); (3) it avoids the use of a limiting slope. These advantages are illustrated in Figure 2, which is based on Equation 41.

With sedimentation equilibrium experiments at different speeds one can also use

$$\Delta(\sum_i c_i M_i^2) = c(M_w M_z)^{eq} - c(M_w M_z)^0 = 3 M_A M_B K c_A c_B \tag{42}$$

for an A + B ⇄ AB association (*14*). Methods have also been reported for evaluating n and m if they are not known *a priori* (*13, 14*); we will not discuss these details here.

Now let us illustrate a somewhat more complicated problem. Suppose the mixed association is described by Equation 2; how do we do the analysis? Here we note that

$$cM_w^{eq} = c_A M_A + c_B M_B + K_{AB} c_A c_B M_{AB} + 2K_{A_2} c_A^2 M_A \tag{43}$$

and

$$\Delta(cM_w) = cM_w^{eq} - cM_w^0 = \frac{2K_{AB} M_A M_B c_A c_B}{M_{AB}} + K_{A_2} c_A^2 M_A \tag{44}$$

Since $\beta = c_A^0/c_B^0$,

$$\lim_{\substack{c \to 0 \\ \beta = \text{const}}} \frac{\Delta(cM_w)}{c_A^0 c_B^0} = \frac{2K_{AB} M_A M_B}{M_{AB}} + \beta K_{A_2} M_A = \chi \tag{45}$$

Note that

$$\lim_{\substack{c \to 0 \\ \beta = \text{const}}} \frac{c_A^2}{c_A^0 c_B^0} = \lim_{\substack{c_A^0 \to 0 \\ \beta = \text{const}}} \frac{\beta c_A^2}{(c_A^0)^2} = \beta \tag{46}$$

Thus by doing these experiments at two or more values of β, one can obtain two or more values of χ, which can be used to obtain K_{AB} and K_{A_2}. In this situation one must use a limiting intercept; there is no simple plot available as there was from Equation 41.

Analysis of Mixed Associations When $\bar{v}_A \neq \bar{v}_B$ and $\Psi_A \neq \Psi_B$. So far we have assumed that $\bar{v}_A = \bar{v}_B = \bar{v}$ and $\Psi_A = \Psi_B = \Psi$; under these circumstances one could evaluate and use M_w^{eq}. Now suppose we wished to study the interaction of a bacterial hapten which was a polysaccharide with an antibody which was a protein. Why should one expect the partial specific volumes or the specific refractive index increments to be the same for these two reactants? Does this negate the analysis? The answer is no, and this is how we overcome these minor complications.

Note that the mass balance equations become

$$\tilde{n}_A{}^0 = n_A{}^0 - n_0 = \Psi'_A c_A{}^0 = \Psi'_A\left(c_A + \frac{nKc_A{}^n c_B{}^m}{M_{A_nB_m}} M_A\right) \tag{47}$$

$$\text{and } \tilde{n}_B{}^0 = n_B{}^0 - n_0 = \Psi'_B c_B{}^0 = \Psi'_B\left(c_B + \frac{mKc_A{}^n c_B{}^m}{M_{A_nB_m}} M_B\right) \tag{47a}$$

for a mixed association described by the first equation. Furthermore, note that

$$\tilde{n} = \tilde{n}_A{}^0 + \tilde{n}_B{}^0 = n - n_0 = \Psi'_A c_A + \Psi'_B c_B + \frac{Kc_A{}^n c_B{}^m}{M_{A_nB_m}}[n\Psi'_A M_A + m\Psi'_B M_B] = \Psi'_A c_A + \Psi'_B c_B + \Psi'_{AB} K c_A{}^n c_B{}^m \tag{48}$$

This means that $\Psi_{A_nB_m}$, the refractive index increment of A_nB_m, is defined by

$$\Psi'_{A_nB_m} = \frac{nM_A\Psi'_A + mM_B\Psi'_B}{M_{A_nB_m}} \tag{49}$$

Similarly we will define $\bar{v}_{A_nB_m}$

$$\bar{v}_{A_nB_m} = \frac{nM_A\bar{v}_A + mM_B\bar{v}_B}{M_{A_nB_m}} \tag{50}$$

Analysis of Mixed Associations from Conventional Sedimentation Equilibrium Experiments. In these experiments one measures a quantity $\overline{M}_1{}^{eq}$ (*14, 28*) instead of $M_w{}^{eq}$. The basic sedimentation equilibrium equation for each reactant is

$$\frac{d \ln \tilde{n}_i}{d(r)^2} = L_i M_i \tag{51}$$

$$\tilde{n}_i = n_i - n_0 = \Psi'_i c_i$$

This can be integrated to give

$$\tilde{n}_{ir} = \tilde{n}_{ir_m} \exp[L_i M_i(r^2 - r_m{}^2)] = \tilde{n}_{ir_m} \exp[\hat{\phi}_{ir}] \tag{52}$$

where $\hat{\phi}_{ir} = L_i M_i (r^2 - r_m^2)$. The Rayleigh optical system gives us information proportional to $n = \sum_i n_i = n - n_0$; the number of fringes J is related to $n - n_0$ by

$$J = \frac{h}{\lambda}(n - n_0) \tag{53}$$

Here h is the thickness of the ultracentrifuge centerpiece (usually 12 mm), $n - n_0$ is the refractive index difference between the solution (n) and the solvent (n_0), and λ is the wavelength of light used (generally 546 μ). Thus with the Rayleigh optical system one can obtain n, which can be expressed as

$$\begin{aligned} \tilde{n}_r &= \tilde{n}_{\mathrm{A}r} + \tilde{n}_{\mathrm{B}r} + \Psi'_{\mathrm{A}_n\mathrm{B}_m} K c_{\mathrm{A}}{}^n c_{\mathrm{B}}{}^m \\ &= \Psi'_{\mathrm{A}} c_{\mathrm{A}r_m} e^{\hat{\phi}_{\mathrm{A}r}} + \Psi'_{\mathrm{B}} c_{\mathrm{B}r_m} e^{\hat{\phi}_{\mathrm{B}r}} + \Psi'_{\mathrm{A}_n\mathrm{B}_m} K c_{\mathrm{A}r_m}^{\;n} c_{\mathrm{B}r_m}^{\;m} e^{n\hat{\phi}_{\mathrm{A}r} + m\hat{\phi}_{\mathrm{B}r}} \end{aligned} \tag{54}$$

Equation 54 is analogous to Equation 28, and one can take three values of $\tilde{n}_r$ at different radial positions to solve for $c_{\mathrm{A}r_m}$, $c_{\mathrm{B}r_m}$, and $K c_{\mathrm{A}r_m} c_{\mathrm{B}r_m}$, and from this evaluate K. With the schlieren optics one obtains

$$\begin{aligned} \frac{d\tilde{n}}{d(r^2)} &= L_{\mathrm{A}} \tilde{n}_{\mathrm{A}} M_{\mathrm{A}} + L_{\mathrm{B}} \tilde{n}_{\mathrm{B}} M_{\mathrm{B}} + L_{\mathrm{A}_n\mathrm{B}_m} \tilde{n}_{\mathrm{A}_n\mathrm{B}_m} M_{\mathrm{A}_n\mathrm{B}_m} \\ &= \tilde{n} \overline{M}_1{}^{\mathrm{eq}} \frac{(1 - \bar{v}_{\mathrm{w}}\rho)\omega^2}{2RT} \end{aligned} \tag{55}$$

$$\text{Here, } \overline{M}_1{}^{\mathrm{eq}} = \frac{\sum_i \tilde{n}_i M_i (1 - \bar{v}_i \rho)}{\sum_i \tilde{n}_i (1 - \bar{v}_i \rho)} = \frac{\sum_i \tilde{n}_i M_i (1 - \bar{v}_i \rho)}{\tilde{n}(1 - \bar{v}_{\mathrm{w}} \rho)} \tag{56}$$

$$i = \mathrm{A, B, or A}_n\mathrm{B}_m$$

$$\tilde{n} = \sum \tilde{n}_i$$

$$(1 - \bar{v}_{\mathrm{w}}\rho) = \sum_i \tilde{n}_i (1 - \bar{v}_i \rho) / \sum_i \tilde{n}_i$$

With the aid of Equation 48, Equation 56 can be written as

$$\overline{M}_1{}^{\mathrm{eq}} = \frac{\sum_j c_j M_j (1 - \bar{v}_j \rho)(\partial \tilde{n}/\partial c_j)_{T,\, c_{i \neq j}}}{\tilde{n}(1 - \bar{v}_{\mathrm{w}}\rho)} \tag{56a}$$

$$j = \mathrm{A \text{ or } B}$$

One can use Equation 55 to obtain an analog Equation 31 which can be used to obtain K; the details for doing this will be found in Adams' paper (*14*).

Analysis from Archibald Experiments or from Sedimentation Equilibrium Experiments at Different Speeds. This discussion here is illustrated by the association described by Equation 1. When $\bar{v}_A \neq \bar{v}_B$ and $\Psi_A \neq \Psi_B$, analogs of Equations 17 and 18 will yield $\overline{M}_1^{eq}$ when extrapolated to zero time. Similarly, $\overline{M}_1^{eq}$ will also be obtained from the analog of Equation 13 when values of $\overline{M}_{1,t}$ are extrapolated to zero time. For this situation

$$\tilde{n}\overline{M}_1^{eq} = \Psi'_A c_A M_A + \Psi'_B c_B M_B + \Psi'_{A_nB_m} \cdot \frac{K c_A{}^n c_B{}^m}{M_{A_nB_m}} [n^2 M_A{}^2 + 2nm M_A M_B + m^2 M_B{}^2] \tag{57}$$

$$\text{whereas } \tilde{n}\overline{M}_1{}^0 = \Psi'_A c_A{}^0 M_A + \Psi'_B c_B{}^0 M_B = \Psi'_A c_A M_A + \Psi'_B c_B M_B + \Psi'_{A_nB_m} \frac{K c_A{}^n c_B{}^m}{M_{A_nB_m}} [n M_A{}^2 + m M_B{}^2] \tag{58}$$

$$\text{Thus, } \Delta(\tilde{n}\overline{M}_1{}^0) = \frac{K c_A{}^n c_B{}^m}{M_{A_nB_m}} \{M_A{}^2 (n^2 \Psi'_{A_nB_m} - n\Psi'_A) + 2mn M_A M_B \Psi'_{A_nB_m} + M_B{}^2 (m^2 \Psi'_{A_nB_m} - m\Psi'_B)\} = \frac{K c_A{}^n c_B{}^m}{M_{A_nB_m}} \hat{Q}$$

$$\text{or } \Delta(n\overline{M}_1{}^0)/\hat{Q} = K c_A{}^n c_B{}^m / M_{A_nB_m} \tag{59a}$$

The analysis from this step on is similar to that described following Equation 40, and it is possible to set up an equation analogous to Equation 41 and make a plot from it (*cf.* Figure 2).

Analysis from Light Scattering Experiments. When $\Psi_A \neq \Psi_B$, then the light scattering equation becomes

$$1000\Delta R(0) = \sum_i \Psi'^2_i c_i M_i = \sum_j \Psi'_j c_j M_j \, (\partial n/\partial c_j)_{T, c_{k \neq j}} \tag{60}$$

$$i = \text{A, B, or } A_nB_m$$

$$j = \text{A or B}$$

Note that we get neither M_w^{eq} nor $\overline{M}_1^{eq}$ here but an entirely different relation. The analysis will be illustrated with two examples.

Case I. $A + B \rightleftarrows AB$. For this case one notes that

$$\left(\sum_i \Psi'^2_i c_i M_i\right)^{eq} = c_A M_A \Psi'^2_A + c_B M_B \Psi'^2_B + K c_A c_B M_{AB} \Psi'^2_{AB} \tag{61}$$

With the aid of Equation 61, the mass balance equations (34 and 35), and the quantity $(\sum_i \Psi_i^2 c_i M_i)^0$ one can obtain

$$\Delta(\sum_i \Psi'^2_i c_i M_i) \equiv (\sum_i \Psi'^2_i c_i M_i)^{eq} - (\sum_i \Psi'^2_i c_i M_i)^0 \tag{62}$$
$$= \frac{2K c_A c_B M_A M_B \Psi'_A \Psi'_B}{M_{AB}}$$

Using the appropriate mass balance equations (*cf.* 34 and 35) for c_A^0 and c_B^0, one notes that Equation 62 can be rewritten as

$$\Delta(\sum_i \Psi'^2_i c_i M_i) = \frac{2KM_A M_B \Psi'_A \Psi'_B}{M_{AB}} \left[c_A^{\circ} - \frac{\Delta(\sum_i \Psi'^2_i c_i M_i)}{2M_B \Psi'_A \Psi'_B} \right] \tag{63}$$
$$\left[c_B^0 - \frac{\Delta(\sum_i \Psi'^2_i c_i M_i)}{2M_A \Psi'_A \Psi'_B} \right]$$

As in Equation 41, one can plot the quantity $\Delta(\sum_i \Psi_i^2 c_i M_i)$ against the product on the right side; the slope of this plot is $2KM_A M_B \Psi_A \Psi_B / M_{AB}$. Since all quantities making up the slope except K are known, one can easily obtain K. The plot based on Equation 62 is analogous to that shown in Figure 2 and has all the advantages of this plot.

Case II. A + B ⇄ AB and 2A ⇄ A₂. Here one has

$$(\sum_i \Psi'^2_i c_i M_i)^{eq} = c_A M_A \Psi'^2_A + c_B M_B \Psi'^2_B + K_{AB} c_A c_B M_{AB} \Psi'^2_{AB} \tag{64}$$
$$+ 2K_{A_2} c_A^2 \Psi'^2_A M_A$$

The quantity $\Delta(\sum_i c_i M_i \Psi_i^2)$ can be expressed as

$$\Delta(\sum_i \Psi'^2_i c_i M_i) = \frac{2K_{AB} \Psi'_A \Psi'_B M_A M_B c_A c_B}{M_{AB}} + K_{A_2} c_A^2 M_A \Psi'^2_A \tag{65}$$

At constant β ($\beta = c_A^0 / c_B^0$) one can obtain

$$\lim_{\substack{c \to 0 \\ \beta = \text{const}}} \frac{\Delta(\sum_i \Psi'^2_i c_i M_i)}{c_A^0 c_B^0} = \frac{2K_{AB} \Psi'_A \Psi'_B M_A M_B}{M_{AB}} + \beta K_{A_2} \Psi'^2_A M_A \tag{66}$$

By doing a series of experiments at two or more values of β, one can evaluate K_{AB} and K_{A_2}.

Nonideal Solutions

Conventions Used to Define Nonideality. The equations developed in the preceeding section for the ideal case have really crystallized our operational definition of ideal conditions, since they require (1) that $y_{A_nB_m}/y_A{}^n y_B{}^m = 1$ at *all* concentrations, and (2) that the experimental techniques give us a measure of $cM_w{}^{eq}$ or its analogs, such as $(\sum_i \Psi_i{}^2 c_i M_i)^{eq}$ or $\tilde{n}\overline{M}_1{}^{eq}$. For the nonideal case it will be assumed that the natural logarithm of the activity coefficient, y_i, on the c-scale (gram/liter) can be represented as

$$\ln y_A = B_{AA}c_A{}^0 + B_{AB}c_B{}^0 \tag{67}$$

$$\text{and } \ln y_B = B_{BA}c_A{}^0 + B_{BB}c_B{}^0 \tag{68}$$

for the reactants A and B. The reason for writing $\ln y_i$ (i = A or B) in this manner is that $c_i{}^0$ (i = A or B) represents the *total* concentration of each reactant that is present in the solution. Furthermore, if no mixed association is present then c_i and $c_i{}^0$ are one and the same. Now for an association described by Equation 1 one could write

$$\frac{y_{A_nB_m}}{y_A{}^n y_B{}^m} = 1 + \alpha_1 c_A{}^0 + \alpha_2 c_B{}^0 + \ldots \tag{69}$$

since this ratio must approach one at infinite dilution of the reactants. This would mean that the mass balance equation (*cf.* 34 and 35) would become

$$\begin{aligned} c_A{}^0 &= c_A + \frac{Kc_A{}^n c_B{}^m}{1 + \alpha_1 c_A{}^0 + \alpha_2 c_{B0} + \ldots} \\ &= c_A + Kc_A{}^n c_B{}^m \,(y_A{}^n y_B{}^m / y_{A_nB_m}) \end{aligned} \tag{70}$$

for reactant A; a similar equation applies for reactant B. Now these relations (Equations 67–70) could bc inserted into the appropriate equations for cM_{wa} or its analogs. However, do we really have enough information or precision to evaluate these quantities? To simplify matters we shall define a nonideal system as one in which Equations 67 and 68 apply and in which

$$y_{A_nB_m}/y_A{}^n y_B{}^m = 1 \tag{71}$$

Some reasons for doing this are as follows: (1) it simplifies the treatment and gives us a practical method for evaluating K and B_{AB} (or B_{BA}); (2)

real systems approach this model at very low concentrations; (3) it is analogous to the treatment used for the analysis of self-associating systems (*29, 30, 31, 32, 33*). It will not always be necessary to assume that Equation 71 applies; for a special circumstance in the sedimentation equilibrium experiment one can assume Equation 69 to apply. This is discussed later.

Analysis by Conventional Sedimentation Equilibrium Experiments. The analysis of nonideal, mixed associations from conventional sedimentation equilibrium experiments is a very difficult matter, and at present seems to be an impasse. These difficulties arise because the redistribution of the reactants is combined with nonideal behavior. We can illustrate the difficulties with the following example. The equation for reactant A can be written as

$$d\ln c_{\mathrm{A}} + \left(\frac{\partial \ln y_{\mathrm{A}}}{\partial c_{\mathrm{A}}}\right)_{T,c_{\mathrm{B}}} dc_{\mathrm{A}} + \left(\frac{\partial \ln y_{\mathrm{A}}}{\partial c_{\mathrm{B}}}\right)_{T,c_{\mathrm{B}}} dc_{\mathrm{B}} = LM_{\mathrm{A}} d(r^2) \qquad (72)$$

This can be integrated between r_m and r, and then rewritten in exponential form to give

$$\begin{aligned} c_{\mathrm{A}r} = c_{\mathrm{A}r_m} \exp \Big\{ LM_{\mathrm{A}} (r^2 - r_m^2) - \int_{c_{\mathrm{A}r_m}}^{c_{\mathrm{A}r}} \left(\frac{\partial \ln y_{\mathrm{A}}}{\partial c_{\mathrm{A}}}\right)_{T,c_{\mathrm{B}}} dc_{\mathrm{A}} \\ - \int_{c_{\mathrm{B}r_m}}^{c_{\mathrm{B}r}} \left(\frac{\partial \ln y_{\mathrm{A}}}{\partial c_{\mathrm{B}}}\right)_{T,c_{\mathrm{A}}} dc_{\mathrm{B}} \Big\} = c_{\mathrm{A}r_m} \exp \Big\{ LM_{\mathrm{A}} (r^2 - r_m^2) \qquad (73) \\ - \sum_i \int_{c_{ir_m}}^{c_{ir}} \left(\frac{\partial \ln y_{\mathrm{A}}}{\partial c_i}\right)_{T,c_j} dc_i \Big\} = c_{\mathrm{A}r_m} \exp \hat{\Phi}_{ir} \end{aligned}$$

A similar equation results for $c_{\mathrm{B}r_m}$. Now *a priori* we do not know ($\partial \ln y_{\mathrm{A}}/\partial c_{\mathrm{B}}$), $c_{\mathrm{A}r}$, $c_{\mathrm{B}r}$, or K; we do not know B_{AA} and B_{BB} (*see* Equations 67 and 68) from measurements on pure A and pure B. We have no way of knowing $\hat{\Phi}_{ir}$ *a priori,* so we cannot use equations similar to 28 or 31. Similar considerations apply if $\bar{v}_{\mathrm{A}} \neq \bar{v}_{\mathrm{B}}$ and $\Psi_{\mathrm{A}} \neq \Psi_{\mathrm{B}}$. Thus, it appears at present that we are forced to use the Archibald experiment or sedimentation equilibrium experiments at different speeds to analyze mixed associations when one is restricted to Rayleigh and/or schlieren optics.

There is one exception to this situation. If one has an ultracentrifuge equipped with a photoelectric scanner (*34, 35*) as well as Rayleigh and schlieren optics, one might be able to evaluate $c_{\mathrm{A}r}$, $c_{\mathrm{B}r}$, and hence $Kc_{\mathrm{A}r}^{n}c_{\mathrm{B}r}^{m}$ for a mixed association under the following circumstance. Suppose A and B each have a different absorption spectrum in the range of 220 to 560 μ. Now if there is no change in the absorption spectrum of A or B on forming the complex, then one can calculate cM_{w}^{0} at every r.

From Rayleigh and schlieren optics, one could calculate cM_{wa}, and hence one could calculate $\Delta(cM_{wa})$ at every r in the solution column of the ultracentrifuge cell and use the analysis used with the other methods (*see* Equation 76 and the discussion following it). This could also be done even if only one of the reactants had an absorption spectrum, since $c = c_A{}^0 + c_B{}^0$. The other possibility is that the complex A_nB_m has a different absorption spectrum from the reactants, or that in the formation of A_nB_m some chromophoric groups are buried so that the absorbance of A and B represents a quantity proportional to c_A and c_B. In either case one could then calculate

$$K_{\text{app}} = \frac{c_{A_nB_m}}{c_A{}^n c_B{}^m} = K\,\frac{y_{A_nB_m}}{y_A{}^n y_B{}^m} = K\,(1 + \alpha_1 c_A{}^0 + \alpha_2 c_B{}^0 + \ldots) \quad (74)$$

It would then be possible to obtain K as well as α_1 and α_2. The dye toluidene blue has been reported to change color when it is bound to heparin (*36*); undoubtedly other examples are known. Here the sedimentation equilibrium experiment could be used to study dye binding; such studies have been reported by Steinberg and Schachman (*37*).

Analysis by the Archibald Method or by Sedimentation Equilibrium Experiments at Different Speeds. Instead of using $M_w{}^{eq}$ here, one uses M_{wa}, the apparent weight-average molecular weight. For the Archibald experiment one obtains $M_{wa,t}$ at r_m or r_b by the application of Equations 13–16. The extrapolation of $M_{wa,t}$ to zero time gives M_{wa}. For sedimentation equilibrium experiments at different speeds, one can evaluate M_{wa} by two different methods; here one uses either Equations 17 or 18. For a mixed association such as A + B ⇄ AB, the basic sedimentation equilibrium equation can be written as

$$\begin{aligned} dc = {} & L_0 \sum_i c_i M_i\, (\partial c/\partial c_i)_{T,c_j}\, d(r^2) \\ & - \sum_i c_i\, (\partial c/\partial c_i)_{T,c_j} \sum_j c_j M_j \left(\frac{\partial \ln y_i}{\partial c_j}\right)_{T,P,c_k} L_0 d\,(r^2) \\ & - \sum_{ij}\sum c_i c_j\, (\partial c/\partial c_i)_{c_k} \left(\frac{\bar{v}_i}{1000}\right) M_i L_0 d\,(r^2) - \ldots \end{aligned} \quad (75)$$

The terms $L_0 = (1 - \bar{v}\rho_0)\omega^2/2RT$, $\Lambda_0 = L_0(r_b{}^2 - r_m{}^2)$, and $\sum_{ij}\sum \bar{v} c_i c_j (\partial c/\partial c_i) M_i/1000$ arise from the fact that the quantity $(1 - \bar{v}\rho)$ can be written as follows (*28*):

$$\left(1 - \bar{v}\rho\right) = (1 - \bar{v}\rho_0)\left(1 - \frac{\bar{v}c}{1000}\right)$$

Here ρ_0 is the density of solvent. Also we assume (*18*) that

$$\left(\frac{\partial \ln y_i}{\partial c_j}\right)_{c_k} dc_j \simeq \left(\frac{\partial \ln y_i}{\partial c_j}\right)_{c_k} L_0 c_j M_j d(r^2)$$

It then follows that

$$\lim_{\Lambda_0 \to 0} \frac{c_{r_b} - c_{r_m}}{\Lambda_0} = c_0 M_{wa} \tag{76}$$

$$= c_0 M_w^{eq} - \sum_{ij}\sum c_i c_j \left(\frac{\partial c}{\partial c_i}\right) M_j \left(\frac{\partial \ln y_i}{\partial c_j}\right) - \sum_{ij}\sum \left(\frac{\bar{v}}{1000}\right) c_i c_j \left(\frac{\partial c}{\partial c_i}\right) M_i - \ldots$$

$$= c_A M_A + c_B M_B + K c_A c_B M_{AB} - B_{AA} M_A (c_A^2 + 2K c_A^2 c_B + [K c_A c_B]^2)$$

$$- B_{BB} M_B (c_B^2 + 2K c_A c_B^2 + [K c_A c_B]^2)$$

$$- 2B_{AB} M_B (c_A c_B + K c_A c_B^2 + K c_A^2 c_B + [K c_A c_B]^2) - \frac{\bar{v} c^2 M_w^{eq}}{1000}$$

Here we have used Equations 34, 35, 67, and 68, and we have made use of a relation (*38*)

$$B_{AB} M_B = B_{BA} M_A$$

which is obtained from the relation $a_{ij} = a_{ji}$ ($a_{ij} = \partial \ln a_i / \partial m_j$) on the molality scale. When this relation is converted to the c-scale one can show that $B_{AB} M_B = B_{BA} M_A$. One can also show that

$$c_0 \lim_{\Lambda_0 \to 0} \frac{\ln c_{r_b}/c_{r_m}}{\Lambda_0} = c_0 M_{wa} \tag{77}$$

This expression for cM_{wa} is quite similar to the expression obtained in light scattering experiments. When $\Psi_A = \Psi_B$ (and $\bar{v}_A = \bar{v}_B$), the light scattering equation becomes

$$1000 \left[\frac{\Delta R(0)}{K\Psi'^2}\right] = \sum_i c_i M_{ia} (\partial c/\partial c_i)_{T,c_j} = cM_{wa} = \sum_i c_i M_i (\partial c/\partial c_i)_{T,cj} \tag{78}$$

$$- \sum_{ij}\sum c_i c_j (\partial c/\partial c_i)_{T,cj} M_j \left(\frac{\partial \ln y_i}{\partial c_j}\right)_{T,P,c_k} - \sum_i c_i^2 \bar{v}_i M_i / 1000$$

Here the term $\sum_i c_i^2 \bar{v}_i M_i / 1000$ arises when one converts from the molality scale to the grams/liter concentration scale; the details for doing this, as well as problems encountered in studying multicomponent systems, are discussed by Casassa and Eisenberg (*39*).

In writing the equations for the three component system (solvent, solutes A and B) we have treated the analysis as if the solutes were neutral. When working with proteins or other macromolecules which can ionize, it will be assumed that the solutes A and B have each been dialyzed against the same buffer or solvent, that the working stock is made up from dialyzed solutions of A and B, and that dilutions (at constant β) are made with the solvent (buffer) in dialysis equilibrium with the concentrated solutions of A and B. Then it will be assumed that this procedure allows us to use equations applicable to three component systems. This problem is discussed in great detail by Casassa and Eisenberg (*39*).

The analysis for an A + B $\rightleftarrows$ AB association (or for any mixed association containing the AB species) is complicated since one obtains K and B_{AB} (or B_{BA}) together in one term. At this point let us define the quantity cM_{wa}^*, where

$$cM_{wa}^* = cM_{wa} + B_{AA}(c_A^0)^2M_A + B_{BB}(c_B^0)^2M_B + \frac{\bar{v}c^2M_w^0}{1000} \tag{79}$$

for the Archibald experiment or the sedimentation equilibrium experiment at different speeds. For the light scattering experiment one would define cM_{wa}^* by

$$cM_{wa}^* = cM_{wa} + B_{AA}M_A(c_A^0)^2 + B_{BB}M_B(c_B^0)^2 + \frac{\bar{v}(c_A^0)^2M_A}{1000} + \frac{\bar{v}(c_B^0)^2M_B}{1000} \tag{79a}$$

Using Equation 79 one notes that

$$\begin{aligned}\Delta(cM_{wa}^*) \equiv cM_{wa}^* - cM_w^0 = {} & (2Kc_Ac_BM_AM_B/M_{AB}) \\ & - B_{AA}M_A\left[2Kc_A^2c_B\left(1 - \frac{M_A}{M_{AB}}\right) + (Kc_Ac_B)^2\left(1 - \frac{M_A^2}{M_{AB}^2}\right)\right] \\ & - B_{BB}M_B\left[2Kc_Ac_B^2\left(1 - \frac{M_B}{M_{AB}}\right) + (Kc_Ac_B)^2\left(1 - \frac{M_B^2}{M_{AB}^2}\right)\right] \\ & - 2B_{BA}M_A\left[c_Ac_B + Kc_Ac_B^2 + Kc_A^2c_B + (Kc_Ac_B)^2\right] \\ & - \frac{\bar{v}c}{1000}\left[\frac{2Kc_Ac_BM_AM_B}{M_{AB}}\right]\end{aligned} \tag{80}$$

At constant β one notes the following relations:

$$\lim_{\substack{c\to 0 \\ \beta = \text{const}}} \frac{\Delta(cM_{wa}^*)}{c_A^0c_B^0} = 2M_AM_B\left(\frac{K}{M_{AB}} - \frac{B_{BA}}{M_B}\right) \tag{81}$$

The term $K/M_{AB} - B_{BA}/M_B$ (or $K/M_{AB} - B_{AB}/M_A$) is also the limiting term obtained in osmotic pressure experiments (*8*). To estimate K, one notes that

$$\lim_{c\to 0}\left(\frac{\partial}{\partial c}\right)_{T,\beta}\left[\frac{\Delta(cM_{wa}^*)}{c_A{}^0 c_B{}^0}\right] \tag{82}$$

$$= \frac{\beta}{1+\beta}\lim_{c_A{}^0\to 0}\left(\frac{\partial}{\partial c_A{}^0}\right)_{T,\beta}\left[\frac{\Delta(cM_{wa}^*)}{c_A{}^0 c_B{}^0}\right]$$

$$= \frac{\beta}{1+\beta}\left\{\frac{-2K}{M_{AB}}\left(M_B + \frac{M_A}{\beta}\right)\left(\frac{KM_AM_B}{M_{AB}} - B_{AB}M_B\right)\right.$$

$$- 2B_{AA}M_AK\left(1 - \frac{M_A}{M_{AB}}\right)$$

$$+ \frac{2B_{BB}M_BK}{\beta}\left(1 - \frac{M_B}{M_{AB}}\right)$$

$$\left.- 2B_{AA}M_AK\left(\frac{\beta+1}{\beta}\right) - \frac{2\bar{v}KM_AM_B}{1000M_{AB}}\left(\frac{\beta+1}{\beta}\right)\right\}$$

This gives us a second equation containing K and B_{AB}, which can be used to evaluate these quantities. In obtaining Equation 82 we have used the mass balance equations (34 and 35) and the quantity β to obtain the following relations:

$$\left(\frac{\partial c_A{}^0}{\partial c}\right)_{T,\beta} - \frac{\beta}{1+\beta} \tag{83}$$

$$\lim_{c\to 0}\left(\frac{\partial c_A{}^0}{\partial c_A}\right)_{T,\beta} = 1 \tag{84}$$

$$\lim_{c\to 0}\left(\frac{\partial c_B}{\partial c_A{}^0}\right)_{T,\beta} = \frac{1}{\beta} \tag{85}$$

For an association of the type $2A + B \rightleftarrows A_2B$ (or related associations not involving AB) one can get K and B_{AB} directly. Here one notes that Equations 79 or 79a can be used to obtain

$$\lim_{\substack{c\to 0\\ \beta=\text{const}}}\frac{\Delta(cM_{wa}^*)}{c_A{}^0 c_B{}^0} = -2B_{BA}M_A \tag{86}$$

and
$$\lim_{c\to 0}\left(\frac{\partial}{\partial c}\right)_{T,\beta}\left[\frac{\Delta(cM_{wa}^*)}{c_A{}^\circ c_B{}^\circ}\right] \tag{87}$$

$$= \frac{\beta}{1+\beta}\left[\frac{K}{M_{A_2B}}(2M_A{}^2 + 4M_AM_B)\right]$$

For the sedimentation equilibrium experiments at different speeds one could use $\Delta(cM_wM_z)_{app}$, the nonideal analog of $\Delta(cM_wM_z)$ or $\Delta(\sum_i c_iM_i^2)$, in the analysis; this is not discussed further here.

Analysis When $\bar{v}_A \neq \bar{v}_B$ and $\Psi_A \neq \Psi_B$. For the sedimentation equilibrium experiments at different speeds or the Archibald method one measures $\widetilde{n}\overline{M}_{1a}$; thus one would use $\Delta(\widetilde{n}M_{1a}^*)$, which is the analog of the quantity $\Delta(cM_{wa}^*)$, and proceed in a manner similar to that described above. We will not elaborate further on this procedure here.

In the light scattering experiment one measures $\sum_i \Psi_i c_i(\partial\widetilde{n}/\partial c_i)_{T,c_j}M_{ia}$ under nonideal circumstances. To do the analysis one defines $\sum_i \Psi_i c_i(\partial\widetilde{n}/\partial c_i)_{T,c_j}M_{ia}^*$, the analog of Equation 79a, first. Then one calculates

$$\Delta(\sum_i \Psi_i^2 c_i M_{ia}^*) = \sum_i \Psi_i c_i \left(\frac{\partial\widetilde{n}}{\partial c_i}\right)_{T,c_j} M_{ia}^* - (\sum_i \Psi_i^2 c_i M_i)^0$$

and this quantity is used in the analysis. For an association of the type $2A + B \rightleftarrows A_2B$, one can show

$$\lim_{\substack{c\to 0\\ \beta=\text{const}}} \frac{\Delta\left(\sum_i \Psi_i^2 c_i M_{ia}^*\right)}{c_A^0 c_B^0} = -B_{AB}M_B(\Psi_A^2 + \Psi_B^2) \qquad (88)$$

$$\text{and} \lim_{c\to 0} \left(\frac{\partial}{\partial c}\right)_{T,\beta} \frac{\Delta\left(\sum_i \Psi_i^2 c_i M_{ia}^*\right)}{c_A^0 c_B^0} \qquad (89)$$

$$= \frac{\beta}{1+\beta} \lim_{c_A^0\to 0} \left(\frac{\partial}{\partial c_A^0}\right)_{T,\beta} \frac{\Delta\left(\sum_i \Psi_i^2 c_i M_{ia}^*\right)}{c_A^0 c_B^0}$$

$$= \frac{\beta}{1+\beta}\left\{\frac{K}{M_{A_2B}}[2\Psi_A^2 M_A^2 + 4\Psi_A\Psi_B M_A M_B]\right\}$$

Note that $\overline{M}_{1a}$ and M_{ia} refer to the apparent values of $\overline{M}_1^{eq}$ and M_i, respectively.

Discussion

We have shown some of the ways that mixed associations could be analyzed by light scattering, Archibald, or sedimentation equilibrium experiments. It might be appropriate at this point to make a comparison of some advantages and disadvantages of each technique.

The light scattering experiment is potentially the fastest way to do these experiments. It does appear to require more solution; 2 ml of solution seems to be the smallest amount of liquid that can be used. One serious difficulty appears to be the removal of dust from aqueous solutions, although Millipore filtering or other forms of filtering can be used to remove dust. For slow or moderately fast reactions one could use light scattering experiments to follow the kinetics of a reaction described by Equation 1, since one could measure M_w (or cM_w) at various time intervals. This should be applicable under the same conditions that stopped flow experiments are applicable. It would require special cells, but one could also use this to see if the active form of an enzyme is an aggregate (dimer or higher aggregate). Here one could use rapid mixing techniques to blend substrate with the enzyme and follow changes in M_w. One could also study pressure effects on chemical equilibria between macromolecules, using special cells designed for this purpose; this has been done very elegantly by Payens and Heremans (*40*) on their studies of pressure effects on the self-association of β-casein.

The sedimentation equilibrium experiment requires much smaller volumes of solution, about 0.15 ml. With six-hole rotors and multichannel centerpieces (*41*) it is potentially possible to do fifteen experiments at the same time. For situations where the photoelectric scanner can be used one might (depending on the extinct coefficients) be able to go to much lower concentrations. Dust is no problem since the centrifugal field causes it to go to the cell bottom. For conventional sedimentation equilibrium experiments, the analysis of mixed associations under nonideal conditions may be virtually impossible. Also, sedimentation equilibrium experiments take time, although methods are available to reduce this somewhat (*42, 43*). For certain situations the combination of optical systems available to the ultracentrifuge may allow for the most precise analysis of a mixed association. The Archibald experiment may suffer some loss in precision since one must extrapolate the data to the cell extremes (r_m and r_b) to obtain $M_{wa,t}$, which must then be extrapolated to zero time. Nevertheless, all three methods indicate that it is quite possible to study mixed associations. We have indicated some approaches that could be used to overcome problems of nonideality, unequal refractive index increments, and unequal partial specific volumes.

In this paper we have used the quantity $(1 - \bar{v}\rho_0)$ in writing equations for sedimentation equilibrium experiments. Some workers prefer to use the density increment, $1000(\partial\rho/\partial c)_{T,\mu}$, instead when dealing with solutions containing ionizing macromolecules. This procedure was first advocated by Vrij (*44*), and its advantages are discussed by Casassa and Eisenberg (*39*). Nichol and Ogston (*13*) have used the density increment in their analysis of mixed associations. The subscript μ means that all of the diffusible solutes are at constant chemical potential in the buffer solution (the solvent) and in the macromolecular solution; this condition is achieved by dialysis.

Literature Cited

1. Steiner, R. F., *Arch. Biochem. Biophys.* (1954) **49,** 71.
2. Steiner, R. F., *Biopolymers* (1970) **9,** 1465.
3. Timasheff, S. N., Kirkwood, J. G., *J. Amer. Chem. Soc.* (1953) **75,** 3124.
4. Singer, S. J., in "The Proteins," 2nd ed., Vol. 3, H. Neurath, Ed., Academic Press, New York, 1965, p. 270.
5. Steiner, R. F., *Arch. Biochem. Biophys.* (1955) **55,** 235.
6. Timasheff, S. N., in "Electromagnetic Scattering," M. Kerker, Ed., MacMillan, New York, 1963, p. 337.
7. Nichol, L. W., Bethune, J. L., Kegeles, G., Hess, E. L., in "The Proteins," 2nd ed., Vol. 2, H. Neurath, Ed., Academic Press, New York, p. 305.
8. Adams, E. T., Jr., Pekar, A. H., Soucek, D. A., Tang, L. H., Barlow, G., Armstrong, J. L., *Biopolymers* (1969) **7,** 5.
9. Steiner, R. F., *Biochemistry* (1968) **7,** 2201; (1970) **9,** 4268.
10. Ackers, G. K., *Advan. Protein Chem.* (1970) **24,** 343.
11. Winzor, D. J., in "Physical Principles and Techniques of Protein Chemistry," S. J. Leach, Ed., Academic Press, New York, 1969, Part A, p. 451.
12. Timasheff, S. N., Townsend, R., in "Physical Principles and Techniques of Protein Chemistry," S. J. Leach, Ed., Academic Press, New York, 1970, Part B, p. 147.
13. Nichol, L. W., Ogston, A. G., *J. Phys. Chem.* (1965) **69,** 4365.
14. Adams, E. T., Jr., *Ann. N.Y. Acad. Sci.* (1969) **164,** 226.
15. Chun, P. W., Kim, S. J., *J. Phys. Chem.* (1970) **74,** 599.
16. Klainer, S. M., Kegeles, G., *J. Phys. Chem.* (1955) **59,** 952.
17. LaBar, F. E., *Biochemistry* (1966) **5,** 2362, 2368.
18. Albright, D. A., Williams, J. W., *J. Phys. Chem.* (1967) **71,** 2780.
19. Archibald, W. J., *J. Phys. Colloid Chem.* (1947) **51,** 1204.
20. Kegeles, G., Klainer, S. M., Salem, W. J., *J. Phys. Chem.* (1957) **61,** 1286.
21. Kegeles, G., Rao, M. S. N., *J. Amer. Chem. Soc.* (1958) **80,** 5721.
22. Fujita, H., Inagaki, H., Kotaka, T., Utimaya, H., *J. Phys. Chem.* (1962) **66,** 4.
23. Adams, E. T., Jr., *Proc. Nat. Acad. Sci. U.S.* (1964) **51,** 109.
24. Kerker, M., "The Scattering of Light and Other Electromagnetic Radiation," Academic Press, New York, 1969.
25. Zimm, B. H., *J. Chem. Phys.* (1948) **16,** 1093, 1099.
26. Tanford, C., "Physical Chemistry of Macromolecules," John Wiley & Sons, New York, 1961, pp. 302–305.
27. Van Holde, K. E., "Physical Biochemistry," Prentice-Hall, Englewood Cliffs, N. J., 1971, pp. 192–197.
28. Van Holde, K. E., Baldwin, R. L., *J. Phys. Chem.* (1958) **62,** 734.
29. Adams, E. T., Jr., Fujita, H., in "Ultracentrifugal Analysis in Theory and Experiment," J. W. Williams, Ed., Academic Press, New York, 1963, p. 119.
30. Adams, E. T., Jr., Williams, J. W., *J. Amer. Chem. Soc.* (1964) **86,** 3454.
31. Adams, E. T., Jr., *Fractions* (1967) No. 3.
32. Albright, D. A., Williams, J. W., *Biochemistry* (1968) **7,** 67.
33. Van Holde, K. E., Rossetti, G. P., Dyson, R. D., *Ann. N.Y. Acad. Sci.* (1969) **164,** 279.
34. Schachman, H. K., *Biochemistry* (1963) **2,** 887.
35. Chervenka, C. H., *Fractions* (1971) No. 1.
36. Whistler, R. L., Smart, C. L., "Polysaccharide Chemistry," Academic Press, New York, 1953, p. 412.

37. Steinberg, I. Z., Schachman, H. K., *Biochemistry* (1967) **5,** 3728.
38. Kirkwood, J. G., Goldberg, R. J., *J. Chem. Phys.* (1949) **18,** 54.
39. Casassa, E. F., Eisenberg, H., *Advan. Protein Chem.* (1964) **19,** 287.
40. Payens, T. A. J., Heremans, K., *Biopolymers* (1969) **8,** 335.
41. Yphantis, D. A., *Biochemistry* (1964) **3,** 297.
42. Hexner, P. E., Radford, L. E., Beams, J. W., *Proc. Nat. Acad. Sci. U.S.* (1961) **47,** 1848.
43. Griffith, O. M., *Anal. Biochem.* (1967) **19,** 243.
44. Vrij, A., Ph.D. Dissertation, Univ. of Utrecht, 1959.

RECEIVED January 17, 1972. Work supported by grants GM 15551 and GM 17611, National Institutes of Health. It is based on the MS theses submitted by Allen H. Pekar (nonideal, sedimentation equilibrium experiments and Archibald method) and by Peter J. Wan (light scattering) to the Graduate School of the Illinois Institute of Technology.

23

The Use of Thin Film Dialysis and High Resolution NMR to Study Conformation and Association Phenomena

L. C. CRAIG, H. C. CHEN, and W. A. GIBBONS

The Rockefeller University, New York, N. Y. 10021

Further progress in our understanding of the nature of large polymeric substances will require the development of methods with greater discrimination in the study of molecular size, conformation, molecular interactions, and other solution properties. It is postulated that this can be accomplished only by the use of a variety of approaches depending on different parameters. A short review of the present stage of development of two approaches, thin-film dialysis and high-resolution NMR, is given. The former is shown to be a sensitive tool for estimation of Stokes radius and for detection of molecular interactions and changes in conformation while the latter gives detailed information concerning the relative positions of the atoms within the molecule. They are shown to have complimentary value by the use of well characterized naturally occurring antibiotic polypeptides and hormones as models.

In spite of intensive research by countless numbers of capable scientists for the past 30 years our methods for precise separation and characterization of larger molecules do not have the discrimination needed to solve many of the problems now posed by what we have already learned about the behavior of such substances. This is true even with naturally occurring biopolymers where the amazing selectivity of the synthetic processes of living tissues permits the isolation of preparations which show the behavior expected if the individual molecules of the preparation should be really all of the same size, shape, and composition. Research with biopolymers can therefore be considered in a different category from that so important for industrial polymers since in many cases it relates to single molecular species. The importance of the latter point depends on the purpose of the research but it would appear to be all important from the standpoint of precise separation methods development, structure determination, and non-covalent molecular interactions.

Present-day literature dealing with the structure and conformation of large molecules presents a confusing array of approaches. Some of them are well supported by experimental data and model building such as the x-ray diffraction studies with crystalline material. Other approaches are theoretical; they require extensive computer calculation, involve assumptions which have not been rigorously proved, and at best allow only a rough correlation with overall existing data. The substances studied are often of such great complexity as to make rigorous experimental verification virtually impossible.

Nature and the synthetic organic chemist have provided an amazing array of substances of molecular weights ranging in the hundreds; these substances can be separated and documented now with great precision with modern methods which include x-ray crystallography, UV, IR, NMR, gas chromatography, and mass spectroscopy. Nature has provided a corresponding array of substances of molecular weights ranging in the tens of thousands, *e.g.*, the proteins. On the other hand pure substances of intermediate molecular size seem to have offered more difficulty and not to have been so plentiful. The separation and characterization methods so successful for the smaller molecules have for the most part not been applicable nor have those developed for the proteins been applicable or shown sufficient resolution.

For some years an objective of this laboratory has been the development of methods for isolating and characterizing the last-mentioned size of natural product. Countercurrent distribution has been a chief tool but more recently we have been developing a membrane-diffusion method which we have called "thin-film dialysis" (*1*). It has considerable potential for studying molecular size, conformation, molecular interactions, and other solution behavior, particularly when combined with the rapidly developing possibilities suggested by high resolution NMR.

Our approach has been essentially empirical in nature with less emphasis on the theoretical. We have isolated single substances, proved their purity, and determined their covalent structure by classical methods of organic chemistry; we have then used these substances of molecular weight ranging from 1,000 to 14,000 as model solutes for the study of conformation and intermolecular interaction. Solutes of special interest have been gramicidin SA (*2*), bacitracin A (*3*), polymyxin B, and the tyrocidines A, B, and C (*4*). All are cyclic antibiotic polypeptides. The first three behave in aqueous solution as reasonably ideal solutes and do not associate, but the tyrocidines associate strongly and are interesting models for the study of association phenomena. Other model solutes of similar nature are the hormones vasopressin (*5*) and oxytocin (*6*) and their synthetic analogs (*7*). Studies with these solutes which are also cyclic polypeptides have been carried out in other laboratories (*8*).

The presence of a covalently linked ring structure reduces the numbers of theoretically possible conformational forms. This fact makes such small polymeric solutes desirable models for establishing certain concepts regarding the structural features responsible for measurable properties which can be detected by ORD, CD, NMR, diffusability, tritium-proton exchange kinetics, etc. Gramicidin SA (*2*) is a good example since it was the first solute of this class whose complete structure, both covalent and conformation in solution, was determined. It was determined to have an antiparallel β pleated sheet structure and to have four amide-hydrogen bonds. NMR studies were of great importance in confirming this structure which had been proposed earlier by Schwyzer (*9*) as the most probable one on the basis of the reasoning of organic chemistry. Its correctness, however, had been questioned in several laboratories (*10, 11*) because we had found in 1965 (*12, 13*) that it gave the precise type of ORD spectrum that had become widely accepted as specific for the conformation of the α helix. Our confirmation of the antiparallel pleated sheet structure by NMR, tritium exchange, and other techniques therefore was a contribution to the study of polymers in that it showed that ORD and CD data alone could not be taken as a reliable measure of helicity. Other conformations can give the same type of pattern. It is interesting that in spite of this now generally known uncertainty ORD and CD are still widely used to measure helicity in polymeric solutes without any mention of the uncertainty.

A number of excellent publications have now appeared dealing with NMR studies of oxytocin (*14, 15*), vasopressin (*16, 17*), and their analogs. These follow the earlier pattern set in the gramicidin SA study and are of great value in forming ground rules for interpretation of NMR data.

A logical further step in this general study should deal with linear molecules lacking the covalently bonded ring which are of the so-called "random-coil" type. Excellent models are to be found in the angiotensin hormones (*7*) and their synthetic analogs, bradykinin and analogs, ACTH hormones (*18*), the calcitonins, glucagon, and protamines. We have had many of these under study to some degree with the methods outlined above but much of the study is only preliminary. The data, however, do permit a few conclusions. The angiotensins appear to have a definite tightly coiled conformation (or possibly two interconvertible forms) depending on the pH and solution environment (*7*). This had been shown by thin-film dialysis and the presence of slow exchangeable amide protons (*19*). In fact all of these random-coil peptides appear to have a diffusional size in aqueous solution indicative of rather tightly coiled spheres depending on the ionic strength and pH (*20*). Urea and guanidinium chloride have the expected effect of increasing the measured diffusional size except where there is molecular association.

The tyrocidines associate strongly and have been thoroughly investigated (*4*) from this standpoint. Extensive study has shown the interaction to be primarily hydrophobic in nature but requiring the rigid fixed-ring structure for expression. Solutes such as urea and guanidinium chloride which increase the diffusional size of linear peptides make the tyrocidines more diffusable. This is true also with insulin (*21*) and glucagon (*18*). The latter was first shown by thin-film dialysis to associate. Association, with the tyrocidines, could be detected by NMR through line broadening as might be expected (*22*).

As experience with the different models has accumulated it has become possible to refine certain techniques and make more meaningful interpretations. One of these has concerned the thin-film dialysis technique. For some years we have considered that the fixed charge on Visking "dialysis" membranes was sufficiently low (*23*) so that ion-exchange effects could be neglected. Actually, in calibrating the membrane if a change in ionic strength significantly altered the half-escape time with a test solute such as bacitracin A, that roll of casing was discarded.

Recently in working with the random-coil peptides such as ACTH which carry a high charge density, it has been found that a higher degree of freedom from fixed charge on the membrane is required in order to achieve reproducible results. This stimulated a search for a method to remove fixed charge completely. Membranes treated with glycine amide and a cyclohexyl dimide coupling agent (1-cyclohexyl-3-(2-morpholinoethyl)-carbodiimidometho-*p*-toluene sulfonate, Aldrich Chemical Co.) proved to be completely satisfactory for highly charged solutes (*20*). They also appeared to show less absorptive properties for certain solutes and gave lower blanks for ultraviolet absorbance measurements in the 220 to 230 mμ range.

With these membranes we were able to show that in thin-film dialysis neither the escape rate into the diffusate nor diffusion across the membrane itself is the controlling rate expressed by the escape plot. Therefore the escape plot is a true measure of the probability of the solute finding its way into the membrane. It was suggested 15 years ago (*24*) that thin-film dialysis could be performed in such a way that it would be a sensitive measure of diffusional size. This has since been confirmed with many solutes (*25*) at a given temperature.

Interpretation of temperature effects, however, remained uncertain. With polypeptides in particular the effect of a change in temperature was unpredictable. Certain ones followed rather closely the Stokes-Einstein relationship for free diffusion; others deviated strongly. It was thought this was caused either by a change in hydration or a change in conformation or both.

If the escape pattern is a true reflection of the probability of the solute entering the pores of the membrane, it should show close adherence to the Stokes-Einstein (*26*) relationship. That is to say the activation energy of this particular type of dialysis should agree closely with the activation energy of free diffusion. For a number of rigid solutes such as gramicidin SA we have found this to be true over the temperature range 10° to 70°C. On the other hand the slope of a plot of the reciprocal of the half-escape time against the reciprocal of the absolute temperature for random-coil peptides deviates either positively or negatively from the slope required by the Stokes-Einstein relationship.

These findings strengthen our confidence in the reliability of the thin-film dialysis method for determining diffusional size provided it is carried out with the precautions and care now shown experimentally to be required. It then becomes a direct measure of the probability of a solute finding its way into a pore from the solution side. Thus it can be a measure of true diffusional size, and the basis of the sensitivity or selectivity is a function of the limiting pore size relative to the diffusional size of the solute molecule. The selectivity or probability is dictated by the relationship

$$A = A_o (1 - r/R)^2$$

where A_o is the cross-sectional area of the pore, r is the effective radius of the diffusing particle, and R is the radius of the cross-sectional area of the pore. This is a relationship suggested many years ago in ultrafiltration (*28*). Since we now know from nuclear magnetic relaxation measurements that solutes of the size we are dealing with are tumbling at the rate approximating 10^{10} times per second, the longest cross-sectional dimension of the molecule is the most important one in determining diffusional size.

Thin-film dialysis can thus be considered as a way to estimate Stokes radius with a precision approximating ±3% (*24*) when suitable models of known dimensions are available for comparison. In addition to this it offers an excellent approach to the study of molecular interactions, self-association, and binding (*29, 30*). It therefore is excellent for supplementing the type of information which can be derived from high-resolution NMR study which in many cases together with model building and the restrictions of steric hindrance can reveal the precise positions certain atoms in a molecule occupy in space relative to each other. Many times however with NMR, a single unique position cannot be extracted from the data but instead only several possibilities. Here supplementary knowl-

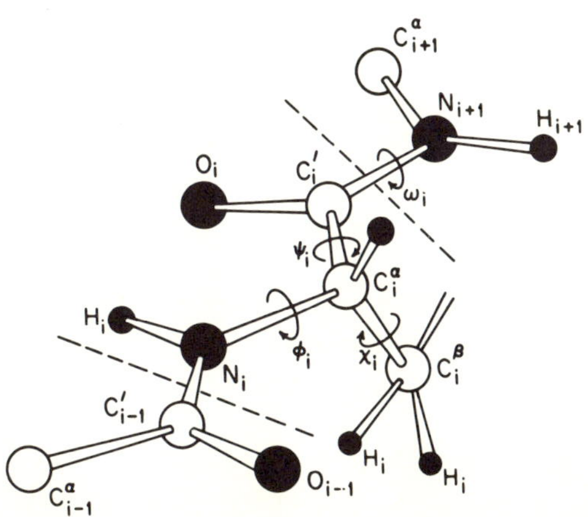

Figure 1. Drawing illustrating IUPAC terminology describing the conformation of a peptide unit

edge of the overall size can be helpful. Reliable interpretation of some of the NMR data also requires a knowledge of the effect of the solvent and whether or not self-association is occurring.

The general overall approach of high-resolution proton NMR to problems of conformation involves the following procedures: 1) Obtain the spectrum as a function of magnetic field strength, solvent, temperature, and pH. 2) Integrate to obtain a proton count. 3) Analyze the NMR spectra obtained in 1, principally by spin decoupling, INDOR, and isotopic labeling, to give coupling constants and chemical shifts of each proton in the molecule. 4) Study the amide proton-hydrogen bonding by deuterium exchange, double resonance, relaxation measurements, and perturbations by changes of solvent, temperature, pH, and paramagnetic shift reagents. 5) Correlate NMR parameters to molecular parameters such as dihedral angles. 6) Compute or model build to find the conformation most consistent with the above data and relevant data from sources other than NMR.

The conformation of a polypeptide is determined by the allowed or most favorable rotational position of each single bond connecting each atom in the molecule. The bonds fixing the backbone structure are given in Figure 1. Since the C–N bond is known to be essentially fixed and trans with respect to the carbonyl and amide proton in practically all peptides, the degrees of freedom are limited to rotation around the other two bonds called ϕ and ψ according to IUPAC terminology (*31*). The dihedral angle ϕ influences the coupling constants of the two vicinal protons designated $^3J_{Nc}$. The single proton on the α carbon atom is also coupled to the proton or protons on the β carbon to give $^3J_{\alpha\beta}$. Determination of these coupling constants furnishes data which limits the stereochemistry around these atoms according to the Karplus relationship (*32*) but does not provide a unique stereochemistry (*33*). Four possibilities remain to be considered which can be reduced to two by the use of conformational energy maps (*34*). For the cyclic polypeptides, model building and strong evidence for hydrogen bonding of specific protons provided the further limitation expressed in the structural formula for gramicidin SA shown in Figure 2.

In gramicidin SA four protons were shown to have very slow deuterium-proton exchange rates which could be assigned to specific protons.

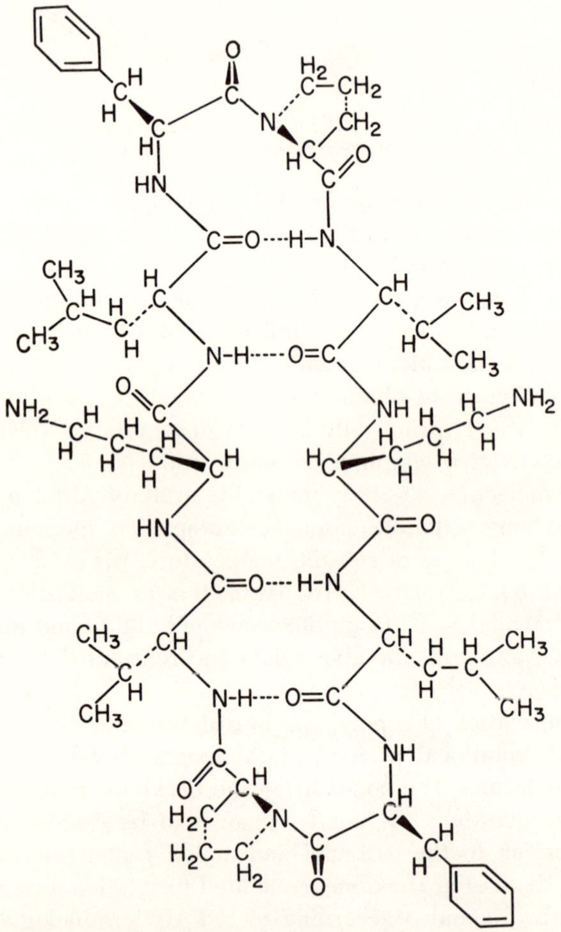

Figure 2. Schematic drawing of model with ω angles all 0 except Leu = 10°. ϕ angles of Val, Orn, and Leu = 30°, Phe = 150°; ψ angles of Val, Orn, and Leu = 0, Phe = 150° and Pro = 130°. Dashed lines O – – – H indicate hydrogen bonds.

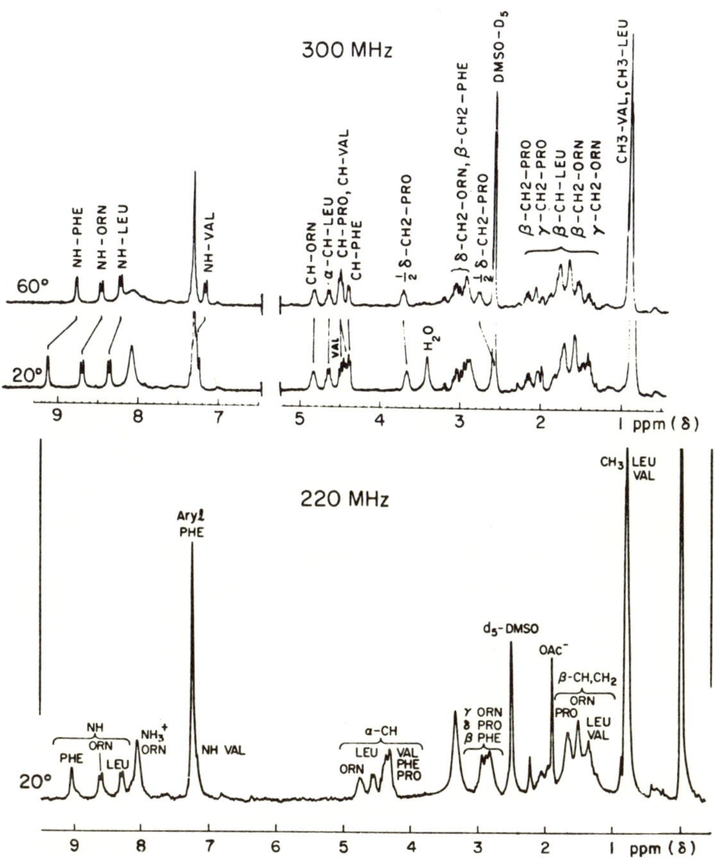

Figure 3. High-resolution NMR spectra of gramicidin SA

The rates were confirmed by tritium-exchange studies (*35*). The very slow rate of exchange for this peptide must result from hydrogen bonding, a conclusion that is not so certain with other peptides with protons whose rate of exchange are only somewhat slower than rates known for simple amides.

The effect of a change in temperature or solvent on chemical shift is a parameter which must be carefully studied. On the basis of several examples it was proposed that evidence for hydrogen bonding (*36*) could be derived from the failure of an amide-proton chemical shift to be sensitive to a change in temperature. Later observations from this laboratory have shown that the chemical shift of certain amide protons not hydrogen bonded also may not be sensitive to a change in temperature.

The chemical shift may reveal the position of one atomic grouping with respect to another by anisotropic shielding. This is an effect likely to be sensitive to temperature and solvent changes as seen in the valine amide-proton resonance in Figure 3.

Gramicidin SA was a particularly suitable model for establishing and testing certain ground rules for the use of NMR in conformation studies. The challenge now is the extension of this type of study to larger molecules. Here the resolution of individual resonances rapidly becomes a problem. The development of spectrometers with stronger magnetic fields (*37*) and equipped with Fourier transform is a promising approach which is being vigorously pursued and will be particularly important for the study of more dilute solutions. Simplification of overlapping resonances by an INDOR technique is also a promising newer approach (*38*).

Regardless of how much improvement can be made in NMR techniques, conformation problems will always require the support of other techniques such as x-ray diffraction, CD, ORD, UV, proton exchange, thin-film dialysis, model building, and energy-minimization calculations.

In this short review it is not possible to cover more than a bare outline of the possibilities presented by only two approaches to the study of conformation. These were chosen because of our own current interest and because they supplement each other well. From the practical standpoint high-resolution NMR is a very expensive and intricate approach while thin-film dialysis is very simple and inexpensive.

Examples of the way these two entirely different approaches compliment each other are developing in conformation studies with the hormones angiotensin, oxytocin, and vasopressin. In 1964 Craig, Harfenist, and Paladini (*7*) published the comparative half-escape times shown in Table I in 0.01*N* acetic acid and another series in Table II using a different membrane less porous and not as selective as the first. Oxytocin and vasopressin are cyclic octapeptides with an S–S linkage closing the ring at the 1–6 positions (*5, 6*). Their size is thus limited except for the side

Table I. Comparative Dialysis Rates of Selected Solutes of Known Structure

Solute	*Molecular Weight*	*T/2 hr*
Carbowax 1000	950-1050	2.4
Angiotensin I	1282	0.95
Tyrocidine B (ring split)	1348	2.1
Cyclohexaamylose	972	2.1
Polymyxin B_1	1208	5.6
Ring fragment of polymyxin B_1	762	1.4
Gramicidin S	1142	2.4
Bacitracin	1420	1.8

Table II. Comparative Escape Rates of Oxytocin and Vasopressin Analogs

Peptide	*T/2 hr*	*No. of Separate Runs*
Angiotensinamide	3.6	3
Oxytocin	3.2	5
8-Lysine vasopressin	5.3	3
8-Histidine vasopressin	4.5	1
8-Lysine vasotocin	5.2	1
1-Deamino-8-lysine vasopressin	3.1	3
1-Deamino oxytocin	3.5	2
1-Acetyl-8-lysine vasopressin	4.1	2
1-*N*-Methyl oxytocin	4.1	1
4-Decarboxamido oxytocin	3.8	1
D-Leucine oxytocin	3.7	2
4-Deamido oxytocin	2.6	2
Oxytocin dimer	9.8	1

chain containing three amino acid residues. From the data in Table II it is seen that 8-lysine vasopressin, 8-histidine vasopressin, and 8-lysine vasotocin all have half-escape times definitely longer than oxytocin and deaminooxytocin. Removal of the basic charge on oxytocin did not appreciably change the half-escape time but with 8-lysine vasopressin the time was reduced to about that found for oxytocin. It was therefore postulated that the side chain of the vasopressins was more extended than the oxytocins and that the oxytocins were as compact as possible. This now appears to be confirmed by NMR studies (*15, 17*), one of which postulated (*15*) a hydrogen bond between the terminal glycineamide proton and a cystine amide carbonyl. If the latter is true, the hydrogen bond would not be possible in 9-sarcosine oxytocin and the side chain would be more extended. The data in Table I indicate a larger diffusional size for 9-sarcosine oxytocin. The suggestion of the more extended conformation of 8-lysine vasopressin also appears to have been confirmed by NMR studies (*17*).

It was postulated from the comparative data in Tables I and II and other observations that angiotensin had a definite compact conformational structure even though it is a linear peptide and should be a random coil. Tritium-exchange studies (*39*), CD (*40*), and NMR evidence (*41, 42*) now give support to this view. NMR studies with angiotensin, oxytocin, and vasopressin thus far have not indicated conformational restriction on the benzene rings, but the thin-film dialysis data are inconsistent with any conformation in which these bulky groups are extended from the otherwise compact conformation.

Literature Cited

1. Craig, L. C., Chen, H. C., Printz, M., Taylor, W. I., in "Characterization of Macromolecular Structure," *Natl. Acad. Sci., U.S., Publ.* **1573**, 315 (1968).
2. Stern, A., Gibbons, W. A., Craig, L. C., *Proc. Natl. Acad. Sci., U.S.* (1968) **61**, 734.
3. Craig, L. C., Phillips, W. F., Burachik, M., *Biochemistry* (1969) **8**, 2348.
4. Burachik, M., Craig, L. C., Chang, J., *Biochemistry* (1970) **9**, 3293.
5. Du Vigneaud, V., Lawler, H. C., Popenoe, E. A., *J. Amer. Chem. Soc.* (1953) **75**, 4880.
6. Du Vigneaud, V., Ressler, C., Tripett, S., *J. Biol. Chem.* (1953) **205**, 949.
7. Craig, L. C., Harfenist, E. J., Paladini, A. C., *Biochemistry* (1964) **3**, 764.
8. Bovey, F. A., Brewster, A. I., Patel, D. J., Tonelli, A. E., Torchia, D. A., *Acc. Chem. Res.* (1972) **5**, 193.
9. Schwyzer, R., in "Amino Acids and Peptides with Antimetabolic Activity," G. E. W. Wolstenholme, Ed., p. 171, Churchill, London, 1958.
10. Liquori, A. M., Conti, F., *Nature* (1968) **217**, 635.
11. Vanderkooi, G., Leach, S. J., Nemethy, G., Scott, R. A., Scherago, H. A., *Biochemistry* (1966) **5**, 2991.
12. Ruttenberg, M. A., King, T. P., Craig, L. C., *J. Amer. Chem. Soc.* (1965) **87**, 4196.
13. Craig, L. C., *Proc. Natl. Acad. Sci., U.S.* (1968) **61**, 152.
14. Urry, D. W., Walter, R., *Proc. Natl. Acad. Sci., U.S.* (1971) **68**, 956.
15. Walter, R., in "Structure Activity Relationships of Protein and Polypeptide Hormones," Part 1, *Proc. Intern. Symp., 2nd, Liege, Sept. 1971*, Excerpta Medica, Amsterdam, p. 181.
16. von Dreele, P. H., Brewster, A. I., Scheraga, H. A., Ferger, M. F., du Vigneaud, V., *Proc. Natl. Acad. Sci., U.S.* (1971) **68**, 1028.
17. Glickson, J. D., Urry, D. W., Walter, R., *Proc. Natl. Acad. Sci. U.S.* (1972) **69**, 2566.
18. Craig, L. C., Fisher, J. D., King, T. P., *Biochemistry* (1965) **4**, 311.
19. Printz, M. P., Williams, H. P., Craig, L. C., *Proc. Natl. Acad. Sci. U.S.* (1972) **69**, 378.
20. Craig, L. C., Kac, H., Chen, H. C., Printz, M. P., in "Structure Activity Relationships of Protein and Polypeptide Hormones," Part 1, *Proc. 2nd Intern. Symposium, Liege*, Excerpta Medica, Amsterdam, p. 176, Sept. 28, 1971.
21. Craig, L. C., King, T. P., Konigsberg, W., *Ann. N.Y. Acad. Sci.* (1960) **88**, 571.
22. Stern, A., Gibbons, W. A., Craig, L. C., *J. Amer. Chem. Soc.* (1969) **91**, 2794.
23. Craig, L. C., Ansevin, A., *Biochemistry* (1963) **2**, 1268.
24. Craig, L. C., Pulley, A. O., *Biochemistry* (1962) **1**, 89.
25. Craig, L. C., in "Methods of Enzymology," Vol. XI, C. H. W. Hirs, Ed., p. 870, Academic Press, New York, 1967.
26. Longsworth, L. G., in "Electrochemistry in Biology and Medicine," T. Shedlovsky, Ed., p. 225, John Wiley & Sons, Inc., New York, 1955.
27. Craig, L. C., Chen, H. C., *Proc. Natl. Acad. Sci. U.S.* (1972) **69**, 702.
28. Ferry, J. D., *J. Gen. Physiol.* (1936) **20**, 95.
29. Stouffer, J. E., Hsu, J. S., *Biochemistry* (1966) **5**, 1195.
30. Chen, H. C., Craig, L. C., *Bioorg. Chem.* (1971) **1**, 51.
31. Edsall, J. T., Flory, J. P., Kendrew, J. C., Liquori, A. M., Nemethy, G., Ramachandran, G. N., Scheraga, H. A., *J. Biol. Chem.* (1966) **241**, 1004.
32. Karplus, M., *J. Amer. Chem. Soc.* (1963) **85**, 2870.
33. Gibbons, W. A., Nemethy, G., Stern, A., Craig, L. C., *Proc. Natl. Acad. Sci. U.S.* (1970) **67**, 239.

34. Ramachandran, G. N., Sasisekharan, C., in "Advances in Protein Chemistry," Vol. 21, Academic Press, New York, 1968.
35. Laiken, S. L., Printz, M. P., Craig, L. C., *Biochemistry* (1969) **8**, 519.
36. Ohnishi, M., Urry, D. W., *Biochem. Biophys. Res. Commun.* (1969) **36**, 194.
37. Johnson, L. F., *Anal. Chem.* (1971) **43**, 28A.
38. Gibbons, W. A., Alms, H., Bockman, R. S., Wyssbrod, H. R., *Biochemistry* (1972) **11**, 1721.
39. Printz, M. P., Williams, H. P., Craig, L. C., *Proc. Natl. Acad. Sci. U.S.* (1972) **69**, 378.
40. Fermandjian, S., Morgat, J.-L., Fromageot, P., *Eur. J. Biochem.* (1971) **24**, 252–258.
41. Weinkam, R. J., Jorgensen, E. C., *J. Amer. Chem. Soc.* (1971) **93**, 7048.
42. Bleich, H., unpublished data from this laboratory.

RECEIVED January 17, 1972. Work supported in part by U.S. Public Health Service Grant No. A.M. 02493.

24

Theoretical Model for Determining Monomer–Polymer Reaction Stoichiometry from Equilibrium Gel Partition

BRUCE F. CAMERON

Papanicolaou Cancer Research Institute, 1155 Northwest 14th St., Miami, Fla. 33136, and Department of Medicine, Division of Hematology, University of Miami School of Medicine, POB 875—Biscayne Annex, Miami, Fla. 33152

ALAN D. ADLER

New England Institute, Grove St., Ridgefield, Conn. 06877

A theoretical analysis of an equilibrium gel partition system for determination of association reaction stoichiometry is described. Equations are generated for generalized monomer–polymer equilibrium, including sequential equilibria. Numerical solutions of the generated polynomial equations were obtained for $pM \rightleftharpoons P$, *p up to 6, and the slope of the graph of the ratio of monomer equivalents external to the gel phase to total monomer equivalents as a function of total initial concentration was shown to be a strong function of the stoichiometric order. A numerical analysis of slopes and values of this curve in the region of the inflection yielded a parameter independent of specific equilibrium constants and characteristic of the stoichiometric coefficient for simple generalized polymerization.*

A relatively common feature of many problems involving molecular weight determination of biopolymers is that of association–dissociation equilibrium. Subunit structure of enzyme proteins is well recognized (*1*), and methods of dissociation of subunits to obtain monomer molecular weight are widely utilized (*2*). A previous paper described the application of an equilibrium gel partition method to the analysis of macromolecular association in a monomer–dimer case (*3*). The experimental parameters in a system utilizing the Sephadex series of gel filtration materials were described and optimized, and the method was applied to the problem of the dissociation of the hemoglobin tetramer (*3*, *4*). The present work is a theoretical extension of the mathematical model to generalized association stoichiometry, for like and unlike monomers, including sequential association with detectable intermediates for the system containing like interactants.

Theory

Case Ia. Multiple Association of Like Interactants. For the reaction

$$p\mathbf{M} \rightleftharpoons \mathbf{P}, \text{ where } \mathbf{P} \equiv \mathbf{M}_p$$

the gel partition system is essentially

$$\overset{K_\alpha}{p\mathbf{M} \rightleftharpoons \mathbf{P}} \qquad\qquad \overset{K_\beta}{p\mathbf{M} \rightleftharpoons \mathbf{P}}$$

$$\overset{K_\mathbf{M}}{\mathbf{M} \rightleftharpoons \mathbf{M}}$$

$$\overset{K_\mathbf{P}}{\mathbf{P} \rightleftharpoons \mathbf{P}}$$

where the external phase is represented by α (volume $= V_\alpha$), and the internal phase is represented by β (volume $= V_\beta$). The relevant equilibrium constants are defined in terms of the number of moles of species **M** and **P** and concentrations [**P**] and [**M**] as

$$K_\alpha = [\mathbf{P}_\alpha]/[\mathbf{M}_\alpha]^p = \frac{\mathbf{P}_\alpha}{\mathbf{M}_\alpha{}^p} V_\alpha{}^{(p-1)}$$

$$K_\mathbf{M} = [\mathbf{M}_\beta]/[\mathbf{M}_\alpha] = \frac{\mathbf{M}_\beta}{\mathbf{M}_\alpha} \frac{V_\alpha}{V_\beta}$$

$$K_\mathbf{P} = [\mathbf{P}_\beta]/[\mathbf{P}_\alpha] = \frac{\mathbf{P}_\beta}{\mathbf{P}_\alpha} \frac{V_\alpha}{V_\beta}$$

Note that K_β is also determined by the above equations since

$$K_\beta = K_\alpha\{K_\mathbf{P}/(K_\mathbf{M})^p\}$$

Now define

$$k_\alpha = K_\alpha/V_\alpha{}^{(p-1)} = \frac{\mathbf{P}_\alpha}{\mathbf{M}_\alpha{}^p} \tag{1}$$

$$k_\mathbf{M} = K_\mathbf{M}/\frac{V_\beta}{V_\alpha} = \frac{\mathbf{M}_\beta}{\mathbf{M}_\alpha} \tag{2}$$

$$k_{\mathbf{P}} = K_{\mathbf{P}} \cdot \frac{V_\beta}{V_\alpha} = \frac{\mathbf{P}_\beta}{\mathbf{P}_\alpha} \tag{3}$$

$$\mu = 1 + k_{\mathbf{M}} \tag{4}$$

$$\pi = 1 + k_{\mathbf{P}} \tag{5}$$

Conservation of mass requires that

$$\mathbf{M}_{\text{tot}} = (\mathbf{M}_\alpha + \mathbf{M}_\beta) + p(\mathbf{P}_\alpha + \mathbf{P}_\beta) \tag{6}$$

where $\mathbf{M}_{\text{tot}}$ is the total equivalents of reactant in the entire reaction volume. Substitution in Equation 6, first for $\mathbf{P}_\beta$ and $\mathbf{M}_\beta$ from Equations 3 and 2, and then for $\mathbf{P}_\alpha$ from Equation 1 and simplification using Equations 4 and 5 yields the following expression for $\mathbf{M}_{\text{tot}}$

$$\mathbf{M}_{\text{tot}} = \mu\mathbf{M}_\alpha + p\pi k_\alpha \mathbf{M}_\alpha{}^p \tag{7}$$

Defining

$$\mathbf{M}_0 = \mathbf{M}_{\text{tot}}/p\pi k_\alpha \tag{8}$$

$$\kappa = \mu/p\pi k_\alpha \tag{9}$$

Substitution of Equations 8 and 9 into Equation 7 and rearrangement gives the following functional relationship of $\mathbf{M}_\alpha$ to $\mathbf{M}_0$ (and hence $\mathbf{M}_{\text{tot}}$)

$$\mathbf{M}_\alpha{}^p + \kappa\mathbf{M}_\alpha - \mathbf{M}_0 = 0 \tag{10}$$

By Descartes' rule (on alternation of signs), this general equation can have only one positive real root (5) which then must be the root of physical significance and must lie between 0 and $\mathbf{M}_{\text{tot}}$. In application to a real system, use of the equation in this form requires a method for measuring $\mathbf{M}_\alpha$, and hence it is necessary to distinguish experimentally between $\mathbf{M}_\alpha$ and $\mathbf{P}_\alpha$. Often such a measurement is not possible, and what can be measured is $\mathbf{M}_\alpha{}^0$, the total number of monomer equivalents in the external phase. By definition

$$\mathbf{M}_\alpha{}^0 = \mathbf{M}_\alpha + p\mathbf{P}_\alpha \tag{11}$$

Combining this with Equation 1

$$\mathbf{M}_\alpha{}^0 = \mathbf{M}_\alpha + pk_\alpha \mathbf{M}_\alpha{}^p \tag{12}$$

and

$$\mathbf{M}_\alpha{}^0 = R + pk_\alpha R^p \tag{13}$$

where R is the particular root of Equation 10 found by analysis. It is the variation of $\mathbf{M}_\alpha{}^0$, or the dimensionless quantity $\mathbf{M}_\alpha{}^0/\mathbf{M}_{tot}$, that may be analyzed as a function, $f(\mathbf{M}_{tot})$, to determine stoichiometry (*i.e.*, the value of p) and the various equilibrium constants.

Case Ib. Multiple Association of Unlike Interactants. For the reaction

$$\mathbf{M}_1 + \mathbf{M}_2 + \ldots + \mathbf{M}_p \rightleftarrows \mathbf{P}$$

the extension from association of like to association of unlike particles is essentially similar to that described in a previous analysis of colligative properties (6). However, because of the individual distribution equilibria of the species $\mathbf{M}_1$, $\mathbf{M}_2$, ..., $\mathbf{M}_p$, it is not possible to guarantee that $\mathbf{M}_{1\alpha} = \mathbf{M}_{2\alpha} = \ldots = \mathbf{M}_{p\alpha}$ as in the colligative case; hence, no generalized solution is possible, and any individual case must be treated specifically according to the scheme outlined.

If it could be arranged that $\mathbf{M}_{1\alpha} = \mathbf{M}_{2\alpha} = \ldots = \mathbf{M}_{p\alpha}$, and that all $K_\mathbf{M}$ values be equal throughout the concentration range of interest, then the generalized equation is identical with that for the homogeneous equilibrium since

$$K_\alpha = [\mathbf{P}_\alpha]/[\mathbf{M}_{1\alpha}]\,[\mathbf{M}_{2\alpha}] \ldots [\mathbf{M}_{p\alpha}] = \mathbf{P}_\alpha/[\mathbf{M}_{1\alpha}]^p$$

Such a condition might hold, for example, in the mixed association of unlike monomers of identical molecular weight in gel permeation.

In such a case, the condition that the polymer is formed by a mixed association of monomer units would be reflected in the numerical value of the association equilibrium constant. The value for unlike association differs from that for like association by a factor p^p since the K_{eq} is referred to $\mathbf{M}_\alpha$ moles of monomer for like particle association and to $\mathbf{M}_\alpha/p$ moles of each monomer for unlike particle association, while the measurements are referred to systems both containing $\mathbf{M}_\alpha$ moles of monomer. Alternately it can be shown that this factor arises from a statistical entropy contribution.

For certain classes of polymerization, such as simple copolymerization, a similar generalized analysis could be derived from considerations of monomer reactivity ratios. However, in the completely general case (such as assembly of a protein from subunits) it would require knowledge of stoichiometries to specify these ratios, and it is just these stoichiometries that one is attempting to obtain from this analysis.

Case II. Sequential Association of Like Interactants. For the reaction

$$2\mathbf{M} \rightleftharpoons \mathbf{P}_2 \overset{\mathbf{M}}{\rightleftharpoons} \mathbf{P}_3 \overset{\mathbf{M}}{\rightleftharpoons} \ldots \overset{\mathbf{M}}{\rightleftharpoons} \mathbf{P}_{(p-1)} \overset{\mathbf{M}}{\rightleftharpoons} \mathbf{P}_p$$

The equivalent gel partition system is essentially

$$p\mathbf{M} \overset{K_{p\alpha}}{\rightleftharpoons} \mathbf{P}_p \qquad\qquad p\mathbf{M} \overset{K_{p\beta}}{\rightleftharpoons} \mathbf{P}_p$$

$$(p-1)\mathbf{M} \overset{K_{(p-1)\alpha}}{\rightleftharpoons} \mathbf{P}_{(p-1)} \qquad\qquad (p-1)\mathbf{M} \overset{K_{(p-1)\beta}}{\rightleftharpoons} \mathbf{P}_{(p-1)}$$

$$\vdots \qquad\qquad \vdots$$

$$2\mathbf{M} \overset{K_{2\alpha}}{\rightleftharpoons} \mathbf{P}_2 \qquad\qquad 2\mathbf{M} \overset{K_{2\beta}}{\rightleftharpoons} \mathbf{P}_2$$

$$\mathbf{M} \overset{K_{\mathbf{M}}}{\rightleftharpoons} \mathbf{M}$$

$$\mathbf{P}_2 \overset{K_{2\mathbf{P}}}{\rightleftharpoons} \mathbf{P}_2$$

$$\vdots$$

$$\mathbf{P}_{(p-1)} \overset{K_{(p-1)\mathbf{P}}}{\rightleftharpoons} \mathbf{P}_{(p-1)}$$

$$\mathbf{P}_p \overset{K_{p\mathbf{P}}}{\rightleftharpoons} \mathbf{P}_p$$

where the external phase is represented by α (volume $= V_\alpha$), and the internal phase is represented by β (volume $= V_\beta$). Note that for analysis the sequential reaction equilibrium constants are reparametized, *i.e.*,

$$K_i = \prod_{j=2}^{i} K'_j \qquad 2 \leqslant i \leqslant p$$

where $K'_j = [\mathbf{P}_j]/[\mathbf{P}_{j-1}][\mathbf{M}]$, $j \geqslant 2$, and $[\mathbf{P}_1] \equiv [\mathbf{M}]$. In analogy with Equations 1–6, define

$$k_{i\alpha} = K_{i\alpha}/V_\alpha^{(i-1)} = \frac{\mathbf{P}_{i\alpha}}{\mathbf{M}_\alpha{}^i} \qquad 2 \leqslant i \leqslant p \tag{14}$$

$$k_{\mathbf{M}} = K_{\mathbf{M}}\frac{V_\beta}{V_\alpha} = \frac{\mathbf{M}_\beta}{\mathbf{M}_\alpha} \tag{15}$$

$$k_{i\mathbf{P}} = K_{i\mathbf{P}}\frac{V_\beta}{V_\alpha} = \frac{\mathbf{P}_{i\beta}}{\mathbf{P}_{i\alpha}} \qquad 2 \leqslant i \leqslant p \tag{16}$$

$$\mu = 1 + k_{\mathbf{M}} \tag{17}$$

$$\pi_i = 1 + k_{i\mathbf{P}} \qquad 2 \leqslant i \leqslant p \tag{18}$$

By conservation of mass

$$\mathbf{M}_{\text{tot}} = \mathbf{M}_\alpha + \mathbf{M}_\beta + \sum_{i=2}^{p} i(\mathbf{P}_{i\alpha} + \mathbf{P}_{i\beta}) \tag{19}$$

Substituting into Equation 19 from Equations 14–16 and simplifying using Equations 17 and 18

$$\mathbf{M}_{\mathrm{tot}} = \mu\mathbf{M}_\alpha + \sum_{i=2}^{p} i\pi_i k_{i\alpha}\mathbf{M}_\alpha{}^i \tag{20}$$

Define

$$\mathbf{M}_0 = \mathbf{M}_{\mathrm{tot}}/p\pi_p k_{p\alpha} \tag{21}$$

$$\kappa_i = i\pi_i k_{i\alpha}/p\pi_p k_{p\alpha} \qquad 2 \leqslant i \leqslant (\mathrm{p} - 1) \tag{22}$$

$$\kappa = \mu/p\pi_p k_{p\alpha} \tag{23}$$

Substitution of Equations 21–23 into Equation 20 and rearrangement yields

$$\mathbf{M}_\alpha{}^p + \sum_{i=p-1}^{2} \kappa_i\mathbf{M}_\alpha{}^i + \kappa\mathbf{M}_\alpha - \mathbf{M}_0 = 0 \tag{24}$$

Again, since there is only one alternation of sign there is only one real positive root (5), which must lie between 0 and $\mathbf{M}_{\mathrm{tot}}$, and this must be the root of physical significance.

In terms of the measured variable, the total number of monomer equivalents in phase α is, by definition

$$\mathbf{M}_\alpha{}^0 = \mathbf{M}_\alpha + \sum_{i=2}^{p} i\mathbf{P}_{i\alpha} \tag{25}$$

For a particular solution, where R is the particular root of Equation 24 found by analysis, and the respective $\mathbf{P}_{i\alpha}$'s are substituted from Equation 14

$$\mathbf{M}_\alpha{}^0 = R + \sum_{i=2}^{p} ik_{i\alpha}R^i \tag{26}$$

Note that for a sequential association with fewer species, the general equation is modified so as to eliminate those terms for which the specific equilibrium does not exist. For example, for

$$\mathbf{M} + \mathbf{M} \rightleftharpoons \mathbf{P}_2$$
$$\mathbf{P}_2 + \mathbf{M} + \mathbf{M} \rightleftharpoons \mathbf{P}_4$$

Equation 24 would become

$$\mathbf{M}_\alpha{}^4 + \kappa_2\mathbf{M}_\alpha{}^2 + \kappa\mathbf{M}_\alpha - \mathbf{M}_0 = 0$$

and Equation 26

$$\mathbf{M}_\alpha{}^0 = R + 2k_{2\alpha}R^2 + 4k_{4\alpha}R^4$$

A similar solution may be written down for any arbitrary sequential association.

Results

Computer simulation was carried out on an IBM 370/155 computer, using double-precision calculation and the double-precision version of the Bairstow method polynomial roots subroutine in the IBM scientific subroutines package. For the monomer–dimer case, the quadratic can be and was solved exactly—*e.g.*, see Ref. *3*. Estimates of propagated error are provided in the tables where needed. The analysis given here reduces to that given previously (*3*) when the system concerned is a monomer–dimer system.

Take Equation 10 for the specific case of dimerization

$$\mathbf{M}_\alpha{}^2 + \kappa\mathbf{M}_\alpha - \mathbf{M}_0 = 0$$

Now substitute here in terms of the variables as defined in the earlier study (*3*), where N_0 is the total equivalents of monomer—*i.e.*, let

$$\mathbf{M}_0 = \frac{N_0 V_\alpha}{2\delta K_\alpha}$$

and

$$\kappa = \frac{\mu V_\alpha}{2\delta K_\alpha}$$

These substitutions yield the equation

$$\frac{2\delta K_\alpha}{V_\alpha}\mathbf{M}_\alpha{}^2 + \mu\mathbf{M}_\alpha - N_0 = 0$$

which is identical with Equation 5 of ref. *3*.

Representative plots of $\mathbf{M}_\alpha{}^0/\mathbf{M}_{tot}$ for the dimerization reaction with varying values of the distribution parameters K_M and K_P are given in Figure 1 as a semilogarithmic plot. Values of the distribution parameters were chosen here in relation to a gel partition model, where the only factor influencing distribution is a molecular size function; that is, K_X

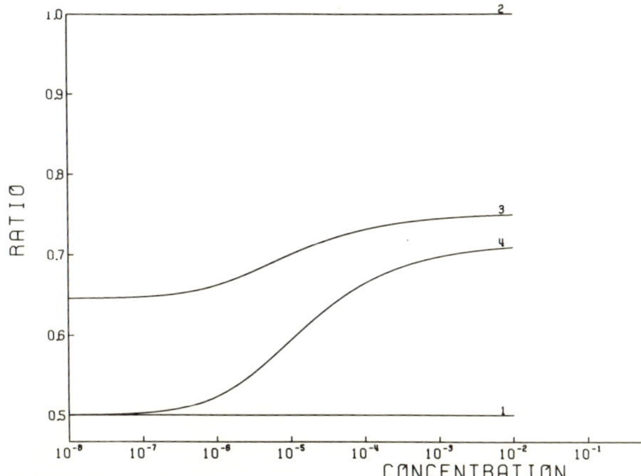

Figure 1. Concentration dependence of $M_\alpha{}^0/M_{tot}$ for the monomer–dimer reaction. Concentration: $M_{tot}/(V_\alpha + V_\beta)$. Ratio: $M_\alpha{}^0/M_{tot}$. All curves, $V_\alpha = V_\beta = 1.00$, $K_\alpha = 1 \times 10^5$. (1) $K_M = K_P = 1.00$; (2) $K_M = K_P = 0.00$; (3) $K_M = 0.55$, $K_P = 0.33$; (4) $K_M = 1.00$, $K_M = 0.50$.

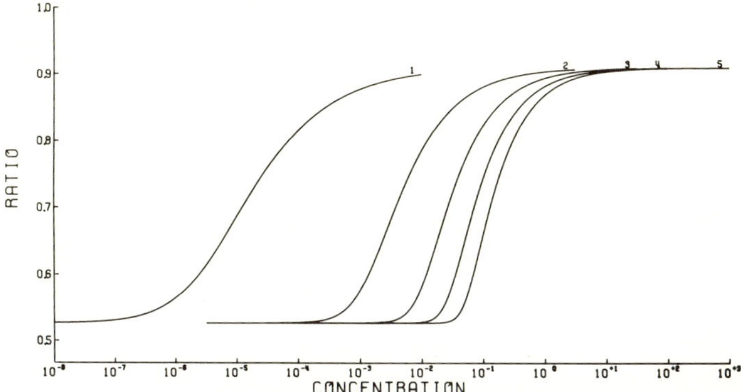

Figure 2. System and axes as in Figure 1, calculated for various stoichiometries. All curves, $V_\alpha = V_\beta = 1.0$, $K_\alpha = 1 \times 10^5$, $K_M = 0.90$, $K_P = 0.10$. (1) Monomer–dimer, (2) monomer–trimer, (3) monomer–tetramer, (4) monomer–pentamer, (5) monomer–hexamer.

varies from zero for a completely excluded molecule to one for a completely penetrant molecule.

Such limits for K_X are not required. The mathematical model requires only that K_X be positive, a necessity for the constant to have physical meaning. The mechanism of the interaction of the gel with the experimental molecule is unspecified, and may be adsorption, ion exchange, or any other process representable by an "equilibrium constant" defined as a concentration ratio of "gel-associated" to "gel-non-associated" material. Multiple interactions are also allowed, and the same mathematical generalization holds. The apparent K_X is then a combination of individual interaction parameters. Thus the analysis can be extended to multiple-phase equilibria in general.

The same function is plotted in Figures 2 and 3 for the monomer–dimer through monomer–hexamer stoichiometry, for a fixed association equilibrium constant of 1×10^5. The distribution coefficients for Figure 2 are chosen for K_M constant and K_P the same irrespective of the size of the polymer while for Figure 3 K_M is constant, and K_P is that value predicted for the polymer using a gel permeation model in which the square root of molecular weight is a linear function of the cube root of the partition constant. This functional relationship is one suggested for gel permeation on theoretical (7) and experimental (8) grounds in the limit of zero flow (9).

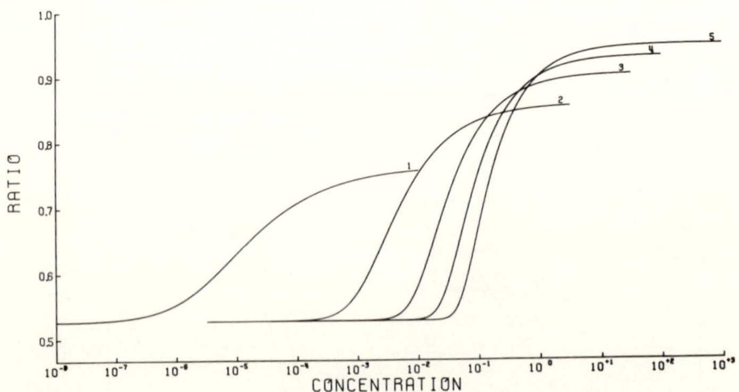

Figure 3. The respective curves are identical with those of Figure 2 except that K_P is chosen to correspond to a theoretical gel permeation experiment (see text). All curves, $K_M = 0.90$. (1) $K_P = 0.32$, (2) $K_P = 0.17$, (3) $K_P = 0.11$, (4) $K_P = 0.08$, (5) $K_P = 0.06$ (for stoichiometries monomer–dimer through monomer–hexamer, respectively).

The slope of the plot is a function of the reaction stoichiometry although the functional dependence is complicated. In addition, the "thermodynamic point," *i.e.*, the point which corresponds to K_{eq} (which is at 1×10^{-5} for all curves given here) does not correspond to the point of inflection, and even for the simplest case, dimerization, the point of inflection does not correspond to the midpoint. In addition, the curves are not symmetrical about the midpoint. Thus, no simple analysis of curve shape based on midpoint value or inflection point can give the thermodynamic or stoichiometric numbers.

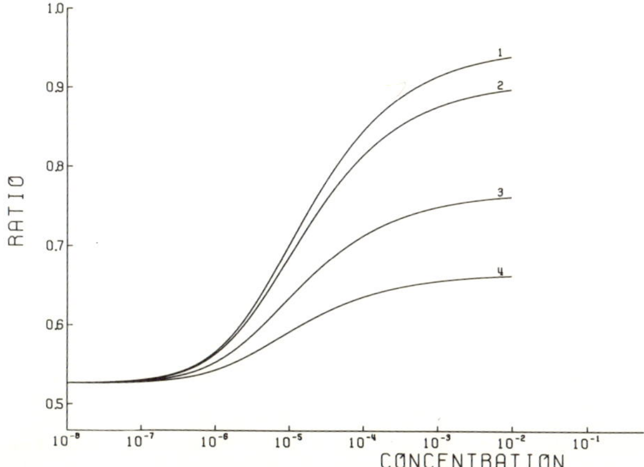

Figure 4. System and axes as in Figure 1, monomer–dimer reaction, varying K_D. All curves, $V_\alpha = V_\beta = 1.0$, $K_\alpha = 1 \times 10^5$, $K_M = 0.90$. (1) $K_D = 0.05$, (2) $K_D = 0.10$, (3) $K_D = 0.30$, (4) $K_D = 0.50$.

These points are emphasized by Figure 4, which is a series of plots of $\mathbf{M}_\alpha{}^0/\mathbf{M}_{tot}$ *vs.* $\mathbf{M}_{tot}$ in the same semilogarithmic scale for the dimerization stoichiometry with varying $K_\mathbf{D}$ (partition coefficient for the dimer), $K_\mathbf{M}$ held constant. Data points from Figures 2 and 4 are collected in Table I with respect to slopes and values at the inflection point.

The slope at the inflection is a function of the partition coefficients, and if stoichiometric information is to be readily obtained, some transformation of variable must be applied to eliminate this dependency. Several reasonable transformations, based on the functional form of the graphed curve, were tried, and in Table II one is displayed which, for the dimerization reaction, satisfies this constraint. The ratio of the slope at the inflection with respect to the ordinate was found by division by the value of $\mathbf{M}_\alpha{}^0/\mathbf{M}_{tot}$ at the inflection. $\mathbf{M}_\alpha{}^0/\mathbf{M}_{tot}$ is essentially an inverse weighted partition coefficient

$$\frac{\mathbf{M}_{tot}}{\mathbf{M}_\alpha{}^0} = \frac{\mathbf{M}_\alpha + p\mathbf{P}_\alpha}{\mathbf{M}_\alpha + p\mathbf{P}_\alpha} + \frac{\mathbf{M}_\beta + p\mathbf{P}_\beta}{\mathbf{M}_\alpha + p\mathbf{P}_\alpha}$$

and

$$\frac{\mathbf{M}_{tot}}{\mathbf{M}_\alpha{}^0} = 1 + k_\mathbf{X}$$

where **X** represents equivalents of monomer. This still does not yield a constant independent of $K_\mathbf{M}$; however, recognizing that the original equa-

Table I. Curve Parameters for the Monomer–*n*-mer System[a]

p	K_P	*Slope*	*Value*
2	0.05	0.1683	0.7035
2	0.10	0.1512	0.6855
2	0.30	0.0960	0.6271
2	0.50	0.0555	0.5847
3	0.10	0.2362	0.6673
4	0.10	0.2938	0.6553
5	0.10	0.3366	0.6461
6	0.10	0.3703	0.6390

[a] Slope = slope at the point of inflection, value = ordinate value at the point of inflection. Conditions: $V_\alpha = V_\beta = 1.0$, $K_\alpha = 1 \times 10^5$, $K_M = 0.90$.

Table II. Value of ξ[a] as a Function of K_X for the Monomer–Dimer Reaction[b]

K_M	K_D	$\xi \pm 0.002$
0.90	0.05	0.929
0.90	0.10	0.929
0.90	0.30	0.928
0.90	0.50	0.925

[a] ξ has been defined operationally in the text.
[b] Conditions: $V_\alpha = V_\beta = 1.0$, $K_\alpha = 1 \times 10^5$.

Table III. Value of ξ as a Function of Stoichiometry (p)[a]

p	$\xi \pm 0.004$
2	0.929
3	1.491
4	1.889
5	2.195
6	2.442

[a] Conditions: $V_\alpha = V_\beta = 1.0$, $K_\alpha = 1 \times 10^5$, $K_M = 0.90$, $K_P = 0.10$.

tion contains terms of the form $(1 + k_M)$ and $(1 + k_P)$, and that the abscissa is a logarithmic axis, division of this ratio by log $[(1 + k_M)/(1 + k_P)]$ was attempted. The result is a number which is independent of the partition coefficients. A series of simulations with varying K_α and V_α indicated that this parameter, which is designated ξ, is independent of all parameters of the analysis except stoichiometric coefficient. The value of ξ for the various stoichiometries plotted in Figure 2 is given in Table III.

It would be difficult in practice to use the value of ξ to determine stoichiometry since it requires not only the slope but also the value at the inflection point, which may be very difficult if not impossible to recognize with sufficient accuracy in real error-prone data. Also, a knowledge of the values of K_M and K_P is required. However, it has been shown that curve shape can give stoichiometric data in and of itself; further numerical modeling is required to define practicable curve parameters for analysis of experimental data. For stoichiometry, some experimental uncertainty is tolerable since ξ is a discontinuous function, and to infer a stoichiometry one requires only that ξ be approximately equal to one of the calculated values for a given set of considered stoichiometries.

Literature Cited

1. Darnall, D. W., Klotz, I. M., *Arch. Biochem. Biophys.* (1972) **149,** 1.
2. Shapiro, A. L., Vinuela, E., Maizel, J. V., *Biochem. Biophys. Res. Commun.* (1967) **28,** 815.
3. Cameron, B. F., Sklar, L., Greenfield, V., Adler, A. D., *Separ. Sci.* (1971) **6,** 217.
4. Kellett, G. L., *J. Mol. Biol.* (1971) **59,** 401.
5. Mostowski, A., Stark, M., "Introduction to Higher Algebra," p. 285, MacMillan, New York, 1964.
6. Adler, A. D., O'Malley, J. A., Herr, A. J., Jr., *J. Phys. Chem.* (1967) **71,** 2896.
7. Porath, J., *Pure Appl. Chem.* (1963) **6,** 233.
8. Andrews, P., *Biochem. J.* (1964) **91,** 222.
9. Guttman, C. M., DiMarzio, E. A., *Macromolecules* (1970) **3,** 681.

RECEIVED January 17, 1972. Work supported in part by U. S. Public Health Grant AM-09 001.

25

Polypeptide Chain Molecular Weight Determination by Gel Permeation Studies on Agarose Columns in 6*M* Guanidinium Chloride

KENNETH G. MANN and DAVID N. FASS

Department of Biochemistry, University of Minnesota, St. Paul, Minn. 55455

WAYNE W. FISH

Department of Biochemistry, Medical University of South Carolina, Charleston, S. C.

Gel permeation chromatography of protein linear random coils in guanidinium chloride allows simultaneous resolution and molecular weight analysis of polypeptide components. Column calibration results are expressed in terms of a log M *vs.* K_d *plot or of effective hydrodynamic radius* (R_e)*. For linear polypeptide random coils in 6M GuHCl,* R_e *is proportional to* $M^{0.555}$*, and* $M^{0.555}$ *or* R_e *may be used interchangeably. Similarly, calibration data may be interpreted in terms of* $N^{0.555}$ *(N is the number of amino acid residues in the polypeptide chain), probably the most appropriate calibration term provided sequence data are available for standards.* R_e *for randomly coiled peptide heteropolymers is insensitive to amino acid residue side-chain composition, permitting incorporation of chromophoric, radioactive, and fluorescent substituents to enhance detection sensitivity.*

The partitioning of a solute between the stationary and mobile phases of a gel permeation column is a function of the molecular size and shape of the solute and the size distribution of gel pores separating the two phases. For a gel permeation column operating under conditions in which an equilibrium distribution of solute between the phases is obtained, the relative elution position of a solute component is a function of its effective hydrodynamic radius, R_e (*1*). If polymer solutes of similar chemical composition and shape are compared, the elution positions of a series of solutes will be related to their molecular weights since under the circumstance of identical shape molecular weight and R_e are related functions.

Since the actual structures of the pores of the insoluble polymer supports used as gel permeation chromatography resins are not known, an exact theoretical treatment of the technique is not possible, and the gel permeation characteristics of a solute can only be compared in a relative fashion. (In alluding to molecular weights determined by this method, we shall use the term M_{app} since values obtained by this method are approximate.) All estimates of polymer dimensions based on gel chromatography are therefore relative and based on those determined for polymers whose dimensions are known.

Many soluble native proteins are compact, essentially spherical structures with frictional ratios (f/f_{min}) around 1.25. (The term f/f_{min} represents the ratio of the measured frictional coefficient to the minimal value which could be obtained for the equivalent anhydrous sphere. This

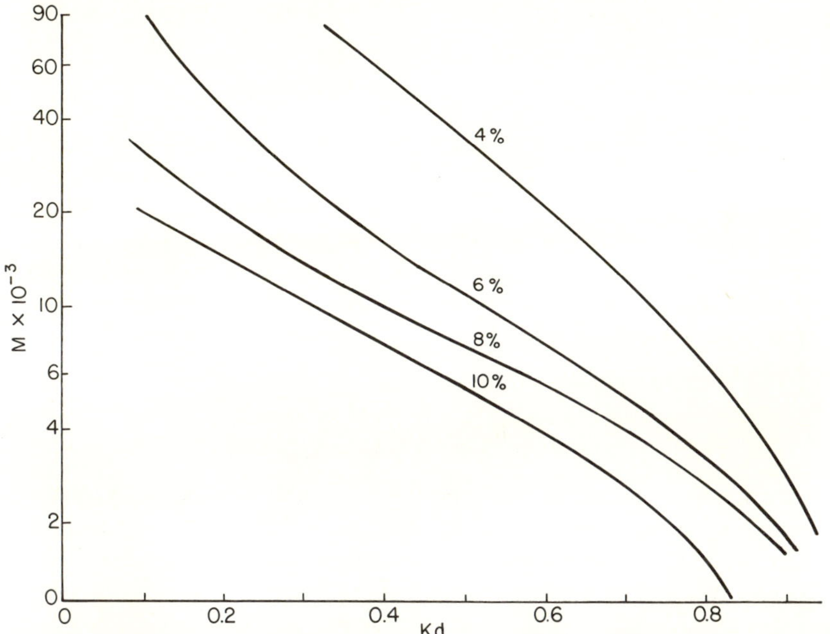

*Figure 1. Gel permeation data for reduced proteins in 6*M *GuHCl on 4, 6, 8, and 10% agarose resins plotted in terms of log* M *vs.* K_d. *Proteins used to calibrate these columns were well-known materials of established electrophoretic purity. Molecular weights used for calibration are from the literature* (5, 7).

observation led to the exploration of gel permeation chromatography as a means of determining the molecular weights of native proteins (2). While comparative gel chromatography has been widely used in biochemistry for molecular weight determinations, this method is inherently dangerous since gel permeation elution position (V_e) can only be related to molecular weight if the "standard" protein used to calibrate a column and the unknown possess identical shapes and there is an absence of noncovalent binding of the protein to the gel filtration medium. The validity of apparent molecular weights (M_{app}) determined in this fashion depend on the assumption regarding solute shape, which can be tested only by measurements whose complexity equals or exceeds that of the classical exact methods of molecular weight determination.

Table I. Useful Limits and Resolution Capabilities of Agarose Resins for Linear Random Coils in 6*M* GuHCl

			ΔK_d			
% Agarose Comp	M *at* K_d = *0.1*	M *at* K_d = *0.9*	*70,000–50,000*	*30,000–20,000*	*10,000–5000*	*5000–1500*
4	250,000	2800[a]	0.075	0.080	0.113	0.115
6	85,000	1850	0.050	0.085	0.180	0.125
8	32,000	1700	—	0.090	0.220	0.295
10	23,000	700[a]	—	—	0.215	0.265

[a] Estimated.

All proteins studied in concentrated solutions of guanidinium chloride (GuHCl) have been demonstrated to behave hydrodynamically as random coils, possessing no detectable gross noncovalent structure (3). If all covalent crosslinks in the polymer chains are ruptured, the resulting structures behave hydrodynamically as linear random coils, and as such they obey the classical relationship between average dimensions of the flexible polymer defined by the radius of gyration (R_g) and the molecular weight (4)

$$R_g^2 = \alpha^2\beta^2\frac{M}{6M_0} \tag{1}$$

in which β is a measure of monomer effective unit length, α is a measure of chain expansion from nonideality, and M_0 is the monomer residue molecular weight. If we assume that proteins may be treated as homopolymers, M_0 and β become constant; however α is a slowly increasing function of molecular weight in real solvents. In 6*M* GuHCl α has been found to vary as (constant)$M^{0.055}$ (5). Therefore, collecting constant terms, R_g may be expressed as

$$R_g = (\text{constant})M^{0.555} \tag{2}$$

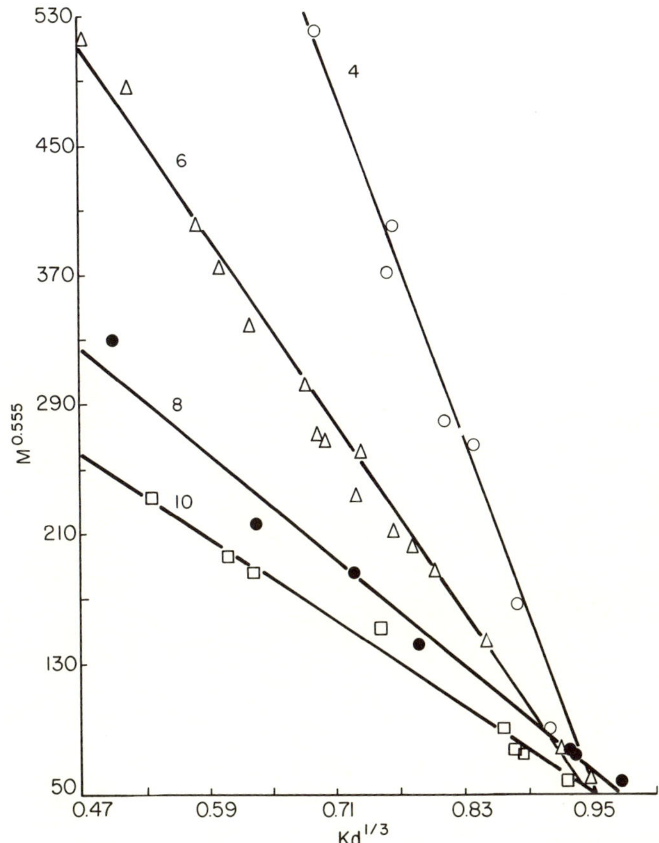

Figure 2. Gel permeation data for polypeptide linear random coils plotted according to the method of Porath (8); $M^{0.555}$ is plotted vs. $K_d^{1/3}$. Lines drawn through the data from each column are lines of best fit determined by linear least-squares analysis. Numerical designation for each curve represents the agarose resin used.

Further, since the term R_e is proportional to R_g, a relationship between the parameter measured by gel filtration of linear random coils and their molecular weights is provided.

Previous studies (5–7) have clearly demonstrated that gel permeation chromatography of reduced linear randomly coiled polypeptide chains in GuHCl provides an accurate, dependable method for molecular weight estimation.

The gel permeation characteristics are best expressed in terms of the distribution coefficient for a solute, K_d, which represents the fraction of the gel internal volume accessible to the solute

$$K_d = \frac{V_e - V_{BD}}{V_{DNP} - V_{BD}}$$

in which V_e is the elution weight of the solvent accumulated prior to the appearance of the maximum concentration of solute in the effluent, V_{BD} the V_e for blue dextran (a high molecular weight polymer which is totally

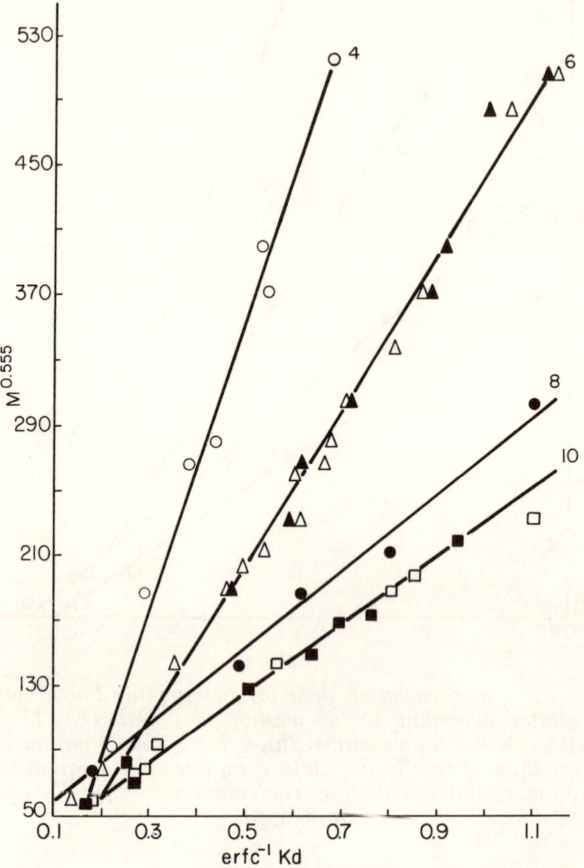

Figure 3. Gel permeation data for linear randomly coiled polypeptides on various agarose resins, plotted according to the method of Ackers (9). $M^{0.555}$ *is plotted vs. the inverse error function complement of* K_d *(erfc*$^{-1}$ K_d*). Lines drawn through the data points represent best fits obtained from linear least-squares analysis of the data. Numerical designation of each curve represents the percent agarose composition for the resin used. Filled triangles on the curve for the 6% resin, and the filled squares on the curve for the 10% resin are points determined using fluorescent proteins. Data for the labeled polypeptides were not included in the least-squares analysis.*

excluded from the gel), and V_{DNP} the V_e for DNP alanine. (The measured parameter V_e is generally expressed in terms of volume. In order to increase precision, we have used effluent weight rather than volume in determining values for K_d. We shall continue, however, to use the standard term V_e and its usual connotations.) The terms $V_{DNP} - V_{BD}$ is a measure of total solvent volume in the gel intersticies, and $V_e - V_{BD}$ is the gel interstitial volume accessible to the solute under investigation. Values of K_d can therefore vary between zero (total exclusion) and one (total inclusion).

Since gel filtration studies are comparative, a set of standard polypeptides of known molecular weight must be used to construct a calibration curve for a given column. Once the K_d's for a number of known polypeptide chain components have been accumulated, a calibration curve can be generated in a number of ways. The most common type of calibration is a plot of the log M *vs.* K_d. Figure 1 presents calibration data for a series of agarose gel columns of identical physical dimensions (1.5×85 cm) but containing resins of different agarose composition. (A complete description of all experimental techniques can be found in Ref. 7.) The resins used were obtained from Pharmacia (6% agarose) and Bio-Rad (4, 8, and 10% agarose), range from 4% to 10% agarose, and cover a useful range of molecular weight determination for linear random coils from approximately 300,000 to 700. The choice of agarose content depends largely on the desired application of the method. As the agarose content increases, the useful range of molecular weight estimation decreases, but the chromatographic resolution inherent in the method increases. The exclusion performance range of each resin can best be expressed in terms of the molecular weights of components which will give observed K_d's of 0.1 and 0.9. On the other hand, resolving performance is best expressed in terms of the ΔK_d observed between pairs of different molecular weight species when they are examined on each resin. Table I presents M at $K_d = 0.1$ and $K_d = 0.9$ as well as the ΔK_d's for a series of molecular weight ranges for each of the agarose resins examined. By maximizing ΔK_d, resolution is enhanced, but the useful range is limited. For most protein work, the 6% agarose presents the best compromise between resolution and range. Examination of Table I would suggest that no obvious advantage would be had by choosing an agarose content greater than 8%. However, the lower limit of the 10% resin is theoretically extended to the range of pentapeptide, although we have not made measurements on this column below the level of dodecapeptide.

It should be noted here that the curves in Figure 1 which were constructed with single lots of each type of agarose are merely representative. If resins of the same nominal agarose content obtained from

Table II. Gel Permeation Characteristics for Crosslinked Random Coils

Protein	M[a]	M_{app} (*cross-linked*)	*% Decrease in* M_{app}	*Crosslinks per 100 Residues*
Serum albumin	69,000	40,000	42.0	2.8
Chymotrypsinogen	25,700	18,000	30.0	2.0
Lysozyme	14,700	8,000	45.6	3.1
Ribonuclease	13,500	9,000	33.3	3.2

[a] Based on value obtained by exact techniques.

different manufacturers are compared, a great deal of variability in calibration curves may be observed (*7*). On the other hand, the calibration curves prepared by different investigators using separate columns but the same lot of agarose are remarkably consistent (*7*).

While the calibration curves presented in Figure 1 provide an adequate treatment of gel permeation calibration data, they are nonlinear in the extremes of the useful calibration regions. Two methods of data treatment may be used to provide linear representations of gel filtration results. These methods substantially increase the usefulness of the technique. The equations for these methods, developed by Porath (*8*) and Ackers (*9*), are as follows:

$$\text{Porath: } K_d^{1/3} = A - BR_e \tag{4}$$

$$\text{Ackers: } R_e = A - B\,\text{erfc}^{-1} K_d \tag{5}$$

While both equations provide linear calibration results, the Porath equation is based on a specific model of the pore dimensions of a gel permeation media while the Acker's treatment is a statistical treatment based on the assumption that for a given resin a Gaussian distribution of pore sizes exists. The terms A and B in each equation have different significance and are adjustable parameters. Since R_e is proportional to $M^{0.555}$ for linear randomly coiled polypeptide chains in 6M GuHCl, these equations can be used to provide linear arrays for the same data shown in Figure 1. Figure 2 provides Porath plots of the gel filtration data for 4% to 10% agarose columns while Figure 3 depicts the same calibration data plotted according to the Acker's method. For some experimental situations ($K_d < 0.1$) neither equation provides for linear representation of the data. This phenomenon can be most easily explained for Acker's treatment since the distribution of pores most likely is not Gaussian when a relatively small number of pores is considered, as at the extreme limits of the pore size distributions.

While the data of Figures 1, 2, and 3 are plotted in terms of M, the parameter actually measured in gel permeation experiments is R_e. The R_e of a random coil can be calculated from intrinsic viscosity data using the expression

$$R_e = \left[\frac{3/4[\eta]M}{2.5N}\right]^{1/3} \tag{6}$$

in which $[\eta]$ is the intrinsic viscosity and N is Avogadro's number. The significance of the term R_e determined from intrinsic viscosity data is exactly equivalent to that used for gel permeation studies; thus column calibration data may be plotted as R_e *vs.* $\text{erfc}^{-1}K_d$, provided intrinsic viscosity data are available for the standards used to calibrate the column. Largely through the efforts of Tanford, Kawahara, and Lapanje (*10*), a sufficient quantity of intrinsic viscosity data for linear and crosslinked randomly coiled polypeptide chains in 6M GuHCl are available. The effect of a crosslinking is to diminish the extent of unfolding of the random coil and hence reduce the observed R_e. While molecular weight estimates of crosslinked random coils using gel filtration in 6M GuHCl are not possible, estimation of R_e for crosslinked coils can be done quite accurately (*7*).

The dependence of M on R_e is valid if and only if the polypeptide chain under investigation is free of covalent (disulfide, etc.) and noncovalent interactions and is a linear random coil. This factor has made gel filtration in 6M GuHCl a potent tool for establishing the absence of crosslinks in denatured proteins. The determination of the quaternary structure of a protein generally involves sedimentation equilibrium studies on the protein under native conditions and conditions which are thought to bring about complete disruption of three-dimensional structure and covalent crosslinks (*e.g.*, 6M GuHCl, 0.1M 2-mercaptoethanol). The latter study involves an assumption rarely tested owing to the difficulty and quantities of material required for intrinsic viscosity measurement. If a protein exists as a linear random coil under reducing conditions in 6M GuHCl, its R_e should be a function of its number of monomer units and hence molecular weight. Equivalency of molecular weight data obtained by an exact technique (*e.g.*, sedimentation equilibrium) and those based on R_e determined by gel filtration in 6M GuHCl provides adequate proof of the attainment of ultimate subunits. An illustration of the dependence of determined M_{app} on the absence of crosslinks can be had by examination of the R_e and M_{app} for lysozyme ($M = 14{,}300$) in 6M GuHCl with crosslinks and as a linear random coil. Crosslinked, randomly coiled lysozyme has an R_e of 26.7 A, and would provide a K_d consistent with an M_{app} of 8000. Reduced lysozyme, on the other hand,

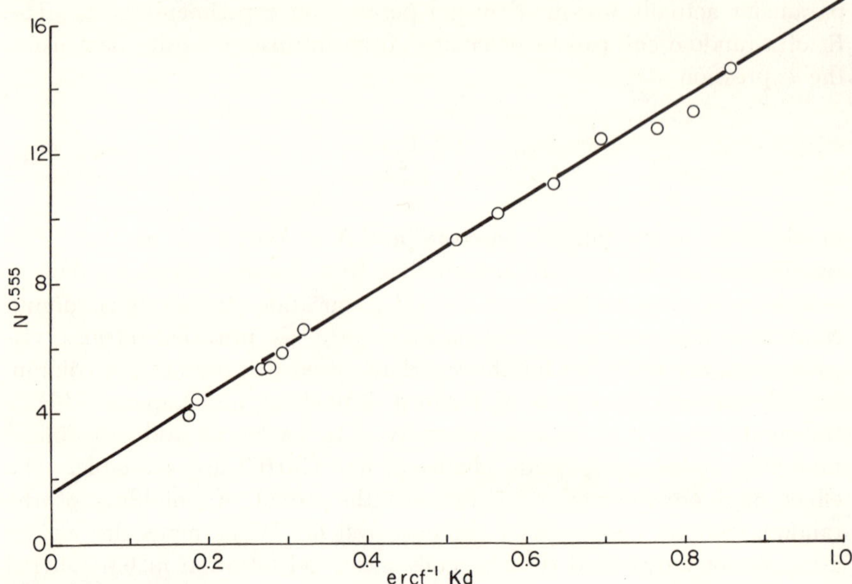

Figure 4. Data for linear randomly coiled polypeptides of known sequence on 10% agarose. $N^{0.555}$ *is plotted* vs. *the inverse error function complement of* K_d *(erfc^{-1} K_d). The line drawn through the points represents a line of best fit determined by least-squares analysis of the data.*

has an R_e of 33.9 A, and would provide a K_d consistent with its true molecular weight when examined on an agarose–GuHCl column.

Table II provides some examples of how the effects of disulfide crosslinks on R_e are reflected in the observed M_{app}'s for a series of proteins of known molecular weight. The divergence between the true M and the M_{app} observed for crosslinked random coils clearly demonstrates the dependence of the method on linear dimensions for the polymer under investigation. It also shows the need for coupling this method with exact methodology for molecular weight determination if incontrovertible data are required.

Besides the need for the assumption that linear random-coil dimensions are fulfilled, a second assumption dealing with protein amino acid composition must be made if molecular weights are to be extracted from gel permeation data. The effective hydrodynamic radius for a polymer is a function of the number of monomer residues in the polymer, and for homopolymers this is obviously a function of M. Although proteins are heteropolymers composed of some 20 different amino acids with residue weights (M_0) between 67.1 grams per mole (glycine) to 186.2 grams per mole (tryptophan), for most proteins the average $\overline{M}_0$ is between 105 and

115 grams per mole, and the assumption that $\overline{M}_0$ is constant does not produce a significant error. This is especially true if one considers that for most of the calibration standards used, the reliability of molecular weight is probably $\pm 5\%$. The studies performed using the 10% agarose column permit a more precise definition of the quantity measured by gel permeation studies of randomly coiled polypeptide heteropolymers. For this column all the standards were polypeptides of known amino acid sequence. Thus, absolute knowledge of the molecular weight and number of amino acid residues is available. Further, since our comparative measurements involve molecular weight only on an average residue basis, it is appropriate to compare a term consistent with that function which the technique actually measures, *viz.*, R_e. The R_e for a linear random coil is a function of the number of residues in the polypeptide chain if any contribution from side-chain steric hindrance can be neglected. Since the term M/M_0 in Equation 1 can be expressed as the number of monomer units in the polymer (N), we can re-express our data in terms of the expression

$$R_e = (\text{constant})N^{0.555} \tag{7}$$

Table III. Number of Amino Acid Residues Determined on 10% Agarose

Protein	*Number of Residues (Sequence)*	*Number of Residues Determined*	*Difference*	*% Error*
Myoglobin	153	144	−9.0	−6.3
Ribonuclease	124	123.2	−0.8	−0.6
Cytochrome-c	104	112.3	8.3	7.0
Myoglobin residues 56–153	97	102.7	5.7	5.6
Lysozyme residues 12–105	93	87.9	−5.1	−5.5
Myoglobin residues 36–131	76	77.6	1.6	2.1
Cytochrome-c residues 1–65	65	64.2	−0.8	−1.2
Myoglobin residues 1–55	55	55.1	0.1	0.2
Insulin B chain	30	28.2	−1.8	−5.8
Lysozyme residues 106–129	24	21.0	−3.0	−12.5
Cytochrome-c residues 81–104	24	25.0	1.0	4.3
Insulin A chain	21	23.4	2.4	11.6
Myoglobin residues 131–153	21	22.5	1.5	11.9
Cytochrome-c residues 66–80	15	14.2	−0.8	−5.3
Lysozyme residues 1–12	12	12.7	0.7	5.7
				Average Error, 5.7%

In this way our initial assumption regarding the constancy of $\overline{M}_0$ for naturally occurring polypeptides is deleted. Figure 4 presents a plot of $\mathrm{erfc}^{-1}K_d$ *vs.* $N^{0.555}$ for polypeptides of known sequence on the 10% agarose column. Over the range of 12–160 residues for a diverse grouping of proteins and peptides, it can be seen that excellent agreement is obtained. Table III presents the number of residues determined for each peptide using the data of Figure 4. Also presented is the difference between the determined value for N using gel permeation chromatography and the true value from the polypeptide sequence data. The average error for determined N is about 5%. This value is most probably a reflection of the experimental error in the technique since neither the assumptions regarding $\overline{M}_0$ nor the errors involving calibration standards whose exact molecular weights are unknown are involved.

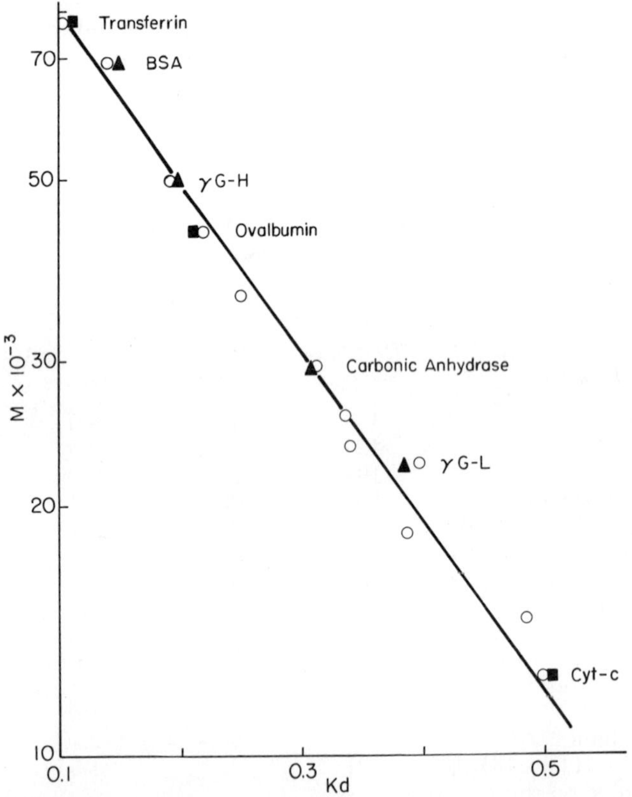

Figure 5. Calibration curve for 6% agarose, prepared with unlabeled polypeptides (open circles), plotted according to the method of Andrews. The log M is plotted vs. K_d*. Superimposed on the curve are data points obtained with fluorescein (filled triangles) and rhodamine (filled squares) labeled proteins.*

The absence of side-chain contributions to gross polymer dimensions in 6*M* GuHCl can be seen by the results obtained with cytochrome-c and the cytochrome-c peptide composed of residues 1–65. Both of these polypeptides contain in addition to a normal complement of amino acids a bulky heme group. However the data of Table III clearly show that even the addition of this large side-chain substituent does not produce anomolous behavior.

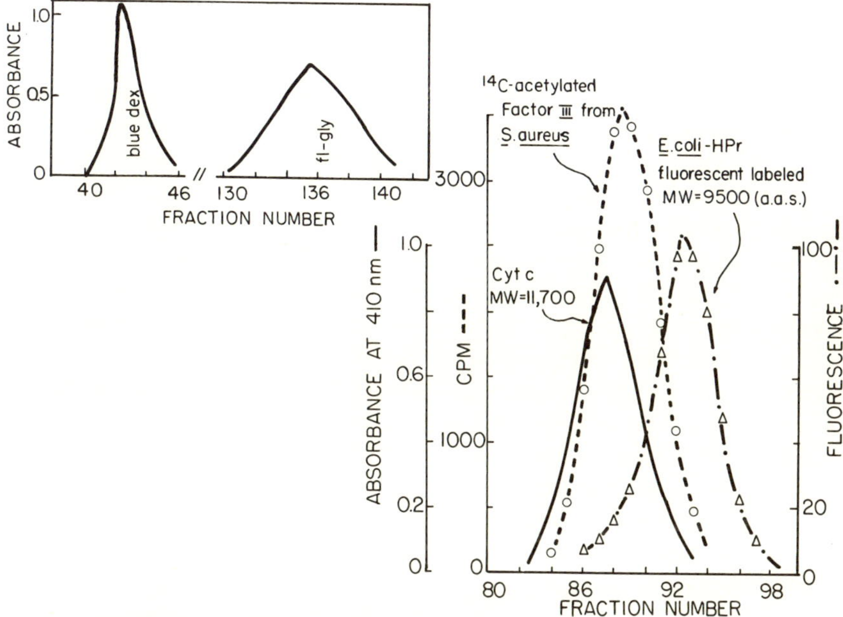

Figure 6. Elution profile obtained on examination of the gel permeation behavior of ^{14}C acetylated S. aureus *Factor III. Cytochrome-c and fluorescein labeled* E. coli *HPr were included in the applied sample as internal standards. Absorbance at 410 nm (cytochrome-c), fluorescein fluorescence (HPr), and radioactivity (Factor III) are plotted* vs. *fraction number. Inset shows the elution profiles for the external (blue dextran) and internal (fluoroglycine) column markers. Data in this figure are from unpublished work* (20).

One of the outstanding characteristics of gel permeation chromatography is the fact that it provides for simultaneous analysis and separation of the components in the system under investigation. The nature of linear polypeptide chains in 6*M* GuHCl is such that the dependence of R_e on M in this solvent is much greater than in solvents which promote native globular configurations. Chymotrypsinogen and lysozyme have R_e's of about 20.6 and 22.5 A in their native globular configurations (*11*). For the same proteins with disulfide bonds reduced in 6*M* GuHCl, the resulting linear random coils have R_e's of 33.9 and 49.0 A, respectively.

Since gel permeation discrimination depends on R_e, it is apparent that dramatically enhanced resolution is obtainable in 6*M* GuHCl. This factor has led to the use of this technique for analysis of such complex mixtures as proteolytic digestion products (*12, 13*) and red cell membrane proteins (*14*). An added dividend of the method is recovery of the isolated polypeptide components for further physical or chemical studies.

Besides the need for accurate molecular weight determination and high resolution, the practicality of gel permeation studies in 6*M* GuHCl to specific problems of protein chemistry depends also on the detection sensitivity of the method. Our initial work made use of either protein turbidity in trichloroacetic acid or the intrinsic chromophores present in naturally occurring polypeptide chains for detection of components in column effluents. Using these procedures analysis generally required at least 1 mg of the applied sample. Effluent absorbance monitoring required the presence of intrinsic chromophores in the protein. While these requirements might not appear to be restrictive, for many biological systems they would prevent use of the method. Since our earlier studies indicated that polymer side-chain characteristics had little effect on gel permeation parameters, it appeared feasible to increase detection sensitivity by means of incorporation of chromophoric, fluorescent, and radioactive labels into the polypeptide chains. We have explored the use of the compounds fluorescein isothiocyanate and rhodamine B (*15*), dansyl chloride (*16*), picryl sulfonic acid (*17*), ^{14}C-acetic anhydride (*18*), and ^{125}I (*19*) as labels for polypeptide chains. None of these modifications alters the measured parameter (K_d) significantly, and all provide an increase in detection sensitivity of at least 1000-fold. Using these prelabeling techniques, molecular weight determinations can be easily performed on as little as 50 μgrams of protein. Further, the possibility of not detecting a component because of lack of an intrinsic chromophore is practically eliminated. Figure 5 presents a calibration curve prepared with unlabeled proteins (based on the Andrew's method) onto which are superimposed the data points collected using rhodamine B and fluorescein labeled proteins. It can be seen that the labeling technique has not altered the gel permeation behavior of the polypeptide chains or the experimental error which remains at about $\pm 7\%$. (Figure 3 also presents data pertinent to the question of using labeled proteins.) The use of these labels provides several advantages even in cases where detection sensitivity is not a problem. For example, by suitable choices of labels a mixture of standards can be used to provide a single experiment which will completely calibrate a column. The labeling procedures also permit the use of internal standards in individual gel filtration experiments. Figure 6 is an illustration of the use of multiple labels. This figure presents the elution profile obtained on analysis of a purified membrane protein, Factor

Table IV. Cyanogen Bromide Molecular Weight Fingerprint Data from 6% and 10% Agarose–GuHCl Columns

Resin	*Protein*	M^{a}(*sequence*)	M_{app}(*determined*)
6% Agarose	α-Lactalbumin	10,270	10,600
		3,900	4,000
	Myoglobin	8,162	8,400
		6,216	5,600
		2,100	2,400
	Cytochrome-c	7,650	7,300
		2,530	2,500
		1,540	1,500
10% Agarose	Lysozyme	10,281	10,000
		2,866	2,450
		1,389	1,400
	Myoglobin	8,162	8,700
		6,216	6,300
		2,100	2,550
	Cytochrome-c	7,650	7,200
		2,530	2,600
		1,540	1,600

[a] Based on value obtained by exact techniques.

III, which had been radioactively labeled by ^{14}C acetylation. Cytochrome-c and fluorescein labeled *E. coli* HPr, a protein of known sequence, have been added to the sample as internal molecular weight standards. The M_{app} determined for Factor III by gel permeation in 6*M* GuHCl, 11,000 ± 300, is in good agreement with the value obtained from sedimentation equilibrium in 6*M* GuHCl–0.1*M* 2-mercaptoethanol, 12,000 (*20*).

The enhanced detection sensitivity provided by prelabeling techniques has permitted our examination of the concentration dependence of K_d over a range of concentration of applied sample from 0.001% to 5%. Over this range K_d is independent of concentration.

The accuracy and resolution capabilities of the technique, coupled with the enhanced detection sensitivity provided by the labeling techniques, has led us to explore the possibility of providing molecular weight fingerprints of the cyanogen bromide (CNBr) cleavage products of proteins. The reagent CNBr cleaves polypeptide chains specifically at the carboxyl terminal of methionine residues in the primary sequence (*21*). Since the methionine content of protein is generally low, a limited number of relatively large peptides are generally produced. The size of these peptides generally precludes their mapping by conventional paper electrophoresis and chromatography techniques. Table IV provides the molecular weights determined for fluorescein labeled cyanogen bromide peptides for a series of proteins of known sequence on 6% and 10%

agarose columns. The agreement between determined M_{app} and the true molecular weights from sequence is rather good. While 6% agarose is suitable for these types of experiments, the better resolving power of 10% agarose in the low molecular weight range renders it better suited to this type of analysis. An illustration of this resolving power can be seen in Figure 7, which presents the elution profile obtained on analysis of a partially cleaved sample of myoglobin on 10% agarose. Components with molecular weights corresponding to the three normal myoglobin peptides which result from cyanogen bromide cleavage at methionines 55 and 131 (C, D, E) are clearly evident. In addition, components corresponding to uncleaved myoglobin (A) and myoglobin cleaved only at methionine 55 (B) are resolved.

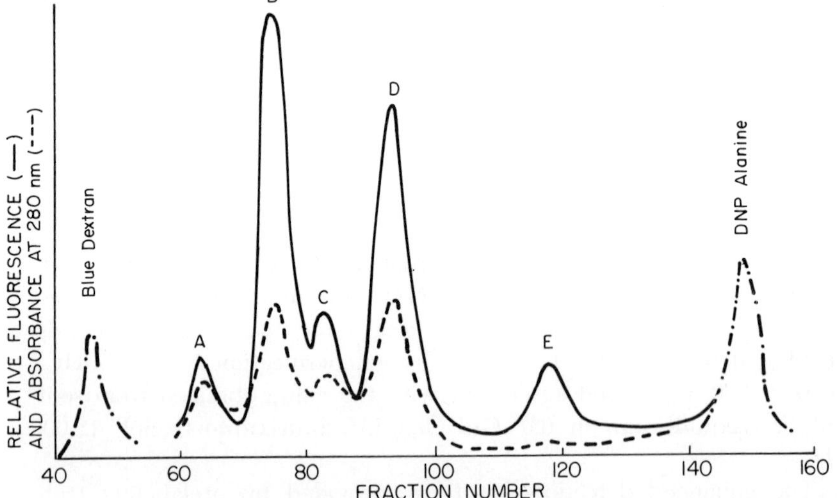

Figure 7. Elution profile for a sample of myoglobin which was partially cleaved with cyanogen bromide and labeled with fluorescein isothiocyanate. Absorbance at 280 μm, (broken line) and fluorescein fluorescence, 493-nm excitation, and 530-nm emission, (solid line) are plotted vs. *fraction number. The identities of components A, B, C, D, and E were inferred on the basis of their* M_{app}*'s (see text).*

The accurate determination of molecular weights in the range of 1400–10,000 indicated by our studies on small naturally occurring peptides and protein CNBr products is an extremely useful aspect of gel permeation studies on agarose–GuHCl columns. Molecular weight determinations by exact techniques are difficult for peptides of low molecular weight in denaturing solvents. Further, the application of the recently developed technique of electrophoresis in sodium dodecyl sulfate (*22*) is excluded from accurate measurements of polypeptides of molecular weight less than 10,000 owing to the shape adopted by polypeptide chains in this solvent (*23*).

Conclusion

Gel permeation studies on agarose–GuHCl columns provide for high resolution and accurate molecular weight determination of linear randomly coiled polypeptide chains. For nonlinear random coils, gel permeation studies provide for accurate determination of the effective hydrodynamic radius of the components.

The technique described here is finding increasingly widespread use for the resolution and molecular weight estimation of naturally occurring polypeptides. Its intended function is to supplement the classical exact techniques of molecular weight analysis and is not intended as an alternative procedure. Although we are aware of no instance in which invalid data have been obtained, the method can only be considered an approximate one since measurement involves calibration with secondary standards, and at present it is not suitable for precise theoretical analysis.

Acknowledgment

The authors acknowledge the technical assistance provided by K. D. Kemp. We also express our appreciation to Val Woodward for his generous support of D. N. Fass. This research was partially supported by a grant from the National Institutes of Health (HE 13923-01), a research grant from the University of Minnesota Graduate School, and a grant from the South Carolina State Appropriation for Research (1970). D. N. Fass was supported by a grant from the National Institutes of Health (GM 01156-08) to the Department of Genetics and Cell Biology, University of Minnesota.

Literature Cited

1. Tanford, C., "Physical Chemistry of Macromolecules," p. 344, Wiley, New York, 1961.
2. Andrews, P., *Biochem. J.* (1965) **96,** 595.
3. Tanford, C., *Advan. Protein Chem.* (1968) **23,** 121.
4. Tanford, C., "Physical Chemistry of Macromolecules," p. 168, Wiley, New York, 1961.
5. Fish, W. W., Mann, K. G., Tanford, C., *J. Biol. Chem.* (1969) **244,** 4989.
6. Davison, P. F., *Science* (1968) **161,** 906.
7. Mann, K. G., Fish, W. W., "Methods in Enzymology," Vol. 26, p. 28, C. H. W. Hirs, S. N. Timasheff, Eds., Academic, New York, 1972.
8. Porath, J., *Pure Appl. Chem.* (1963) **6,** 233.
9. Ackers, G. K., *J. Biol. Chem.* (1967) **242,** 2237.
10. Tanford, C., Kawahara, K., Lapanje, S., *J. Amer. Chem. Soc.* (1967) **89,** 906.
11. Tanford, C., "Physical Chemistry of Macromolecules," p. 359, Wiley, New York, 1961.
12. Mann, K. G., Batt, C. W., *J. Biol. Chem.* (1969) **244,** 6555.

13. Mann, K. G., Heldebrant, C. M., Fass, D. N., *J. Biol. Chem.* (1971) **246,** 6106.
14. Gwynne, J. T., Tanford, C., *J. Biol. Chem.* (1970) **245,** 3269.
15. Rinderkneckt, H., *Sep. Exp.* (1960) **16,** 430.
16. Gray, W. R., "Methods in Enzymology," Vol. 11, Hirs, C. H. W., Ed., p. 139, Academic, New York, 1967.
17. Habeeb, A. F. S. F., *Anal. Biochem.* (1966) **14,** 328.
18. Riordan, J. F., Vallee, B. L., "Methods in Enzymology," Vol. 11, Hirs, C. H. W., Ed., p. 565, Academic, New York, 1967.
19. Greenwood, F. C., Hunter, W. M., Glover, J. S., *Biochem. J.* (1963) **89,** 114.
20. Fish, W. W., Hays, J., Roseman, S., unpublished data.
21. Gross, E., Witkop, B., *J. Biol. Chem.* (1962) **237,** 1856.
22. Weber, K., Osborn, G., *J. Biol. Chem.* (1969) **244,** 4406.
23. Reynolds, J. A., Tanford, C., *J. Biol. Chem.* (1970) **245,** 5161.

RECEIVED January 17, 1972.

26

Small-Angle X-Ray Scattering Measurements of Biopolymer Molecular Weights in Interacting Systems

SERGE N. TIMASHEFF

Graduate Departmenet of Biochemistry, Brandeis University, Waltham, Mass. 02154

The thermodynamic equations for measuring macromolecule molecular weights by small-angle x-ray scattering are derived from fluctuation theory. The multicomponent relation, which must be used in mixed solvents, is shown to be formally identical to those of light scattering and sedimentation equilibrium. Small-angle x-ray scattering measurements of molecular weights in mixed solvents require knowledge of the extent of preferential interactions between the macromolecule and solvent components. Neglect of this contribution may lead to serious errors in reported molecular weights. The theory is illustrated by two systems: β-lactoglobulin A in a water–2-chloroethanol mixture and DNA in concentrated salt solutions.

In recent years it has become increasingly evident that many important biological systems, in particular many enzymes, exist in a stable form as specific aggregates of subunits. Frequently, such intermolecular associations play a key role in determining biological activity and its control (*1, 2, 3*). To arrive at a full understanding of the function of such macromolecules, it is important to know exactly the molecular weights of their subunits.

Since in many systems the intersubunit association forces, even though they are not covalent, are strong enough to resist dissociation by mild procedures (*4*) such as change in pH or salt concentration in dilute buffer, it is necessary to use strong dissociating agents to obtain monomeric species. The materials, most frequently used for this purpose are 6*M* guanidine HCl (GuHCl), concentrated (8*M*) urea, detergents, and organic solvents—*e.g.*, alcohols. While these media usually cause the associated protein species to dissociate fully into subunits, they also denature the proteins—*i.e.*, they disrupt the native globular three-dimensional structure, resulting in a randomly coiled form (in 6*M* GuHCl) (*5, 6*) or in various other conformations, such as rod-like structures (in sodium dodecylsulfate) (*7*) or highly helical forms (when alcohols are used as denaturant) (*8, 9*). While the conformations of the dissociated subunits are different in these cases, they all exhibit strong interactions with the denaturant, which help to stabilize them in solution. Such interactions are seen thermodynamically as a large non-ideality of the system. Since many methods of measuring molecular weights are thermodynamic (*e.g.*, light scattering, small-angle x-ray scattering, sedimentation equilibrium), such interactions can be expected to affect strongly the interpretation of the experimental results. This paper reviews the basic theory of molecular weight measurements by small-angle x-ray scattering and gives examples of the magnitude of the effects which are commonly encountered in multicomponent systems.

Theoretical

When a beam of x-rays strikes an electron, some of the energy is momentarily absorbed, displacing the electron from its unperturbed position. This sets the electron in periodic motion with the same frequency as that of the exciting radiation. As a result the electron radiates an electromagnetic wave in all directions with the same frequency as the exciting radiation. This leads to the experimental observation that the incident radiation is "scattered" by the electron. A theoretical analysis (*10, 11*) of these events leads to the Thompson scattering equation which relates the intensity of x-rays scattered by a single electron, I_e, to that of the incident non-polarized x-radiation, I_o:

$$I_e = \frac{e^4}{m^2\, c^4\, r^2}\, I_o \left(\frac{1 + \cos^2 2\theta}{2}\right) \tag{1}$$

where e is the electronic charge, m is the mass of the electron, c is the velocity of light, r is the distance between the scattering electron and the observer, and $2\,\theta$ is the angle between the directions of the incident and scattered rays. For a particle containing n electrons, the scattering in vacuum (averaged over all orientations), has been shown by Debye (*12*) and Guinier (*13*) to be

$$I\,(s) = n^2\, I_e\, f\,(s) \tag{2}$$

where s is a directional parameter, defined as

$$s = \frac{2 \sin \theta}{\lambda} \tag{3}$$

where λ is the wavelength of the x-rays, and $f(s)$ is the angular dependence of scattering, which is a function of the size, shape, and internal structure of the particle. It is essentially identical with the particle scattering factor, $P(\theta)$, defined by Zimm for light scattering (*14*). For F noninteracting particles, the total scattering is $F\ I(s)$, and if the concentration, C, is expressed in mass (electrons) per volume units, the number of electrons per particle is given by

$$n = \frac{I\ (s)}{I_e\ C\ f\ (s)} \tag{4}$$

For a two-component system (macromolecule, component 2, dissolved in pure solvent, *e.g.*, water, component 1), application of the fluctuation theory of scattering (*15, 16*) shows that in a volume element of volume δV (of the order of s^{-3} at the lowest angles of measurement), the intensity of radiation scattered by the solution in excess of the solvent, $I(s) \equiv I(s)_{\text{solution}} - I(s)_{\text{solvent}}$, is proportional to the fluctuations in the number of electrons, n:

$$\Delta\ I\ (s) = I_e\ \overline{\Delta\ n^2}\ f\ (s) \tag{5}$$

where $\overline{\Delta n^2} \equiv \overline{n^2} - \overline{n}^2$ is the fluctuation in the number of electrons in volume element δV caused by fluctuations in the concentration of the particles by Brownian motion, assuming that the temperature and pressure fluctuations of the solvent and solution are identical (*17*). Defining the electron density, ρ, as the number of electrons per unit volume (usually per cubic Angstrom), we have, for a volume element

$$\rho = \frac{n}{\delta\ V} \tag{6}$$

and substituting into Equation 5,

$$\Delta\ I\ (s) = I_e\ \overline{\Delta \rho^2}\ (\delta V)^2\ f\ (s) \tag{7}$$

For a unit volume, the intensity of scattering will be equal to the total scattering from all ω volume elements which it contains. Since $\omega = 1/\delta V$ and since the fluctuations in electron density are the consequence of fluctuations in concentration of component 2, $\overline{\Delta\ C^2_{e,2}}$, where the concentration $C_{e,i}$ is given as the number of electrons per unit volume (A^3), we have

$$\Delta I(s) = I_e \, \delta V \left(\frac{\partial \rho}{\partial C_{e,2}}\right)^2_{T,p} \overline{\Delta C^2_{e,2}} \, f(s) \tag{8}$$

Introducing into Equation 8 the thermodynamic relation for concentration fluctuation:

$$\delta V \, \overline{\Delta C^2_{e,2}} = - \frac{k \, T \, C_{e,2} \, \overline{V}_1}{\left(\frac{\partial \mu_1}{\partial C_{e,2}}\right)_{T,p}} \tag{9}$$

where k is Boltzmann's constant, T is the thermodynamic temperature, $\overline{V}_1$ is the partial molal volume of solvent, and p is pressure, application of the Gibbs-Duhem equation, $n_1 \, d\mu_1 = n_2 \, d\mu_2$, and defining the chemical potential of the macromolecule, μ_2, in proper concentration units, $g_{e,2}$ (number of electrons of solute per electron of solvent, a molal type of concentration), $\mu_2 = \mu^o{}_2 + RT \ln g_{e,2} + RT \ln \gamma_2$, results in the two-component equation for small-angle x-ray scattering:

$$\frac{I_e \, V g_{e,2} \, f(s)}{N_A \, \Delta I(s)} \left(\frac{\partial \rho}{\partial g_{e,2}}\right)^2_{T,p} = \frac{1}{M_{e,2}} \left[1 + \left(\frac{\partial \ln \gamma_2}{\partial g_{e,2}}\right)_{T,p} g_{e,2}\right] \tag{10}$$

where V is the volume of solution per electron of solvent, $M_{e,2}$ is the molecular weight of the macromolecule in electrons per mole, and N_A is Avogadro's number. From a knowledge of the chemical composition of the macromolecule, $M_{e,2}$ may be related to M_2, its molecular weight, since $M_i = N_A M_{e,i}/q_i$, where q is the number of electrons per gram of macromolecule.

A determination of the molecular weight by small-angle x-ray scattering requires, therefore, knowledge of the scattering intensity as a function of protein concentration and angle of observation, as well as of the electron density increment, $\left(\frac{\partial \rho}{\partial g_{e,2}}\right)_{T,p}$, which replaces the refractive index increment in light scattering. While in light scattering the refractive index increment can be measured directly, the electron density increment must be calculated from a measurement of the partial specific volume, using the relation

$$\left(\frac{\partial \rho}{\partial g_{e,i}}\right)_{T,p} = \frac{1}{V} (1 - \rho_s \psi_i) = \frac{1}{V} \left(1 - \rho_s \frac{10^{24} \, \overline{v}_i}{q_i}\right) \tag{11}$$

where ψ_i is the electron partial specific volume of component i, in units of A^3 per electron, ρ_s is the electron density of the solution, and $\overline{v}_i$ is the partial specific volume of component i in ml/gram. In the more customary mass per volume concentration units, C_e, electrons per A^3, Equations 10 and 11 take the forms:

$$\frac{I_e}{N_A}\frac{C_{e,2}}{\Delta I(s)}\left(\frac{\partial\rho}{\partial C_{e,2}}\right)^2_{T,p} = \frac{1}{M_{e,2}f(s)}\left[1 + \left(\frac{\partial \ln\gamma_2}{\partial C_{e,2}}\right) C_{e,2}(1 - C_{e,2}\psi_2)\right] \times \frac{1}{(1 - C_{e,2}\psi_2)^2} \quad (12)$$

$$\left(\frac{\partial\rho}{\partial C_{e,i}}\right)_{T,p} = \frac{(1 - \rho_s\psi_i)}{(1 - C_{e,i}\psi_i)}$$

When the system becomes multicomponent, such as a protein in a denaturant—*e.g.*, 6*M* GuHCl—the relations become more complicated. For strong interactions, such as complex formation, the fluctuating units evidently consist of protein molecules plus material liganded to them. In a more general sense, when the interactions are weak, the fluctuating unit, in addition to protein, receives contributions from other components of the system, whose freedom of motion (rotation or translation) is affected by the protein molecules. This is reflected in a change in the activity coefficient of the other components arising from the presence of the protein, $(\partial \ln\gamma_i/\partial m_2)_{T,p,mj}$, where m is molal concentration. This means that in the fluctuating units an extensive property, such as the electron density increment, will now have contributions not only from the macromolecule but also from solvent components which interact with it. The extent of these contributions which will be detected is only that arising from the difference between the composition of the solvent in the immediate domain of the macromolecule and that in the bulk (9). Thus, the interactions measured are preferential and not absolute. Formally, when the multicomponent fluctuation theory (*18, 19*) is applied to small-angle x-ray scattering, we obtain the relation

$$\frac{I_eV}{N_A}\frac{g_{e,2}}{\Delta I(s)}\left(\frac{\partial\rho}{\partial g_{e,2}}\right)^2_{T,p,g_{e,3}} = \frac{1}{M_{e,2}(1+D)^2 f(s)}[1 + B' g_{e,2}]$$

$$D = \frac{(\partial\rho/\partial g_{e,3})_{T,p,g_{e,2}}}{(\partial\rho/\partial g_{e,2})_{T,p,g_{e,3}}}\frac{(\partial\mu_3/\partial g_{e,2})_{T,p,g_{e,3}}}{(\partial\mu_3/\partial g_{e,3})_{T,p,g_{e,2}}} \quad (13)$$

$$= \frac{(1 - \rho_s\psi_{3,g_{e,2}})}{(1 - \rho_s\psi_{2,g_{e,3}})}\left(\frac{\partial g_{e,3}}{\partial g_{e,2}}\right)_{T,p,\mu_3}$$

where components 1, 2, and 3 are principal solvent (water), protein, and denaturant, respectively. The symbols have their previous meaning. The subscript $g_{e,i}$ on ψ_j means that the partial specific volume of component j must be measured under conditions at which the molal concentrations of component i are identical in the solution and in the reference solvent. Examination of Equation 13 reveals immediately that in a three-component system, the usual extrapolation to zero macromolecule concentration of small-angle x-ray scattering data does not result in a value of the molecular weight but rather in the product of the molecular weight and a thermodynamic interaction parameter between the macromolecule and the third component. If we pass from the electron concentration units, $g_{e,i}$ to the more familiar g_i units of gram of component i per gram water ($g_{e,i} = g_i\, q_i/q_{H_2O}$), we find that the deviation from the true molecular weight is a measure of the well-known preferential interaction parameter (*20, 21, 22*):

$$(\partial g_3/\partial g_2)_{T,\mu_1,\mu_3} = - g_3\,(\partial g_1/\partial g_2)_{T,\mu_1,\mu_3}\ .$$

Thus, in small-angle x-ray scattering, measurement of the molecular weight of a macromolecule in concentrated solvent requires knowledge of the preferential interaction parameter. This can be measured by techniques such as differential refractometry, densimetry, and isopiestic vapor phase equilibrium measurements. For densimetry,

$$\left(\frac{\partial g_3}{\partial g_2}\right)_{T,\mu_1,\mu_3} = \frac{(\partial d/\partial g_2)_{T,p,\mu_3} - (\partial d/\partial g_2)_{T,p,g_3}}{(\partial d/\partial g_3)_{T,p,g_2}}$$

$$= \frac{(1 - d_s\,\bar{v}_{2,\mu_3}) - (1 - d_s\,\bar{v}_{2,g_3})}{(1 - d_s\,\bar{v}_{3,g_2})} \qquad (14)$$

where d is density in grams/ml. The subscripts indicate that the density measurements on protein solutions must be performed at conditions at which, in turn, the chemical potential and the concentration (in grams per gram water) of the denaturant must be identical in the solution and in the reference solvent. The first condition can be attained to a close approximation (*19*) by dialyzing the protein solution against the solvent of interest and using the dialyzed protein and dialyzate as solution and reference solvent in density measurements. (In measurements of molecular weights by light scattering and sedimentation equilibrium, knowledge of the interaction parameter is not needed since the refractive index increments and partial specific volumes can be measured under conditions which eliminate it. These are conditions at which the solvent components are in chemical equilibrium in the solution and the reference solvent; operationally, this can be attained by dialysis. Such an operation is obviously impossible for small-angle x-ray scattering. This can be shown easily to be true, for example, for light scattering. The light scattering equivalent of Equation 13 can be simplified by:

$$(1 + D)^2 \left(\frac{\partial n}{\partial g_2}\right)^2_{T,p,g_3}$$

$$= \left[\left(\frac{\partial n}{\partial g_2}\right)_{T,p,g_3} + \left(\frac{\partial n}{\partial g_3}\right)_{T,p,g_2}\left(\frac{\partial g_3}{\partial g_2}\right)_{T,p,\mu_3}\right]^2 \qquad (13a)$$

As in Equation 14 for the density increment, the right side of Equation 13a is equal to $(\partial n/\partial g_2)_{T,\mu_1,\mu_3}$, *i.e.*, the refractive index increment measured by using a dialyzed macromolecule solution and the dialyzate as reference solvent.)

Here we present typical examples of systems on which such measurements have been carried out, and we show the magnitude of the contribution of such interactions to apparent molecular weights obtained if the proper correction is not made. The systems examined are β-lactoglobulin in water–2-chloroethanol medium and DNA in high salt concentration.

Results

β-Lactoglobulin A in 40% 2-Chloroethanol. Previous light scattering and differential refractometry measurements (*8, 23*) have shown that β-lactoglobulin exhibits strong preferential interactions with solvent components in the water–2-chloroethanol system. Since the preferential interaction between protein and 2-chloroethanol in this system was found to be maximal at 40% (v/v), the effect of this interaction on the partial specific volume of the protein was determined.

The densities of the solvent and of β-lactoglobulin A (β-Lg) in 40% (v/v) 2-chloroethanol, in the presence of 0.01*M* HCl and 0.02*M* NaCl, were determined, with and without prior dialysis, in a 10-ml pycnometer at 20°C. Solutions were prepared as described previously (*8, 24*). The solutions were filtered through millipore filters in syringe adapters just before the density measurements. Protein concentrations were determined after filtration by ultraviolet absorption at 278 nm. The apparent partial specific volume, $\bar{v}_{\mathrm{app}}$, was calculated from the densities using the standard equation (*21, 25*):

$$\bar{v}_{\mathrm{app}} = \frac{1}{d_o}\left(1 - \frac{d - d_o}{C}\right) \tag{15}$$

where d is the density of the solution in grams/ml, d_o is that of the solvent, and C is the protein concentration in grams/ml.

The results obtained for β-Lg in 40% 2-chloroethanol were:

$$\bar{v}_{2,g_3} = 0.715 \pm 0.003 \text{ ml/gram}$$

$$\bar{v}_{2,\mu_3} = 0.675 \pm 0.002 \text{ ml/gram.}$$

The value of the partial specific volume obtained in 0.01*M* HCl, 0.02*M* NaCl was 0.748 ml/gram, in good agreement with the literature value of 0.751 (*26*).

The significance of such a difference between the partial specific volumes measured at constant chemical potential ($\bar{v}_{2,\mu_3}$) and constant solvent composition ($\bar{v}_{2,m_3}$) can be illustrated as follows. The molecular weight of β-lactoglobulin in 40% 2-chloroethanol was measured by sedimentation equilibrium. With $\bar{v}_{2,\mu_3} = 0.675$, the molecular weight found was 19,900, in fair agreement with the value of 18,800 obtained with the same protein preparation in the absence of denaturant. On the other hand, if $\bar{v}_{2,m_3}$ was used, the apparent molecular weight obtained was 20% too high. In light scattering, where the corresponding parameter is the refractive index increment, use of the dn/dC_2 value measured without prior dialysis resulted in an apparent molecular weight which was 60% too high.

With the measured values of $\bar{v}_2$, the preferential interaction parameter was calculated using Equation 14. With $d_s = 1.092$ grams/ml and $\bar{v}_{3,m_3} = 0.822$ ml/gram (*8*), $(\partial g_3/\partial g_2)_{T,\mu_1,\mu_3}$ was found to be 0.437 gram of 2-chloroethanol per gram of protein. This must be compared with 0.716 gram/gram previously determined by light scattering and differential refractometry (*8, 23*). This agreement can be regarded as acceptable since the value of this parameter is obtained from a small difference between two large numbers. A value of $(\partial g_3/\partial g_2)_{T,\mu_1,\mu_3}$ of 0.437 gram/gram when inserted into Equation 13, with application of the proper conversion parameters, results in a $(1 + D)^2$ contribution in small-angle x-ray scattering of 1.32. Thus, neglect of preferential interaction would result in a reported molecular weight which is 32% too high.

Knowledge of the partial specific volume of a protein in the denaturant, $\bar{v}_{2,g_3}$, which is essential for determining its molecular weight by small-angle x-ray scattering, permits also the calculation of the volume change upon denaturation since

$$\Delta V = M_2\,[\bar{v}_{2,g_3}\text{ (in denaturant)} - \bar{v}_2\text{ (native)}] \quad (16)$$

where ΔV is the molar volume change on denaturation. Since furthermore the extent of preferential interaction with solvent components in the denaturant must also be known to determine the molecular weight accurately, these auxiliary measurements provide important information related to the mechanism of denaturation of the protein.

For the denaturation of β-Lg in 40% 2-chloroethanol, the presently reported partial specific volume measurements result in a ΔV value of −590 ml/mole. This value is very close to that previously reported for the denaturation of this protein by 6.4*M* urea, −610 ml/mole (*27*). Although this similarity of ΔV values is striking, it might be the result of a fortuitous compensation of various effects. The change in volume calculated from the difference in partial specific volumes is the sum of a number of contributions (*28*), such as differences in electrostriction in the two media, changes in the density of solvent components when they interact with protein, as well as the actual difference between the volumes of the native and denatured protein molecules. In 6.4*M* urea, β-Lg is denatured to a disordered conformation while in 40% 2-chloroethanol, the product is close to 50% α-helical (*8*), the side chains being highly exposed to contact with solvent components in both cases. Since the nature of the solvent differs greatly in the two cases, the various protein–solvent interaction effects may be expected to make different contributions to ΔV. The overwhelming effect with β-Lg may nevertheless be one of the annihilation of voids inside the native structure when it unfolds.

Table. I. Small-Angle X-Ray Scattering Measurements of Preferential Interaction of DNA with Solvent Components in Concentrated Salt Solutions

g_3	$(M/L)/(M/L)_o^{\,a}$	D	$(\partial g_{e,3}/\partial g_{e,2})$ (el/el)	$(\partial g_1/\partial g_2)$ $(gram/gram)$
NaCl				
0.012	0.980	−0.010	−0.0072	0.63
0.029	0.968	−0.017	−0.0086	0.32
0.060	0.980	−0.010	−0.0075	0.13
0.122	0.875	−0.063	−0.053	0.47
0.187	0.760	−0.127	−0.120	0.69
NaBr				
0.077	0.893	−0.034	−0.036	0.53
0.115	0.777	−0.073	−0.080	0.80
0.159	0.690	−0.105	−0.119	0.86

[a] Data of Luzzati *et al.* (*29*), recalculated as described in the text.

DNA in Concentrated Salt Solutions. Small-angle x-ray scattering experiments involving a system in which macromolecule–solvent interactions are significant have been reported on DNA in concentrated salt solutions (*29, 30*). Since a DNA molecule is very long compared with the dimensions of a fluctuating volume element while the diameter of a double helix is not, the quantity measured in this case is the mass per unit length, M/L, rather than the molecular weight, M (*31*). Therefore, in Equation 13, $M\,(1 + D)^2$ must be replaced by $(M/L)(1 + D)^2$. Using the recently reported salt concentration dependence of the partial specific volume of DNA (*32*), $\bar{v}_{2,g_3}$, the small-angle x-ray scattering data (*29*) were recalculated and normalized to the theoretical value at zero salt concentration. Analysis of the resulting data in terms of Equation 13 is presented in Table I, where component 1 is water, component 2 is DNA, and component 3 is the salt. In this case, the situation is opposite to that described above for β-Lg in the presence of 2-chloroethanol. In the DNA–salt system, the value of D becomes progressively more negative as the salt concentration increases. Equation 13 shows that a negative value of

D corresponds to a positive value of $(\partial \mu_3/\partial g_2)_{T,p,g_3}$, *i.e.*, to an increase in the chemical potential of the salt when DNA is added. This repulsion of the salt by the nucleic acid in the given system leads to a net preferential hydration of the macromolecule. This is a simple consequence of the fact that at equilibrium the activities of the salt in the domain of the macromolecule and in the bulk solvent must be equal. Since in the presence of the macromolecule the activity coefficient of the salt increases, its concentration must decrease. This deficiency in salt concentration relative to the bulk solvent is given (in electron units) in column 4 of Table I. When expressed as preferential hydration, this effect assumes the values shown in the last column of the table in units of gram of water/gram of DNA. These values are in general agreement with the literature (*33, 34*). The large spread in the extent of preferential hydration reflects simply normal experimental error. For example, when M_{app}/M (where $M_{app} = M(1 + D)^2$—*i.e.*, the apparent molecular weight measured in the given solvent) is of the order of 0.95, a 5% error in the determination of M_{app} results in an error by a factor of two in the preferential interaction parameter.

Discussion

The two systems treated here are good illustrations of the magnitude of preferential interactions between solvent components and macromolecules in biological systems. The results of these interactions are commonly translated in large positive or negative deviations of the apparent molecular weight from the true value. For proteins in guanidine hydrochloride, a medium frequently used for determining the molecular weights of enzyme subunits, the preferential interaction with the denaturant frequently attains values of up to 0.20 gram guanidine hydrochloride per gram of protein (*28, 35*). In terms of molecular weight measurements, this raises the apparent molecular weight by as much as 25% both in sedimentation equilibrium and light scattering experiments (*36, 37, 38*). Therefore, failure to ascertain the contribution of solvent interactions in a molecular weight measurement can easily introduce an uncertainty into the determination of the number of subunits in an associated system when that number is three or greater. In making molecular weight measurements in mixed solvents, it is rarely possible to predict or "guess at" the magnitude, or even sign, of the preferential interaction. In the case of guanidine hydrochloride, $(\partial g_3/\partial g_2)_{T,\mu_1,\mu_3}$ varies between 0 and 0.20 (*28, 35*). A striking example is afforded by ribonuclease in two solvent systems, each consisting of water and an alcohol, *i.e.*, 2-chloroethanol and 2-methyl-2,4-pentanediol. In the first system, the alcohol interacts preferentially with the protein; in the second the protein is preferentially hydrated (*39*).

It seems desirable to consider further the meaning of the solvent interactions which intervene in molecular weight measurements. As shown in Equation 13, the deviation from the true molecular weight, D, is a direct measure only of the change in chemical potential of one component caused by the addition of another component. Within the approximation that $(\partial\mu_3/\partial m_3)_{T,p,m_2} = RT/m_3$, the intercomponent thermodynamic interaction is equal to

$$\left(\frac{\partial \mu_3{}^{(e)}}{\partial m_2}\right)_{T,p,m_3} = \left(\frac{\partial \mu_2{}^{(e)}}{\partial m_3}\right)_{T,p,m_2} = -\frac{RTDM_2}{1000g_3}\alpha$$

$$= -\frac{RTM_2}{1000g_3}\left(\frac{\partial g_3}{\partial g_2}\right)_{T,\mu_1,\mu_3} = \frac{RTM_2}{1000}\left(\frac{\partial g_1}{\partial g_2}\right)_{T,\mu_1,\mu_3} \qquad (17)$$

where α is the ratio of the auxiliary parameters of components 3 and 2, namely refractive index increments in light scattering, electron density increments in small-angle x-ray scattering and buoyancy terms in sedimentation equilibrium, and $\mu_i{}^{(e)} = 1/RT \ln \gamma_i$ is defined by $\mu_i = RT \ln m_i + \mu_i{}^{(e)} + \mu_i{}^0 (T, p)$.

Application of Equation 17 to the two systems discussed here shows that dissolution of β-lactoglobulin in 40% 2-chloroethanol in the final conformation assumed by the protein in this medium stabilizes the system since $(\partial\mu_2{}^{(e)}/\partial m_3) = -6{,}000$ cal/mole protein/mole 2-chloroethanol in 1000 grams of water. This means that in this system the protein has a stronger affinity for the alcohol than for water relative to the bulk solvent composition. As a result, excess alcohol is found in the domain of the protein. Exactly the opposite is true for the DNA in salt system. Here, the obtained values of D result in a destabilization of the system by the salt since $(\partial\mu_2{}^{(e)}/\partial m_3) = 50$–$100$ cal/(mole/A) of DNA per mole of salt in 1000 grams of water. For a double helix of 100 base pairs this amounts to a destabilization of the order of 10 kcal/mole of salt added. As pointed out above, this results in excess water in the domain of the macromolecule —*i.e.*, in preferential hydration of the DNA.

The preferential interaction measured by thermodynamic techniques is, therefore, strictly an activity coefficient effect, which may be expressed in terms of a "preferential binding" parameter, $(\partial g_3/\partial g_2)_{T,\mu_1,\mu_3}$ since

$$\left(\frac{\partial \mu_3}{\partial m_3}\right)_{T,p,m_3} \Big/ \left(\frac{\partial \mu_3}{\partial m_3}\right)_{T,p,m_2} = -\left(\frac{\partial m_3}{\partial m_2}\right)_{T,p,\mu_3} \qquad (18)$$

The "preferential interaction" is, therefore, an expression of the affinity of one component for another in the most general sense, and while the term preferential binding is frequently used to express this effect, binding in this context in no way indicates immobilization of solvent molecules at specific sites on a macromolecule, such as by clathrate formation; it is only an expression of a general change in activity coefficient. Conversely, since the methods of measurement give a value of the interactions which is an average over all the constituent parts of a macromolecule (*e.g.*, over all the amino acid residues of a protein), the results do not preclude different strengths of interaction at various points of the macromolecule or even regions on the macromolecule at which different solvent components interact preferentially. Thus, locally, on a residue level of resolution, the pattern of interactions may be quite complex (*9*, *23*).

In a mixed solvent system a macromolecule may display an overall preferential interaction for one of the solvent components, but this does not eliminate interactions with the other solvent component as well. For example, in the water–2-chloroethanol system, particular regions of the protein molecule, such as ionized side chains, must be interacting with water molecules. Therefore, the extent of preferential interaction observed must be related to the absolute interactions of the protein with the solvent components. In fact, it can be shown (*40*) that:

$$\left(\frac{\partial g_3}{\partial g_2}\right)_{T,\mu_1,\mu_3} = A_3 - g_3 A_1 \tag{19}$$

where A_1 and A_3 are the absolute interactions of components 1 and 3 with component 2 expressed in grams of component i per gram of component 2. Therefore, if the absolute interaction of one component with protein is known and if it is expressed in terms of binding, such as degree of hydration, a measurement of the preferential interaction parameter allows one immediately to calculate the extent to which the other solvent component is interacting with the macromolecule. Using Equation 19 and the assumption that the absolute extent of hydration of a protein changes little on unfolding, the degrees of protein solvation by the denaturant have been calculated for a number of systems (*40*) and related to the extent of denaturation.

A consequence of Equation 19 is that preferential and absolute interactions, when expressed as binding, do not necessarily vary in parallel manner when the solvent composition is changed. The values of A_3 and $(\partial g_3/\partial g_2)_{T,\mu_1,\mu_3}$ may actually assume opposite signs. This is easily illustrated with the help of Figure 1. In the model system depicted, it is assumed that the protein binds a constant amount of water at all solvent compositions; thus A_1 remains constant. It also binds a monotonely increasing amount of the denaturant (A_3 increases) as the proportion of the latter is increased in the medium. Relative to the bulk solvent composition, however, the net observed effect is one of preferential binding of denaturant, when the latter is present in small amounts in the solvent,

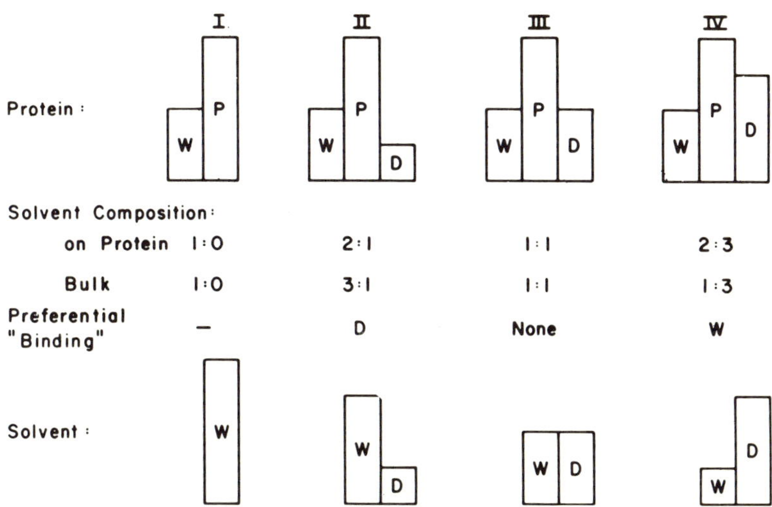

Figure 1. Schematic of preferential interactions. P : protein; W : water; D : denaturant.

followed by a point of zero preferential binding when the solvent composition in the domain of the macromolecule equals that of the bulk solvent; finally a region is attained, at high denaturant concentration, in which the observed preferential interaction is with water. In terms of molecular weight measurements, the observed pattern is one of initially too high apparent molecular weights, followed by a decrease toward the correct value, and finally a region where the measured molecular weights would be too low. Such a situation has been found for a number of proteins in the water–2-chloroethanol system (*9, 23*).

Appendix

Although this paper deals specifically with molecular weight measurements in mixed solvents, it seems worthwhile to consider briefly other types of interactions—namely, those leading to intermacromolecule complexes. In such systems, the stoichiometry and thermodynamics of association may be deduced if the molecular weights are measured as a function of concentration. Small-angle x-ray scattering, however, affords the possibility of measuring, in addition to the molecular weight, a number of other parameters (*13, 31, 41*)—namely, the radius of gyration, R_g, the hydrated volume of the macromolecule, V, the surface-to-volume ratio, S/V, the surface per particle, S/n, and the density in solution of the macromolecule. It is interesting to see what kinds of averages are obtained for these parameters when the system is polydisperse, as is true in associating systems.

If the electron densities of the macromolecules do not change on association, all the parameters may be examined, starting with the knowledge that the intensity of scattering in a mixture is equal to the sum of the scattering intensities of all the components. This leads to the well-known results (*42*) that in small-angle x-ray scattering and light scattering, the molecular weight measured is a weight average while the radius of gyration is an average of type

$$\overline{R^2{}_g} = \frac{\sum_i C_i\, M_i\, R^2{}_{g,i}}{\sum_i C_i\, M_i} \tag{20}$$

Although this average may appear to be a z-average, the situation is complex since the dependence of $R_{g,i}$ on the molecular weight is a function of the shape of the particle (*43*).

A similar analysis has been applied now to the other structural parameters. The results show that the averages of the various structural parameters are of different types.

(1) The hydrated molecular volume of species i, V_i, in A^3, is given by (*31*):

$$V_i = \frac{C_{e,i}\, M_{e,i}\, (1 - \rho_1\, \psi_2)^2}{4\, \pi\, \rho_s \int_0^\infty s^2\, \Delta\, I_i\, (s)\, ds} \tag{21}$$

where all the symbols have their previous meaning. In a polydisperse system, the measured volume is an average of type

$$\langle V \rangle_{\mathrm{Av}} = \frac{\overline{M_{e,w}}\, C_e}{\sum_i \frac{C_{e,i}\, M_{e,i}}{V_i}} \tag{22}$$

where $\overline{M_{e,w}}$ is the weight-average molecular weight, $C_{e,i}$ is the concentration of species i in electrons per A^3, and C_e is the total concentration.

(2) The surface-to-volume ratio, S/V, measured in a mixture is

$$\overline{\left(\frac{S}{V}\right)} = \frac{4\pi^2 \sum_i \lim_{s \to \infty} s^4\, \Delta\, I_i\, (s)}{\sum_i \int_0^\infty s^2\, \Delta\, I_i\, (s)\, ds} = \frac{\sum_i S_i}{\sum_i V_i}$$

$$= \frac{\sum_i \sigma_i\, C_i/M_i}{\sum_i v_i\, C_i/M_i} \tag{23}$$

where σ_i is the surface of a molecule of species i, and v_i is its volume. Therefore, in a polydisperse system, the quantity measured is the ratio of the total surface of the macromolecules to their total volume.

(3) Since the surface measured is the total interface between macromolecules and solvent, in a polydisperse system, the average of the surface per particle is a number average:

$$\overline{\left(\frac{S}{n}\right)} = \frac{\sum_i S_i}{\sum_i C_{e,i}/M_{e,i}} = \frac{\sum_i \sigma_i C_{e,i}/M_i}{\sum_i C_{e,i}/M_i} \tag{24}$$

(4) Finally, the average excess electron density of the hydrated particles over that of the solvent, $\Delta\rho = \rho_2 - \rho_1$, is

$$\overline{\Delta\rho} = \frac{4\pi\, \rho_s \sum_i \int_0^\infty s^2\, \Delta\, I_i\,(s)\, ds}{(1 - \rho_1\, \psi_2) \sum_i C_{e,i}} = \frac{\sum_i C_{e,i}\, \Delta\, \rho_i'}{\sum C_{e,i}} \tag{25}$$

This is a weight average, from which a weight average degree of hydration of the macromolecules may be estimated.

Acknowledgment

The author thanks Katherine S. Rostand for her able performance of the density measurements. This work was supported in part by NIH Grant GM-14603 and NSF Grant GB-12619. Part of this work was performed during the tenure by the author of a John Simon Guggenheim Memorial Foundation Fellowship at the Institut de Biologie Moléculaire, Faculté des Sciences, Paris, France.

Literature Cited

1. Monod, J., Changeux, J. P., Jacob, F., *J. Mol. Biol.* (1963) **6**, 306.
2. Monod, J., Wyman, J., Changeux, J. P., *J. Mol. Biol.* (1965) **12**, 88.
3. Koshland, D. E., Jr., Nemethy, G., Fulmer, D., *Biochemistry* (1966) **5**, 365.
4. Timasheff, S. N., in "Protides of the Biological Fluids," H. Peeters, Ed., Vol. 20, p. 511, Pergamon, New York, 1973.
5. Tanford, C., Kawahara, K., Lapanje, S., *J. Amer. Chem. Soc.* (1967) **89**, 729.
6. Tanford, C., *Adv. Protein Chem.* (1968) **23**, 121.
7. Reynolds, J. A., Tanford, C., *J. Biol. Chem.* (1970) **245**, 5161.
8. Inoue, H., Timasheff, S. N. *J. Amer. Chem. Soc.* (1968) **90**, 1890.
9. Timasheff, S. N., *Accts. Chem. Res.* (1970) **3**, 62.
10. Zachariasen, W. H., "Theory of X-Ray Diffraction in Crystals," Wiley, New York, 1945.
11. Timasheff, S. N., *J. Chem. Educ.* (1964) **41**, 314.
12. Debye, P., *Ann. Phys.* (*Leipzig*) (1915) **46**, 809.
13. Guinier, A., Fournet, J., "Small-Angle Scattering of X-Rays," Wiley, New York, 1955.
14. Zimm, B. H., *J. Chem. Phys.* (1948) **16**, 1099.
15. Tanford, C., "Physical Chemistry of Macromolecules," Wiley, New York, 1961.
16. Kauzmann, W., "Quantum Chemistry," Academic, New York, 1957.

17. Stigter, D., *J. Phys. Chem.* (1960), **64**, 842.
18. Kirkwood, J. G., Goldberg, R. J., *J. Chem. Phys.* (1950) **18**, 54.
19. Stockmayer, W. H., *J. Chem. Phys.* (1950) **18**, 58.
20. Timasheff, S. N., Kronman, M. J., *Arch. Biochem. Biophys.* (1959) **83**, 60.
21. Cassassa, E. F., Eisenberg, H., *Adv. Protein Chem.* (1964) **19**, 287.
22. Timasheff, S. N., in "Electromagnetic Scattering," M. Kerker, Ed., p. 337, Pergamon, New York, 1963.
23. Timasheff, S. N., Inoue, H., *Biochemistry* (1968) **7**, 2501.
24. Pittz, E. P., Lee, C., Badlouzian, B., Townend, R., Timasheff, S. N., in "Methods in Enzymology," C. H. W. Hirs, and S. N. Timasheff, Eds., Vol. 27, Chap. 10, Academic, New York, 1973.
25. Kielley, W. W., Harrington, W. F., *Biochim. Biophys. Acta* (1960) **41**, 401.
26. Pedersen, K. O., *Biochem. J.* (1936) **30**, 961.
27. Christensen, K. L., *Compt. Rend. Trav. Lab. Carlsberg* (1952) **28**, 37.
28. Lee, J. C., Timasheff, S. N., *Biochem.*, in press.
29. Luzzati, V., Nicolaieff, A., Masson, F., *J. Mol. Biol.* (1961) **3**, 185.
30. Luzzati, V., Masson, F., Mathis, A., Saludjian, P., *Biopolymers* (1967) **5**, 491.
31. Luzzati, V., *Acta Cryst.* (1960) **13**, 939.
32. Cohen, G., Eisenberg, H., *Biopolymers* (1968) **6**, 1077.
33. Hearst, J. E., *Biopolymers* (1965) **3**, 57.
34. Hearst, J. E., Vinograd, J., *Proc. Nat. Acad. Sci., U.S.* (1961) **47**, 825.
35. Hade, E. P. K., Tanford, C., *J. Amer. Chem. Soc.* (1967) **89**, 5034
36. Lee, J. C., Frigon, R. P., Timasheff, S. N., *J. Biol. Chem.*, in press.
37. Noelken, M. E., Timasheff, S. N., *J. Biol. Chem.* (1967) **242**, 5080.
38. Timasheff, S. N., Townend, R., in "The Physical Principles and Techniques of Protein Chemistry," Part B, S. Leech, Ed., Part B, p. 147, Academic, New York, 1970.
39. Pittz, E. P., Timasheff, S. N., unpublished data.
40. Inoue, H., Timasheff, S. N., *Biopolymers* (1972) **11**, 737.
41. Pessen, H., Kumosinski, T. F., Timasheff, S. N., *J. Agr. Food Chem.* (1971) **19**, 698.
42. Pessen, H., Kumosinski, T. F., Timasheff, S. N., in "Methods in Enzymology," C. H. W. Hirs, and S. N. Timasheff, Eds., Vol. 27, Chap. 9, Academic, New York, 1973.
43. Geiduschek, E. P., Holtzer, A., *Adv. Biol. Med. Phys.* (1958) **6**, 431.

RECEIVED December 21, 1972. Publication No. 915.

INDEX

D

H

I

L

M

N

O

P

Q

R

S

T

U

V